AF388662

Sustainable Designed Pavement Materials

Sustainable Designed Pavement Materials

Volume 2

Special Issue Editors

Sandra Erkens
Yue Xiao
Mingliang Li
Tao Ma
Xueyan Liu

MDPI • Basel • Beijing • Wuhan • Barcelona • Belgrade • Manchester • Tokyo • Cluj • Tianjin

Special Issue Editors
Sandra Erkens
Delft University of Technology (TUDelft)
The Netherlands

Yue Xiao
Wuhan University of Technology (WUT)
China

Mingliang Li
Research Institute of Highway Ministry of Transport (RIOH)
China

Tao Ma
Southeast University (SEU)
China

Xueyan Liu
Delft University of Technology (TUDelft)
The Netherlands

Editorial Office
MDPI
St. Alban-Anlage 66
4052 Basel, Switzerland

This is a reprint of articles from the Special Issue published online in the open access journal *Materials* (ISSN 1996-1944) (available at: https://www.mdpi.com/journal/materials/special_issues/sdpm).

For citation purposes, cite each article independently as indicated on the article page online and as indicated below:

LastName, A.A.; LastName, B.B.; LastName, C.C. Article Title. *Journal Name* **Year**, *Article Number*, Page Range.

Volume 2
ISBN 978-3-03936-002-4 (Hbk)
ISBN 978-3-03936-003-1 (PDF)

Volume 1-2
ISBN 978-3-03936-004-8 (Hbk)
ISBN 978-3-03936-005-5 (PDF)

Contents

About the Special Issue Editors

Sandra Erkens is the principal specialist in pavement materials and structures at Rijkswaterstaat, the Dutch highway authority. She is a full professor, holding the Chair of Pavement Engineering Practice, at Delft University of Technology. She is an internationally acknowledged expert in pavement materials and structures in general and asphalt concrete in particular. Prof. Erkens was a member of national and international groups involved in developing technical requirements for pavement materials and several (inter)national organizations for the dissemination of research. These include the ISAP technical committee on the Constitutive Modelling of Asphalt Concrete, the organization committee of the two-yearly Dutch conference on Infrastructure (CROW-infradagen) and the organizing committee of the 4th International Chinese European Workshop on Functional Pavement Design. She has been involved in road engineering research since 1997, has published over a hundred papers on her work and is a regular reviewer for conferences and journals.

Yue Xiao is a full-time research professor of the State Key Lab of Silicate Materials for Architectures in the Wuhan University of Technology. He was awarded the Fok Ying Tung Outstanding Young Teacher award by the Ministry of Education of China in 2018. He received the title of CHUTIAN Scholar in material science and engineering from the Hubei provincial department of education in 2014. Prof. Xiao received his Ph.D. degree in Road and Railway Engineering from Delft University of Technology, The Netherlands. He then joined Wuhan University of Technology in 2013. His research interests are in asphalt pavement surfacing, road and pavement materials, asphalt pavement design. Dr. Xiao is now conducting three innovative projects funded by the National Natural Science Foundation of China (NSFC), and projects supported by provincial departments as well. Since 2011, Dr. Xiao has published 62 SCI peer-reviewed journal papers.

Mingliang Li is a full-time associate researcher at the Road Research Center, Research Institute of Highway Ministry of Transport (RIOH) in China. Dr. Li received his Ph.D. degree in Road and Railway Engineering from Delft University of Technology, The Netherlands. His research interests are in functional pavement material and technology, pavement maintenance, asphalt materials, and recycling technology. He has participated in and completed more than 10 national and provincial scientific research projects, such as the National Key R & D plan of the Ministry of Science and Technology, research projects from the Ministry of Transport, etc. He was in charge of more than 20 design and consultant projects, including porous asphalt pavement, Sponge City permeable pavement and in-place recycling, etc. He has published one monograph, participated in the writing of six national and local standards, published more than 30 journal papers, and obtained 18 national patents.

Tao Ma is a full-time research professor and the vice dean of the School of Transportation at Southeast University. He is also the director of Road Engineering Department. His awards include the Young and Middle-Aged Leading Talent of Science And Technology Innovation In Transportation by the Ministry of Transport of China, Fok Ying Tung fund by the Ministry of Education of China, 333 High-Level Talent Project, and the Six Talent Summit Project by Jiangsu Province. He is the deputy director of the National Engineering Laboratory for Advanced Road Materials, deputy director of National Teaching Center of Road Traffic Virtual Simulation, and director of Jiangsu

Key Laboratory for Long-term Service and Safety of Road Infrastructure. His research interests are in asphalt pavement design, road and pavement materials, asphalt pavement maintenance and pavement sustainable development technologies. Dr. Ma is now conducting the science project for outstanding young people founded by the National Natural Science Foundation of China (NSFC), and other innovative projects supported by NSFC and provincial departments. Dr. Ma has published two books, more than 100 technical and journal papers, and has been awarded more than 20 national invention patents.

Xueyan Liu is currently an Associate Professor in the Section of Pavement Engineering of the Faculty of Civil Engineering & Geosciences of TU Delft. He works in the areas of constitutive modelling, numerical modelling, and material experimental characterization. Within the research program of the Section Pavement Engineering, his research topics mostly relate to the development and implementation of constitutive models for the simulation of the static and dynamic response of various pavement engineering materials like soils, asphalt concrete, liner and reinforcing systems, and sustainable development technologies, i.e., multiscale modelling of asphaltic materials, warm/cold asphalt concrete technology, durability of asphalt surfacings on orthotropic steel deck bridge, accelerated pavement test, and pavement continuous monitoring. Dr. Liu was granted his doctoral thesis in 2003. During the same period, Dr. Liu participated in the team that developed the ACRe model for Asphalt Concrete Response currently implemented in the 3D Computer Aided Pavement Analysis system (CAPA-3D). Dr. Liu has published more than 100 technical and journal papers on the mechanics and the finite element modelling of granular, concrete, and asphaltic materials. Dr. Liu is a member of the RILEM Technical Committee of Cracking in Asphalt Pavements WG3 and a member of the Delft Centre for Materials (DCMat). He is also a member of ISAP, AAPT, APSE, and IACMAG. Dr. Liu is an Editorial Board Member of Geomaterials (GM). Dr. Liu was appointed as a Board member of the International Association of Chinese Infrastructure Professionals (IACIP) and a member of the Academic Committee of the Key Laboratory of Road Structure and Materials Transportation Industry of the China Ministry of Transport. He is also actively involved in organizing international and national workshops and conferences and was invited as a Scientific/Technical committee member of several international conferences.

Article

Sustainable Green Pavement Using Bio-Based Polyurethane Binder in Tunnel

Chao Leng [1], Guoyang Lu [2,3], Junling Gao [2], Pengfei Liu [3,*], Xiaoguang Xie [2] and Dawei Wang [2,3,*]

[1]   School of Physical Education, Harbin University, Harbin 150090, China; jwwkc207@hrbu.edu.cn
[2]   School of Transportation Science and Engineering, Harbin Institute of Technology, Harbin 150090, China; lu@isac.rwth-aachen.de (G.L.); 18S132071@stu.hit.edu.cn (J.G.); xxg75@126.com (X.X.)
[3]   Institute of Highway Engineering, RWTH Aachen University, 52074 Aachen, Germany
*   Correspondence: liu@isac.rwth-aachen.de (P.L.); wang@isac.rwth-aachen.de (D.W.);
    Tel.: +49-241-80-20389 (P.L.); +49-241-80-22780 (D.W.)

Received: 25 May 2019; Accepted: 19 June 2019; Published: 21 June 2019

Abstract: As a closed space, the functional requirements of the tunnel pavement are very different from ordinary pavements. In recent years, with the increase of requirements for tunnel pavement safety, comfort and environmental friendliness, asphalt pavement has become more and more widely used in long tunnels, due to its low noise, low dust, easy maintenance, and good comfort. However, conventional tunnel asphalt pavements cause significant safety and environmental concerns. The innovative polyurethane thin overlay (PTO) has been developed for the maintenance of existing roads and constructing new roads. Based on the previous study, the concept of PTO may be a feasible and effective way to enrich the innovative functions of tunnel pavement. In this paper, the research aims to evaluate the functional properties of PTO, such as noise reduction, solar reflection and especially combustion properties. Conventional asphalt (Open-graded Friction Course (OGFC) and Stone Mastic Asphalt (SMA)) and concrete pavement materials were used as control materials. Compared with conventional tunnel pavement materials, significant improvements were observed in functional properties and environmental performance. Therefore, this innovative wearing layer can potentially provide pavements with new eco-friendly functions. This study provides a comprehensive analysis of these environmentally friendly materials, paving the way for the possible application in tunnels, as well as some other fields, such as race tracks in stadiums.

Keywords: tunnel pavement; sustainable pavement material; polyurethane thin overlay; combustion properties; noise reduction; solar reflection

---

1. Introduction

As a closed space, the functional requirements of the tunnel pavement are very different from ordinary pavements, e.g., a higher requirement on the reduction of the pavement noise, proper light reflection to ensure safety and save tunnel lighting energy, closure of the construction process, better pollutant discharge and air purification.

For a long time, cement concrete pavement has been widely used in tunnels, due to its long service life and better lighting effect. However, the long construction period, the high noise during operation, large dust and the fast deterioration rate of the skid resistance at the entrance and exit sections, which significantly influence the tunnel safety and environmental protection, have become a limitation for the further application of the cement concrete on the tunnel construction.

In recent years, with the increase of requirements for tunnel pavement safety, comfort and environmental friendliness, asphalt pavement has become more and more widely used in long tunnels, due to its low noise, low dust, easy maintenance and good comfort [1]. However, conventional tunnel asphalt pavements also cause significant safety and environmental concerns. At high temperatures and under the vertical and horizontal loads during the starting, accelerating, decelerating and braking, shear deformation distresses, such as rutting and upheaval, may occur on the asphalt pavements [2]. In the construction of tunnel asphalt pavement, a large amount of smoke, exhaust gas and dust will be generated. When a fire occurs in a tunnel, the combustion-supporting effect of the asphalt will increase the risk of tunnel fires, especially in long tunnels.

Recently, two main flame retardant methods haven been proposed. The first method is adding flame retardant, which mainly makes asphalt pavement achieve an ideal flame retardant effect. It mainly includes two forms: Directly adding flame retardant to asphalt or flame retardant mineral powder. The second method is gasoline escape, which makes use of the large void characteristics of the asphalt mixtures to make the gasoline escape quickly beyond the pavement, therefore, reduces the amount of gasoline involved in combustion, and achieves the purpose of flame retardant.

At present, most of the research on flame retardant asphalt pavement is to apply mature flame retardants in the field of flame retardant to asphalt. Usually, flame retardants are directly mixed with asphalt at about 160 °C to prepare flame retardant asphalt, and then mixed with coarse and fine aggregates, mineral powder and fibers to form asphalt pavement. Most of the research results of flame retardant asphalts are applied to asphalt felt and asphalt coating. Among them, Jolitz and Kirk [3] used organic bromide, potassium citrate, amine and other flame retardants; Walter [4] used borate flame retardant; Grube and Frankoski [5] used borate flame retardant; Brown et al. [6] used bauxite, brucite, etc.; Graham [7] used halogen flame retardant; Slusher et al. [8] used intumescent flame retardants. In order to determine the flame retardant and mechanism of asphalt synergistic flame retardant by decabromodiphenyl ethane (DBDPE) and Sb_2O_3, Zuo et al. [9] studied the effects of DBDPE and Sb_2O_3 on the flame retardant and thermal decomposition characteristics of SBS asphalt. The test used was based on the results of the limiting oxygen index test, the smoke density test, and the thermal gravimetric and differential thermal test. Cong et al. [10] explains the mechanism of asphalt burning and elucidates the different forms of flame retardants. Xu et al. [11] conducted horizontal burning, limiting oxygen index (LOI) and direct burning tests to evaluate the effects of magnesium hydroxide (MH) on flame retardancy for asphalt. Jia et al. [12] evaluated the effect of three different combinations of flame retardant additives on an asphalt binder flaming using the oxygen index test method. Zhao et al. [13] studied the flame flame-retardant asphalt for a tunnel containing various kinds of flame retardants, including decabromodiphenyl oxide (DBDPO), DBDPE, Sb_2O_3, $ZnBO_3$, $Mg(OH)_2$, and $Al(OH)_3$ under different additive concentrations. Zhang et al. [14] conducted horizontal burning and limiting oxygen index (LOI) to evaluate the effects of mixed decabrombromodiphenyl ethane (DBDPE) and antimony trioxide (Sb_2O_3) on flame retardancy for epoxy asphalt binder. The influence of DBDPE/Sb_2O_3 on the rotational viscosity, the thermal and mechanical properties of epoxy asphalt binder was assessed by thermogravimetric (TG) analysis, differential scanning calorimetry (DSC), and a tensile test. Wu et al. [15] investigated the flame retardant mechanism of hydrated lime (HL) on asphalt mastics via a range of analytical techniques, including the cone calorimeter test, and thermogravimetry and differential scanning calorimetry analysis.

The structural performance of asphalt under heavy traffic loading is a vital aspect for pavement practitioners [16]. However, the effects of flame retardant have an unpropitious impact on the performance of the asphalt, and have not formed a special flame retardant system. There are still many problems, such as toxicity, smoke, and construction difficulties.

In recent years, the demand for functional pavements has kept increasing. New materials and technologies are being developed to improve the functionality of new roads. This innovative polyurethane thin overlay (PTO) was developed for the maintenance of existing roads and constructing new roads. The performance-related mechanical properties of the PTO specimens were evaluated with several preliminary methods in previous studies [17], and the basic strength of the specimen was tested through a uniaxial compression test. From this test, the PTO showed excellent mechanical behavior with roughly 6.6 GPa Youngs' modulus. Furthermore, dynamic stability was evaluated with cyclic compression tests. It presented a superior performance in the long-term durability of the materiel, which was more than ten times higher than conventional porous asphalt. The fatigue resistance was assessed by the indirect tensile strength test. From which test, the strong inherency tensile strength was also examined. Porous PTO can also facilitate stormwater infiltration and increase the driving safety and maneuverability of automobiles [18–21]. Apart from mechanical properties, the hydraulic conductivity and the durability against clogging have also been verified much larger than conventional porous pavement materials, due to excellent void connectivity and pore structures [22].

Based on the previous study, the concept of PTO may be a feasible and effective way to enrich the innovative functions of tunnel pavement. In this paper, the research aims to evaluate the functional properties of PTO. Apart from conventional functionalities, e.g., skid resistance and drainage, some properties, such as noise reduction, solar reflection, and especially combustion properties, have aroused great interest in tunnels of metropolitan cities. A conventional asphalt (open graded friction course (OGFC) and stone mastic asphalt (SMA)) and concrete pavement materials were used as control materials. Compared with conventional tunnel pavement materials, significant improvements were observed in mechanical and functional properties, as well as environmental performance. Therefore, this wearing layer can potentially provide pavements with new eco-friendly functions. This study provides a comprehensive analysis of these environmentally-friendly materials, paving the way for a possible application in tunnels, as well as some other fields, such as in race tracks in stadiums.

2. Methods

2.1. Materials and Preparation of the PTO Specimens

To achieve a high permeability, a void-rich thin overlay, such as OGFC, is currently the most feasible and effective way [23,24]. Although potentially beneficial, permeable pavement materials face some challenges, due to the high-void content (less stone-to-stone contact regions) and the viscous nature of bitumen. In conclusion, the poor mechanical durability [25] and unfavorable clogging behavior [24,26] represent the main obstacle inhibiting a wide application of permeable pavements [27]. The latest research substituted conventional bitumen with bio-based polyurethane (PU) in order to create a sustainable permeable pavement. The bio-based polyurethane consists of various polymers that are synthesized by a poly-addition reaction of a di-isocyanate or a polymeric isocyanate with a polyol. Specifically, among the polyol components, traditional petroleum raw materials are replaced by organic oils. The synthesis is based on the connection of isocyanates and hydroxyl groups that lead to the formation of a urethane group [28]. Polyurethane elastomers consist of the polyol component and the isocyanate component. By modifying the components, a wide range of material properties, ranging from brittle to elastic, can be designed. Preliminary research conducted at RWTH Aachen University (Germany) suggests that PU-bound porous pavement structures exhibit high permeability, high strength, high resistance to permanent deformation and increased fatigue resistance [17,29].

The mechanical and morphological properties of aggregate are essential for the functional and mechanical properties of conventional OGFC [30–32]. Various investigations have focused on using natural or recycled aggregate in pavements [33].

In producing the PTO mixture, the conventional natural aggregate within the particle size range of 2.0–5.6 mm was replaced, and the remaining 30% was filled with natural sand (0–0.2 mm). A 2-component polyurethane product was selected as the binder for the PU specimens. The basic components of PU are shown in Figure 1.

Figure 1. Raw materials for polyurethane (PU) specimens: (**a**) TiO_2-coated aggregate; (**b**) Bio-based polyurethane binder; (**c**) Compaction process; (**d**) PU specimen.

Conventional OGFC, SMA and cement were selected as reference materials in this study. The conventional OGFC and SMA specimens were composed of crushed diabase aggregate, limestone powder, and a polymer modified bitumen binder. The mixtures were prepared by means of Marshall compaction (50 impacts per side). The grain size distribution and detailed mixture design of the PU, OGFC and SMA reference specimens were based on the porosity and the maximum density process, which are given in Figure 2 and Tables 1–3 respectively.

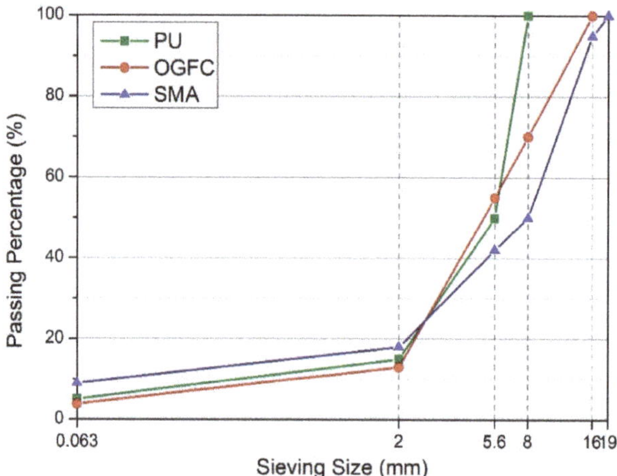

Figure 2. Grain size distribution of PU, open graded friction course (OGFC), and stone mastic asphalt (SMA).

Table 1. Mix design of PU mixture.

PU Materials	Grain Size (mm)	Mass Percentage (%)	Apparent Density (g/cm^3)
Limestone	0–0.063	5.0	2.820
Diabase	0.063–2	15.0	2.850
	2–5.6	52.0	2.850
	5.6–8	100.0	2.850
Polyurethane	2-component polyurethane, 6.5 M.–%		1.09
PU Mixtures	Air void content 28.9 Vol.–%		1.93

Table 2. Mix design of OGFC mixture.

OGFC Materials	Grain Size (mm)	Mass Percentage (%)	Apparent Density (g/cm^3)
Limestone	0–0.075	4.0	2.820
Basalt	0.075–2.36	10.2	2.820
	2.36–9.5	62.3	2.820
	9.5–16	23.5	2.820
Bitumen	Polymer modified bitumen 40/100–65 A, 4.5 M.–%		1.472
OGFC Mixture	Air void content 21.2 Vol.–%		2.090

Table 3. Mix design of SMA mixture.

SMA Materials	Grain Size (mm)	Mass Percentage (%)	Apparent Density (g/cm^3)
Limestone	0–0.075	10.0	2.820
Basalt	0.075–2.36	9.5	2.820
	2.36–9.5	35.5	2.820
	9.5–16	40.0	2.820
	16–19	5.0	
Bitumen	Polymer modified bitumen 40/100–65 A, 5.4 M.–%		1.472
SMA Mixture	Air void content 3.1 Vol.–%		2.471

The preparation of different PU specimens followed a similar procedure to hot-mix asphalt (OGFC and SMA). However, mixing polyurethane can be conducted at room temperature, because the polymerization reaction and viscosity of polyurethane are not strongly affected by temperature. After the two components of polyurethane were thoroughly mixed, the binder is added to the aggregate. The components are mixed for a few minutes to obtain a homogenous mixture in which all surfaces of the aggregate are coated with binder. After mixing, a pre-determined amount of polyurethane-bound mixture is placed into a mold to obtain specimens of the desired bulk density. A heavy roller is used for compaction of the mixture. After approximately 24 h, the hardening process is completed, and the specimens can be removed from the mold.

Another reference specimen had a grade of C40/10 which is conventionally used in PCC pavements in Germany. Table 4 shows the batching proportions for C40/10 used in this study. The design slump values for PCC mixtures were 80.

Table 4. Mix design of the cement (kg/m^3).

Cement	Pulverized Fuel Ash	20 mm	10 mm	Fines	Water	Aggregate Cement Ratio	Water Cement Ratio
330	110	725	345	620	185	3.84	0.42

2.2. Acoustic Performance

Existing studies have shown that tire-road noise mainly happens based on the inner resonance of the tire surface and pavement cavities (Pcavity), air flow around the vehicle body (Vehicle) and tire vibrations (Pvibration) [34]. Pcavity, which determines the noise absorption, is researched by a huge amount of research as the most important factor when evaluating the acoustic performance of the PU specimens [20,35].

In order to examine the sound absorption capacity of pavement material, the impedance tube test according to DIN EN ISO 10534-2 was adopted, which has been widely applied for noise absorption evaluation in previous studies [36,37]. This test method uses an impedance tube, two microphones, an amplifier, and a recorder to evaluate the sound absorption coefficient (see Figure 3). Due to the shape of the tube, the sound waves propagate as flat waves inside the tube. A frequency sweep is generated and played back via the connected amplifier and loudspeakers. The generated sound frequencies are measured with the microphones installed at the tube. The acoustic transfer functions of the two microphone signals are used to calculate the reflection factor and the absorption factor at normal incidence and the impedance ratio of the test material according to DIN EN ISO 10534-2 (https://www.perinorm.com/document.aspx). All test samples, including PU, OGFC, SMA, and cement, were regulated in the same dimension of normal pavement samples with 100 mm diameter and 40 mm height, so that it corresponds to the practical layer thickness. There are three parallel test pieces of each variant; all specimens were measured three times each.

(a)   (b)

Figure 3. Installing the impedance tube for characterizing the noise absorption of PU samples: (**a**) Connecting the sample holders; (**b**) completely installed impedance tube device.

2.3. Reflection Test

The reflection rate is an indicator of how reflective a pavement is of solar radiant. Higher reflection rate means more radiant energy could be reflected back into the air, and pavement will absorb less energy. It has been reported that the reflection rate depends strongly on pavement surface color, service condition, and texture features. In general, the reflection rate of pavement falls in the range of 0.10–0.30 [33], with brighter pavements higher reflection rates. The reflection rate may be not a big problem under a tunnel section, but it is worthy of being investigated for the application of the innovative material in the entrance and exit sections of the tunnel, where is not sealed under the tunnel section. Furthermore, these results can offer comprehensive information and may benefit the readers who would like to apply the innovative material in other practical fields.

In this research, UV/VI/IR Spectrophotometer (see Figure 4), a device commonly applied to detect substance based on the absorption spectrum, was used to test the reflection rate of different samples. Samples were radiated by different lights within the wavelength range of 400–2000 nm with a step length of 5 nm, followed by detection of the reflected energy. Then, the reflection rate of different wavelength light was automatically calculated.

Figure 4. UV/VI/IR Spectrophotometer.

To further test the samples' heat reflection, a specially-designed iodine tungsten lamp device (Figure 5), with a radiant energy of 820 W/m^2, was used to simulate solar radiation. All samples were surrounded by thermal insulation cotton in the bottom and four sides. The temperatures were measured by a sensor placed 2.5 cm in the sample. The setup of this test was defined according to our previous research, which can efficiently test and evaluate the samples' heat reflection in the laboratory. A calibration factor or shift factor which can convert the laboratory results to the field will be studied in the future.

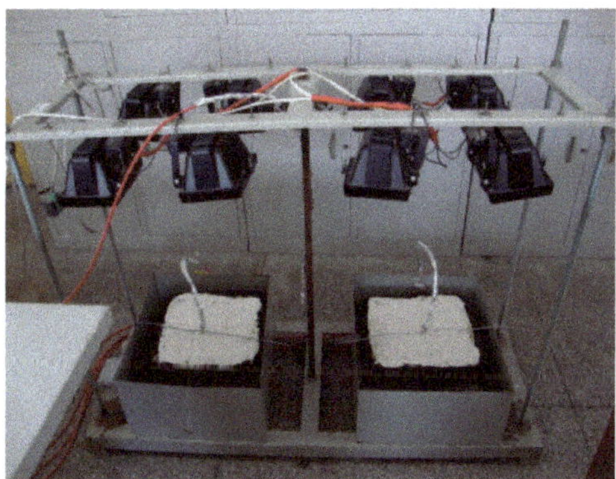

Figure 5. Simulation of solar radiation using an iodine tungsten lamp.

2.4. Flammability Evaluation Method

The ignition of a vehicle is the main cause of the tunnel fire. However, for the pavement engineers, how to reduce the harm caused by the ignition of the pavement materials is the main concern. Especially, the smoke may be extremely harmful to the drivers trapped within the tunnel. Therefore, the combustion performance of the materials was evaluated in this study. The cone calorimeter was designed by Dr. Babrauskas of the National Institute of Standards and Technology (NIST) in 1982 based on the principle of oxygen consumption. It is a vital test instrument for evaluating materials' combustion performance. As shown in Figure 6, the cone calorimeter is mainly composed of a carrier, a combustion chamber, a ventilation system, a flue gas measuring system and a gas analyzer.

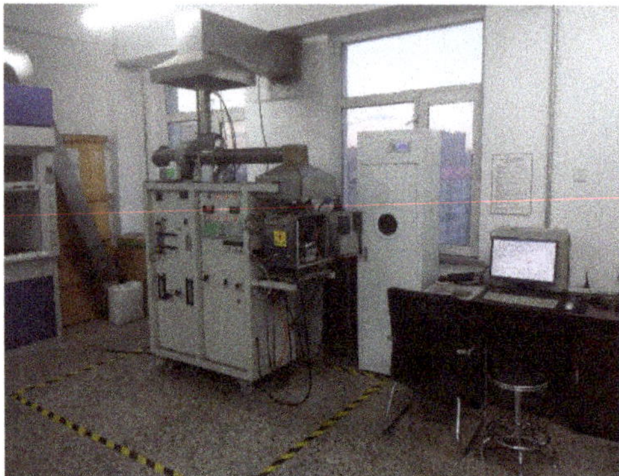

Figure 6. Appearance and components of the cone calorimeter.

The main test parameters it can include the heat release rate (HRR), total heat release (THR), effective combustion heat (EHC), ignition time (TTI), smoke and toxicity parameters, and mass change parameters (MLR). These parameters can be used to evaluate the combustion performance or flame retardancy of materials. Compared with the traditional test method, the cone calorimeter can get more data in one experiment. In addition, the combustion test environment of the cone calorimeter is similar to the real combustion environment, and the test results have a good correlation with the real conflagration, which has a good reference value for the evaluation of the combustion performance of materials.

In this test, it is required that the cross section of the tested specimen be 100 mm × 100 mm square, and the mass should not exceed 200 g. The experimental power of the cone calorimeter is 50 kW/m^2, and the corresponding temperature is 780 °C. Before the test, the side and bottom of the sample were wrapped with aluminum foil, and then the measured sample will be forced to ignite under 50 kW/m^2 thermal radiation intensity. The data obtained during combustion will be collected by a computer.

3. Results and Discussion

3.1. Results of the Acoustic Test

The acoustic absorption coefficients of all specimens are shown in Figure 7. The figure indicates that the acoustic behavior of the four types of material is distinctively different. In general, the acoustic absorption properties of porous pavement material (both PU 8 and OGFC) are far higher than the conventional SMA and concrete specimens. In the relevant frequency range of 800–2500 Hz, PU specimens show a peak absorption coefficient of about 84% at 1400 Hz, whereas the OGFC specimens show a peak absorption coefficient of about 70% at 1000 Hz. The SMA and concrete only show a relatively lower peak noise absorption about 30% and 20% at 800 and 1080 Hz respectively. However, the noise frequency near pavement is usually in the range from 1200 Hz to 1600 Hz if the travel speed around 70 [34,36]. In which case, the PU material can have the highest noise absorption property among all materials. The peak noise absorption frequency is a function of the sample height and the pore structures. During the test, the specimens of PU, OGFC, SMA and concrete were kept in the same height of 4 mm. In this case, the difference can be mostly attributed by the porosity and pore structures. The absorption coefficients of PU samples surpass that of the other samples throughout the entire frequency domain. In comparison with the conventional OGFC, the PU exhibits higher coefficients of absorption across a wider range of frequencies, rendering it a material with far superior acoustic properties. The excellent noise reduction ability is mainly due to the large connective void content within the PU, which can expand the frequency range of absorption. Therefore, the PU material is the most suitable for the application in tunnel pavement, due to the relatively high noise frequency level.

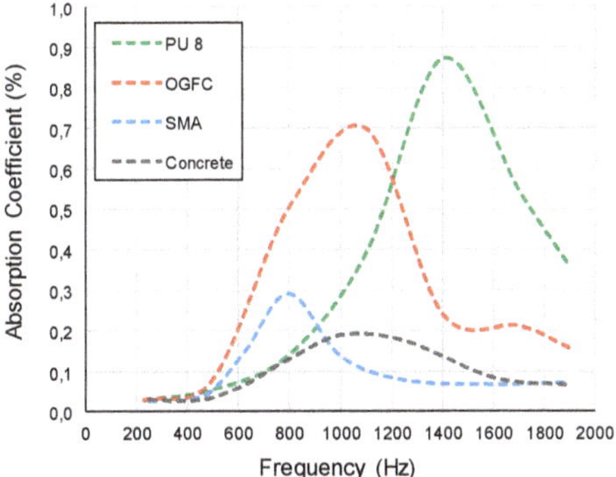

Figure 7. Acoustic absorption-coefficient curves.

3.2. Results of Heat Reflection

Figure 8 presents the results of the heat reflection tests subjected to light in the wavelength of 400 to 2000 nm. It can be observed that the light reflection rate varied significantly among different surfaces. Freshly manufactured asphalt, which is black, has the lowest light reflection in the whole range of wavelength. In other words, the majority of solar radiation will be absorbed by asphalt, thus increasing the road surface temperature. In which case, the OGFC and SMA, that use bitumen as a binder material, exhibit lower heat reflectance values, which in most cases is less than 0.1%. The surface temperature of OGFC and SMA is also relatively high, reaching 74 degrees after 6h solar radiation.

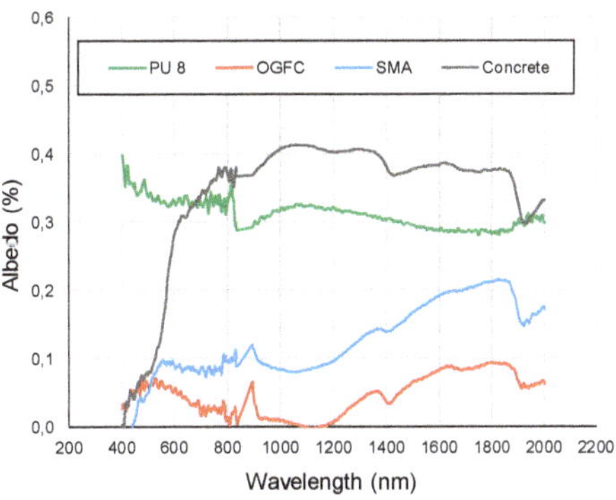

Figure 8. Simulation of solar radiation using iodine tungsten lamp.

In contrast, PU and granite have the highest light reflection rates in wavelength rage of 800 to 2000 nm. PU experiences a continuously upward trend in terms of heat reflection and eventually reaches a peak at 2000 nm, with a reflection rate of around 0.5. Infrared mainly contributes to the most thermal effect of light, especially in practice. The infrared wavelength is in the range of 760–2500 nm, which falls into the highest reflection wavelength range of PU and granite. In this case, PU and granite in the actual application of heating rate are lower.

When exposed to the same simulated solar radiant energy of 820 W/m^2, the surface temperature of PU is lower than that of normal asphalt surface during the testing period (see Figure 9). After about five hours of heating, the temperature of both samples becomes steady. The eventual temperature of PU was 20 °C lower than that of the normal asphalt surface after ten hours of heating. The result indicates that PU, which is brighter, provides higher heat reflection rate in comparison with the black normal asphalt surface, which significantly reduces the pavement surface temperature. Therefore, this innovative pavement surface treatment method can potentially lessen the urban heat island effect.

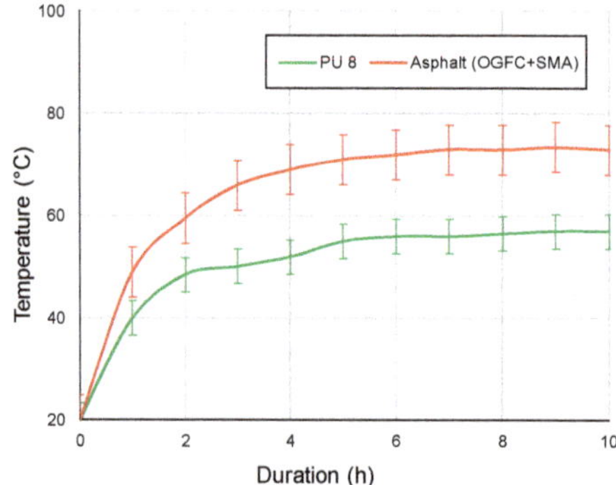

Figure 9. The different reflection results of a heat reflection surface.

3.3. Results of Combustion Tests

3.3.1. Ignition Time (TTI)

TTI is an important parameter for evaluating the combustion properties of materials. It refers to the time spent from heating the surface of materials to continuous combustion at a preset incident heat flux, and its unit is second. It can be used to evaluate and compare the refractory properties of materials. The longer the TTI is, the harder the material is to ignite under the specified experimental conditions.

As shown in Figure 10, the TTI of OGFC is smaller than that of SMA. This is because the air void of OGFC is larger than that of SMA, resulting in the larger exposed asphalt area and air contact area, thus it is easier to ignite. The TTI of porous polyurethane mixture is larger than that of OGFC and SMA, indicating that PU is more difficult to ignite than asphalt. Because there is no obvious ignition phenomenon of cement, it is regarded as incapable of ignition, and its TTI is not shown in the figure.

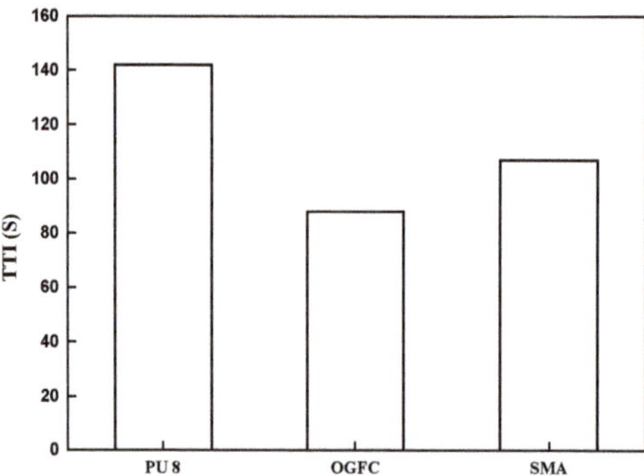

Figure 10. Comparison of ignition time (TTI) among PU, OGFC and SMA.

3.3.2. Heat Release Rate (HRR)

HRR refers to the heat release rate per unit area after the material is ignited under the preset radiation intensity, and its unit is kW/m^2. The maximum value of HRR is the peak value of the heat release rate (pkHRR). The peak value indicates the maximum degree of heat release during combustion.

The greater the HRR and pkHHR are, the greater the heat released from the burning of the material, and the greater the fire hazard.

Figure 11 shows that the heat release rate curve of OGFC, with the highest pkHRR of 120.38 kW/m², is relatively left-centralized, meaning the quickest complete combustion and the greatest risk of fire. The HRR curve of SMA, with the second largest pkHRR, basically encloses that of PU, which indicates that SMA emits more heat per unit time and the heat release of it lasts for a longer time than that of PU. By comparing the curves of PU, SMA and OGFC, it can be seen that PU has better flame retardancy than asphalt. Finally, the HRR curve of cement fluctuates at 0, indicating that the cement will not be ignited.

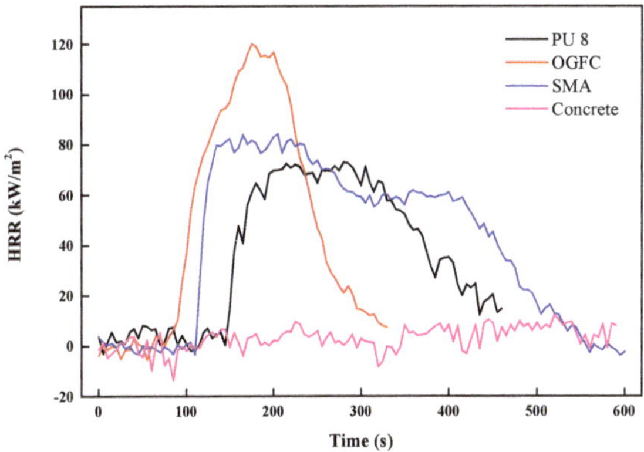

Figure 11. Comparison of heat release rate (HRR) among PU, OGFC and SMA.

3.3.3. Total Heat Release (THR)

THR refers to the sum of heat released by materials from ignition to flame extinction at a preset incident heat flux in the unit of MJ/m². Combining HRR with THR can better evaluate the combustibility and flame retardancy of materials, which has a more objective and comprehensive guiding role for fire research.

Figure 12 shows that SMA has the largest THR of 24.34 MJ/m², because it has the largest asphalt content. Besides, the THR of OGFC is 16.73 MJ/m², close to that of PU of 16.04 MJ/m², but the ignition time of OGFC is small, and the heat release of it is concentrated and intense. Therefore, asphalt pavement is more dangerous than polyurethane pavement when they are on fire. Finally, the total heat released of cement is very small but still exists, because the cement has only seven days age and has not been fully hydrated.

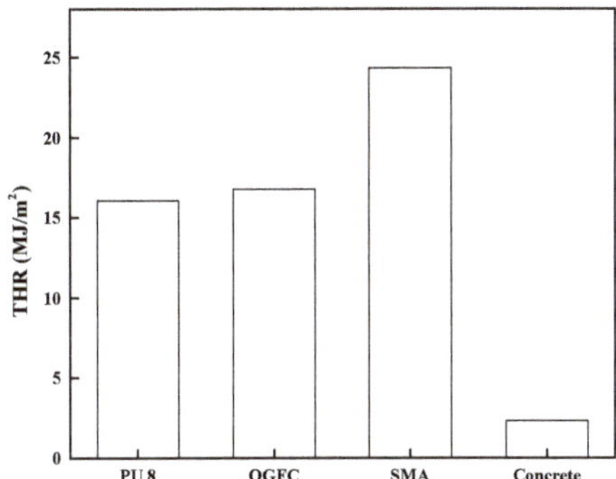

Figure 12. Comparison of total heat release (THR) among PU, OGFC and SMA.

3.3.4. Specific Extinction Area (SEA) and Total Smoke Release (TSR)

SEA is a dynamic parameter to characterize the amount of smoke emitted at every moment in the combustion process, which can reflect the ratio of volatile matter per unit mass to smoke (unit: m^3/kg), while the TSR can reflect the total amount of smoke generation and release per unit area in the fire field (unit: m^3/m^2). These data have a good correlation with the smoke parameters of large-scale experiments.

Figure 13 shows that SMA has the largest average smoke emission and total smoke emission, followed by OGFC, PU and cement. In fires, excessive smoke may lead to asphyxiation, hypoxia and death. Therefore, asphalt pavement is more harmful to the environment and the human body than polyurethane pavement when burning.

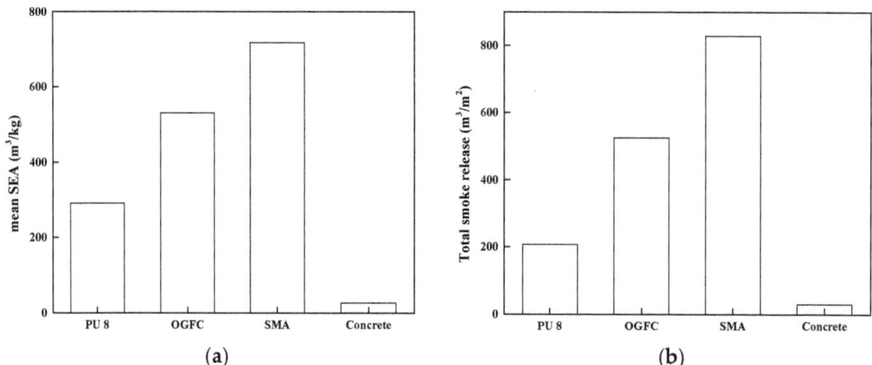

Figure 13. Comparison of the results among PU, OGFC and SMA. (**a**) mean for the specific extinction area (SEA); (**b**) total smoke release (TSR).

3.3.5. Fire Proceeding Index (FPI)

FPI combines ignition time and peak heat release rate, and is the ratio of ignition time to peak heat release rate (unit: $s*m^2/kW$). The larger the FPI value, the stronger the fire resistance.

Figure 14 shows that the FPI of OGFC is smaller than that of SMA, and the FPI of asphalt mixture is smaller than that of polyurethane mixture, which indicates, as previously analyzed, that polyurethane pavement has better flame retardancy than asphalt pavement. Besides, the pkHRR of cement is almost zero, so its FPI value is extraordinary and not shown in the figure.

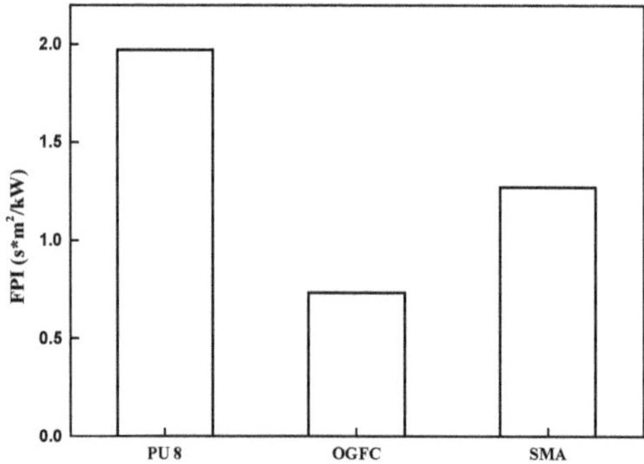

Figure 14. Comparison of fire proceeding index (FPI) among PU, OGFC, and SMA.

4. Summary and Conclusions

In this paper, the functional properties of PTO were evaluated, and compared, with the conventional asphalt (OGFC and SMA) and concrete pavement materials regarding the noise reduction, solar reflection and combustion properties. Significant improvements were observed in functional properties, as well as environmental performance. The following summarizes the main findings of this study:

- Based on the evaluation by the acoustic tube test, both PU and OGFC show superior acoustic properties in comparison with conventional SMA and concrete pavement materials. However, comparing to OGFC, PU has a wider range of noise absorption. Especially for high frequency noise, which more likely exists in the tunnel, PU exhibits a maximum noise absorption coefficient, while other materials have almost no noise absorption within this range. Concluding, PU is more efficient in noise reduction of tunnel pavement.
- The light reflection rate varied significantly among different pavement surfaces. Based on the heat reflection tests, the OGFC and SMA which use bitumen as a binder material, exhibit the lowest heat reflectance values. The concrete presented the highest reflection rate and followed by PU in wavelength it ranged from 800 to 2000 nm. On the other hand, an increase of the radiation time results in a significant increase in the surface temperature of the PU and asphalt material. However, compared to the asphalt material, the increase of temperature on the PU surface is almost 30% less. Therefore, this PU pavement surface can potentially lessen the urban heat island effect. It is meaningful when the PU is applied in the entrance and exit sections of the tunnel, where it is not completely sealed under the tunnel section.
- By comparing the results of the combustion tests, it can be seen that PU has a better flame retardancy than asphalt (OGFC and SMA). Particularly, the TTI of PU is larger than that of OGFC and SMA, indicating that PU is more difficult to ignite than asphalt. Asphalt (OGFC and SMA), with the higher pkHRR than PU, can result in faster combustion and a greater risk of fire. SMA has the largest THR, indicating that the largest amount of heat is released. The THR of OGFC is close to that of PU, but the ignition time of OGFC is small, and thus the heat release is more concentrated and intense. SMA has the largest average smoke emission and total smoke emission, followed by OGFC, PU and cement. FPI of asphalt is smaller than that of PU.

Overall, this study has demonstrated that this innovative wearing layer can potentially provide tunnel pavements with new eco-friendly functions compared with the conventional asphalt materials. In further research, the PTO will be applied and analyzed in the construction of model tunnels. The relative production and laydown direct cost relation for all the mixtures will be investigated in the next step. More comprehensive tests will be carried out to investigate the mechanical and functional properties of the PTO materials, i.e., long-term performance of PTO under repeated dynamic loading (traffic loading in the field) needs to be investigated carefully through dynamic fatigue and fracture tests [38]. The possibility of the PTO applied in race tracks in the stadiums will also be investigated.

Author Contributions: Conceptualization and original draft preparation, C.L. and P.L.; revision, G.L. and X.X.; methodology, D.W.; data curation, J.G.

Funding: This project was carried out at the request of the German Research Foundation (DFG), under research project Grant No. OE 514/4-2 and Grant No. FOR 2089/2 (OE514/1-2).

Conflicts of Interest: The authors declare no conflict of interest.

References

1. Zhang, J.; Wang, M.; Wang, D.; Li, X.; Song, B.; Liu, P. Feasibility study on measurement of a physiological index value with an electrocardiogram tester to evaluate the pavement evenness and driving comfort. *Measurement* **2018**, *117*, 1–7. [CrossRef]
2. Liu, P.; Otto, F.; Wang, D.; Oeser, M.; Balck, H. Measurement and evaluation on deterioration of asphalt pavements by geophones. *Measurement* **2017**, *109*, 223–232. [CrossRef]

3. Jolitz, R.J.; Kirk, D.R. Flame Retardant Asphalt Composition. U.S. Patent 4,804,696, 14 February 1989.

4. Walters, R.B. Flame Retarded Asphalt Blend Composition. U.S. Patent 4,659,381, 21 April 1987.

5. Grube, L.L.; Frankoski, S.P. Flame Retardant Bitumen. U.S. Patent 5,110,674, 5 May 1992.

6. Brown, S.; Mead, N.; Evans, K.; Rodriguez, C.E.; Dayer, A.J. Bauxite Flame-Retardant Fillers for Insulators or Sheathing. U.S. Patent 6,252,173, 26 June 2001.

7. Graham, J. Flame Resistant Ashaltic Compositions. U.S. Patent 4,512,806, 23 April 1985.

8. Slusher, C.C.; Ogren, E.A.; Gorman, W.B.; Thompson, G.S.; Kane, E.G.; Usmani, A.M. Flame Retardant Modified Asphalt-Based Material and Products Therefrom. U.S. Patent 5,516,817, 14 May 1996.

9. Zuo, J.D.; Li, R.X.; Feng, S.H.; Liu, G.Y.; Zhao, J.Q. Flame retardancy and its mechanism of polymers flame retarded by DBDPE/Sb$_2$O$_3$. *J. Cent. South Univ. Technol.* **2008**, *15*, 64–68. [CrossRef]

10. Cong, P.; Yu, J.; Wu, S.; Luo, X. Laboratory investigation of the properties of asphalt and its mixtures modified with flame retardant. *Constr. Build. Mater.* **2008**, *22*, 1037–1042. [CrossRef]

11. Xu, T.; Huang, X.; Zhao, Y. Investigation into the properties of asphalt mixtures containing magnesium hydroxide flame retardant. *Fire Saf. J.* **2011**, *46*, 330–334. [CrossRef]

12. Wu, S.; Cong, P.; Yu, J.; Luo, X.; Mo, L. Experimental investigation of related properties of asphalt binders containing various flame retardants. *Fuel* **2006**, *85*, 1298–1304. [CrossRef]

13. Zhao, H.; Li, H.P.; Liao, K.J. Study on properties of flame retardant asphalt for tunnel. *Pet. Sci. Technol.* **2010**, *28*, 1096–1107. [CrossRef]

14. Zhang, Y.; Pan, X.; Sun, Y.; Xu, W.; Pan, Y.; Xie, H.; Cheng, R. Flame retardancy, thermal, and mechanical properties of mixed flame retardant modified epoxy asphalt binders. *Constr. Build. Mater.* **2014**, *68*, 62–67. [CrossRef]

15. Wu, K.; Zhu, K.; Kang, C.; Wu, B.; Huang, Z. An experimental investigation of flame retardant mechanism of hydrated lime in asphalt mastics. *Mater. Des.* **2016**, *103*, 223–229. [CrossRef]

16. Pitawala, S.; Sounthararajah, A.; Grenfell, J.; Bodin, D.; Kodikara, J. Experimental characterisation of fatigue damage in foamed bitumen stabilised materials using dissipated energy approach. *Constr. Build. Mater.* **2019**, *216*, 1–10. [CrossRef]

17. Lu, G.; Renken, L.; Li, T.; Wang, D.; Li, H.; Oeser, M. Experimental study on the polyurethane-bound pervious mixtures in the application of permeable pavements. *Constr. Build. Mater.* **2019**, *202*, 838–850. [CrossRef]

18. Li, H.; Harvey, J.T.; Holland, T.J.; Kayhanian, M. The use of reflective and permeable pavements as a potential practice for heat island mitigation and stormwater management. *Environ. Res. Lett.* **2013**, *8*, 015023. [CrossRef]

19. Tennis, P.D.; Leming, M.L.; Akers, D.J. *Pervious Concrete Pavements*; No. PCA Serial No. 2828; Portland Cement Association: Skokie, IL, USA, 2004.

20. Lu, G.; Liu, P.; Wang, Y.; Faßbender, S.; Wang, D.; Oeser, M. Development of a sustainable pervious pavement material using recycled ceramic aggregate and bio-based polyurethane binder. *J. Clean. Prod.* **2019**, *220*, 1052–1060. [CrossRef]

21. Liu, P.; Xing, Q.; Wang, D.; Oeser, M. Application of dynamic analysis in semi-analytical finite element method. *Materials* **2017**, *10*, 1010. [CrossRef] [PubMed]

22. Törzs, T.; Lu, G.; Grabe, J.; Wang, D.; Oeser, M. Hydraulic properties of polyurethane-bound permeable pavement materials considering unsaturated flow. *Constr. Build. Mater.* **2019**, *212*, 422–430. [CrossRef]

23. Hill, A.R.; Dawson, A.R.; Mundy, M. Utilisation of aggregate materials in road construction and bulk fill. *Resour. Conserv. Recycl.* **2001**, *32*, 305–320. [CrossRef]

24. Oeser, M.; Hovagimian, P.; Kabitzke, U. Hydraulic and mechanical properties of porous cement-stabilised materials for base courses of PICPs. *Int. J. Pavement Eng.* **2012**, *13*, 68–79. [CrossRef]

25. Mo, L.; Huurman, M.; Wu, S.; Molenaar, A.A.A. Ravelling investigation of porous asphalt concrete based on fatigue characteristics of bitumen–stone adhesion and mortar. *Mater. Des.* **2009**, *30*, 170–179. [CrossRef]

26. Scholz, M.; Grabowiecki, P. Review of permeable pavement systems. *Build. Environ.* **2007**, *42*, 3830–3836. [CrossRef]

27. Zhang, J.; Cui, X.; Li, L.; Huang, D. Sediment transport and pore clogging of a porous pavement under surface runoff. *Road Mater. Pavement Des.* **2017**, *18* (Suppl. 3), 240–248. [CrossRef]

28. Prisacariu, C. *Polyurethane Elastomers: From Morphology to Mechanical Aspects*; Springer Science & Business Media: Berlin/Heidelberg, Germany, 2011.

29. Renken, L.; Oeser, M. Innovative Baustoffkonzepte-Anwendungspotenziale und Charakterisierung von synthetischen Strassenbefestigungen. In *Aachener Mitteilungen Strassenwesen, erd-und Tunnelbau*; Institute of Highway Engineering, RWTH Aachen University: Aachen, Germany, 2014.

30. Xie, X.; Lu, G.; Liu, P.; Wang, D.; Fan, Q.; Oeser, M. Evaluation of morphological characteristics of fine aggregate in asphalt pavement. *Constr. Build. Mater.* **2017**, *139*, 1–8. [CrossRef]

31. Wang, D.; Wang, H.; Bu, Y.; Schulze, C.; Oeser, M. Evaluation of aggregate resistance to wear with Micro-Deval test in combination with aggregate imaging techniques. *Wear* **2015**, *338*, 288–296. [CrossRef]

32. Wang, H.; Wang, D.; Liu, P.; Hu, J.; Schulze, C.; Oeser, M. Development of morphological properties of road surfacing aggregates during the polishing process. *Int. J. Pavement Eng.* **2017**, *18*, 367–380. [CrossRef]

33. Wang, D.; Leng, Z.; Hüben, M.; Oeser, M.; Steinauer, B. Photocatalytic Pavements with Epoxy-Bonded TiO$_2$-Containing Spreading Material. *Constr. Build. Mater.* **2016**, *107*, 44–51. [CrossRef]

34. Sun, W.; Lu, G.; Ye, C.; Chen, S.; Hou, Y.; Wang, D.; Wang, L.; Oeser, M. The state of the art: Application of green technology in sustainable pavement. *Adv. Mater. Sci. Eng.* **2018**, *2018*, 9760464. [CrossRef]

35. Freitas, E.; Mendonça, C.; Santos, J.A.; Murteira, C.; Ferreira, J.P. Traffic noise abatement: How different pavements, vehicle speeds and traffic densities affect annoyance levels. *Transp. Res. Part D Transp. Environ.* **2012**, *17*, 321–326. [CrossRef]

36. Wang, D.; Liu, P.; Leng, Z.; Leng, C.; Lu, G.; Buch, M.; Oeser, M. Suitability of PoroElastic Road Surface (PERS) for urban roads in cold regions: Mechanical and functional performance assessment. *J. Clean. Prod.* **2017**, *165*, 1340–1350. [CrossRef]

37. Wang, D.; Schacht, A.; Leng, Z.; Leng, C.; Kollmann, J.; Oeser, M. Effects of material composition on mechanical and acoustic performance of poroelastic road surface (PERS). *Constr. Build. Mater.* **2017**, *135*, 352–360. [CrossRef]

38. Liu, P.; Chen, J.; Lu, G.; Wang, D.; Oeser, M.; Leischner, S. Numerical Simulation of Crack Propagation in Flexible Asphalt Pavements Based on Cohesive Zone Model Developed from Asphalt Mixtures. *Materials* **2019**, *12*, 1278. [CrossRef]

 *materials* 

Article

Study on Compatibility and Rheological Properties of High-Viscosity Modified Asphalt Prepared from Low-Grade Asphalt

Mingliang Li [1,*], Feng Zeng [2], Ruigang Xu [3,*], Dongwei Cao [4] and Jun Li [1]

1   Road Engineering Research Center, Research Institute of Highway Ministry of Transport, Xitucheng Road, No. 8, Haidian District, Beijing 100088, China; jun.li@rioh.cn
2   China Road & Bridge Corportation, Andingmenwai Street, No. 88, Beijing 100011, China; zengf@crbc.com
3   China Railway Third Bureau Group Co., Ltd. Survey and Design Branch, Jinsong Road, No. 16, Yingze District, Taiyuan 030001, Shanxi Province, China
4   Zhonglugaoke (Beijing) Road Technology Co. Ltd., Xitucheng Road, No. 8, Beijing 100088, China; caodongwei@vip.126.com
*   Correspondence: ml.li@rioh.cn (M.L.); reygandxu@163.com (R.X.); Tel.: +86-158-103-39871 (M.L.)

Received: 11 September 2019; Accepted: 15 November 2019; Published: 17 November 2019

Abstract: High-viscosity modified asphalt is mainly used as a binder for porous asphalt in China and Japan. In order to meet the demand for using porous asphalt under high temperature condition in Africa, high-viscosity asphalt made from low-grade matrix asphalt, which is commonly used in Africa is investigated. Based on simulation of local climate in Africa, the suitable range of high viscosity additive content for different matrix asphalt was obtained by analyzing dynamic viscosity of the asphalt. Through PG high temperature grading, multi-stress repeated creep, accelerated fatigue, temperature sweep and other tests, changes of high temperature, anti-fatigue and anti-shear indicators before and after modification were compared and analyzed and effects of different matrix asphalt were also studied. Finally, considering engineering requirements, mixing and compaction temperatures of various high-viscosity modified asphalt were determined through study of viscosity-temperature characteristics. This research provides a support for preparation of high-viscosity modified asphalt and porous asphalt mixture by using low grade asphalt. The research achievements can help to guide the material design and application of porous asphalt in Africa and other high temperature areas.

Keywords: high-viscosity modified asphalt; low grade asphalt; porous asphalt; rheological property; viscosity-temperature characteristic

---

1. Introduction

Compared with dense-graded asphalt pavement, porous asphalt has the advantages of reducing water mist in rainy days, increasing driving safety, anti-skid, noise reduction and effectively alleviating urban heat island effect because of its large voids content. However, during the application of porous asphalt, performance of the pavement is highly dependent on materials and environment where it is used. When studying the performances of porous asphalt, it is necessary to considering the combined effect of traffic, the climate of the project site, as well as the material properties.

At the beginning of the application of porous asphalt, straight run asphalt or natural asphalt was mainly used as binders. For example, 40/60 straight-run was generally used in the United States. Due to the low viscosity of such binders, porous asphalt was easily damaged, which limited its promotion. And many countries began to use modified asphalt, which greatly improved service life of porous asphalt [1]. Based on improvement and optimization of European experience and according to climate and traffic characteristics of various regions in the country, the Japanese road engineering community focused on study of binders and proposed high-viscosity modified asphalt (HVMA). And it became a key technology for high temperature stability and anti-raveling of porous asphalt in the country [2–4]. At present, HVMA is also use as binder for porous asphalt in China and the dynamic viscosity at 60 °C can be over 100,000 Pa·s [5]. High-viscosity modified asphalt can be prefabricated by blending matrix asphalt and high-viscosity additive (HVA), which is called "wet method" of modification and the technology by putting HVA directly during mixing process of mixture is called "dry method." The HVA is generally a particle shape, with a length between 2~5 mm. It can be melted and dispersed rapidly and evenly in the mixing process of asphalt mixture.

In previous studied, it has been revealed that the main component of HVA is thermoplastic rubber, which forms a polymer network structure between the polymer in the HVA and the asphalt component [6], which generally results a higher viscosity in comparison of SBS modified asphalt. Qin et al. used temperature sweep and frequency sweep tests to investigate the influence of temperature and frequency on anti-rutting performance of high-viscosity modified asphalt and the methods are considered better presents a practical load in the road surface [7]. Tan et al. studied the composite modification of matrix asphalt using a thermoplastic elastomer and SBS polymer modified asphalt. It found that the thermoplastic elastomer particles as high elastic interlocking units are uniformly distributed in the network structure in the modified asphalt, which leads to good high and low temperature performance based on Performance Grade (PG) grading [8]. Cai et al. carried out research on environmental friendly alternative binders for permeable asphalt mixture by recycling engineering wastes including crumb rubber powder and recycled oil to prepare high-viscosity asphalt binders. The results showed that the performance can meet the specification requirements in China [9]. Xu et al. investigated the ageing mechanism of the high-viscosity asphalt and developed rejuvenator material for preventive maintenance [10]. In addition, studies were also performed on the noise reduction properties of porous asphalt using HVMA as a binder [11,12]. However, it showed that the high-viscosity binder mainly improves the mechanical performance of the pavement and has less influence of sound absorption. In existing studies, the high-viscosity asphalt are mainly made from matrix asphalt with grade 70 or composite modification of SBS asphalt. The compatibility of low-grade asphalt, the penetration at 25 °C of which is generally 30~50 (0.1 mm), for preparing high-viscosity asphalt are not studied yet. And there is also little knowledge on performances of HVMA made from low grade asphalt.

This paper focuses on rainy environments in Africa, which is characterized by high temperature in summer and heavy rainfall in spring and summer. When used in such conditions, theporous asphalt mixture should have sufficient resistance to high temperature stability, moisture damage and structural durability. As the average lowest temperature in such areas are higher than 20 °C, the low temperature properties are not taken into account in this study. At the same time, different from existing studies in China and Japan, matrix asphalt used in Africa is mainly low-grade asphalt due to high temperatures in summer. Besides, the performance and chemical composition of matrix asphalt from different sources are also quite different. Therefore, study is carried out for verifying the feasibility of preparing high-viscosity modified asphalt based on low-grade matrix asphalt and the performances of these high-viscosity asphalt used for African highways are investigated. The research will provide a technical support for using HVMA in Africa and other areas with high temperatures in summer and it contributes to promotion and application of porous asphalt based on HVMA.

2. Materials and Methods Study on Optimum Mixing Content of HVA

The mixing proportion of HVA is the main factor affecting the performance of high-viscosity modified asphalt. Therefore, the optimum mixing content of HVA for four low-grade matrix asphalt commonly used in Africa was studied first.

2.1. Matrix Asphalt

The indicators of the four types of matrix asphalt used in West Africa are shown in Table 1. The four types of asphalt are numbered B1, B2, B3 and B4 respectively. Tests were carried out according to European Standard BS EN12591-2009. The chemical analysis results of the four-component tests are shown in Table 2. It can be seen from Tables 1 and 2 that asphaltene content in B1 and B4 asphalt is high, resulting in relatively high softening point but penetration index after aging is relatively low. Asphaltene content in B3 asphalt is the lowest, so its softening point is relatively low but performance declines least after aging.

Table 1. Test results on matrix asphalt.

Items	Unit	Technical Requirement	Test Value			
			B1	B2	B3	B4
Penetration (25 °C, 100 g, 5 s)	0.1 mm	30–45	44	37	40	41.1
Softening point values (TR&B)	°C	52–60	54	53.5	50.8	55.4
Flashpoint	°C	≥240	322	316	308	346
Solubility	%	≥99	99.79	99.92	99.69	99.83
Dynamic viscosity (60 °C)	Pa·s	≥260	485	394	265	566
Residue mass change after TFOT	%	≤0.5				
Residual penetration (25 °C)	0.1 mm	-	31	25.9	32.6	27.5
Penetration ratio	%	≥53	70.5	70	81.5	67

Table 2. Test results of four-component for matrix asphalt.

Asphalt Type	Asphaltene (%)	Saturate (%)	Aromatic (%)	Resin (%)
B1	20.74	13.93	39.20	23.50
B2	16.10	9.80	47.32	25.43
B3	11.28	8.26	51.35	28.84
B4	17.05	14.16	40.03	27.41

2.2. Preparation of HVMA

(1) Test equipment

The equipment required for preparation of high-viscosity asphalt binder includes high-speed shear mixer, heating furnace, thermometer, glass stirring rod. Preparation of high-viscosity asphalt using high-speed shear mixer is shown in Figure 1.

(2) Preparation Process

Preparation of high-viscosity modified asphalt in laboratory is as follows:

(1) Take a certain amount of matrix asphalt and heat the asphalt to about 180 °C or 190 °C (for SBS modified asphalt), then weigh the asphalt. Calculate the amount of high-viscosity additive according to the proportion of the designed content and add it into the asphalt, mix it evenly with glass rod;

(2) Place the sample cup under the high-speed shear machine, set rotation speed to 5000 rpm and shearing shall be continued for 30 min. The temperature during the whole process is controlled between 180 °C and 190 °C;

(3) After shearing and blending, place the prepared HVMA in an oven at 180 °C for 30 min. Then take it out for various tests.

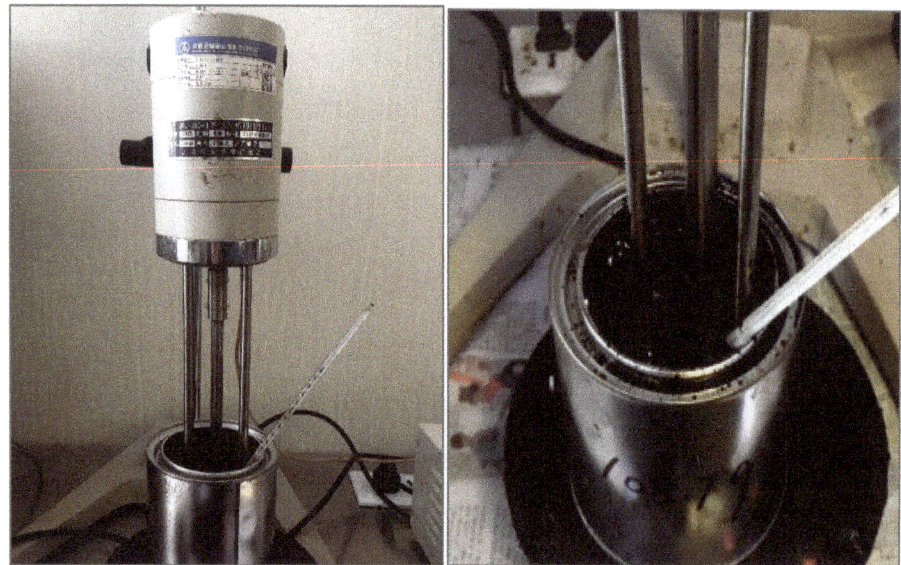

Figure 1. Preparation of HVMA using high-speed shear mixer.

2.3. Selection of Key Indicators for High-Viscosity Asphalt Binder

Existing studies show that [5], with the increase of dynamic viscosity of asphalt at 60 °C, compressive strength, splitting strength and bending strength of porous asphalt mixture are obviously increased. Besides, dynamic stability and other road performance are also significantly improved. A reasonable value of dynamic viscosity at 60 °C ensures porous asphalt to be used under high temperature conditions and without big increase in cost. Therefore, it is necessary to analyze the combined effect of temperature, load and dynamic viscosity on rutting performance of porous asphalt.

Among them, temperature and load are external factors, while dynamic viscosity is an internal factor and dominates the effect. According to existing research [5], the relation between dynamic stability of high-viscosity modified asphalt mixture and dynamic viscosity is as follows:

$$\lg DNS = 0.535 \lg\eta - 0.0653t - 0.63p + 5.592 \quad \left(R^2 = 0.892\right), \tag{1}$$

where,

DNS—Rutting test dynamic stability, times/mm;
η—Asphalt dynamic viscosity at 60 °C, Pa·s;
t—Test temperature, °C;
p—Loading pressure, MPa.

At present, 60 °C is usually chosen as test temperature to evaluate high temperature performance of asphalt mixture in many countries. According to meteorological data of some countries in Africa, the project is located in low latitude area and its climate is humid, hot and rainy, with high temperature in summer. The average maximum temperature for 7 consecutive days can reach 45.5 °C, as shown in Figure 2.

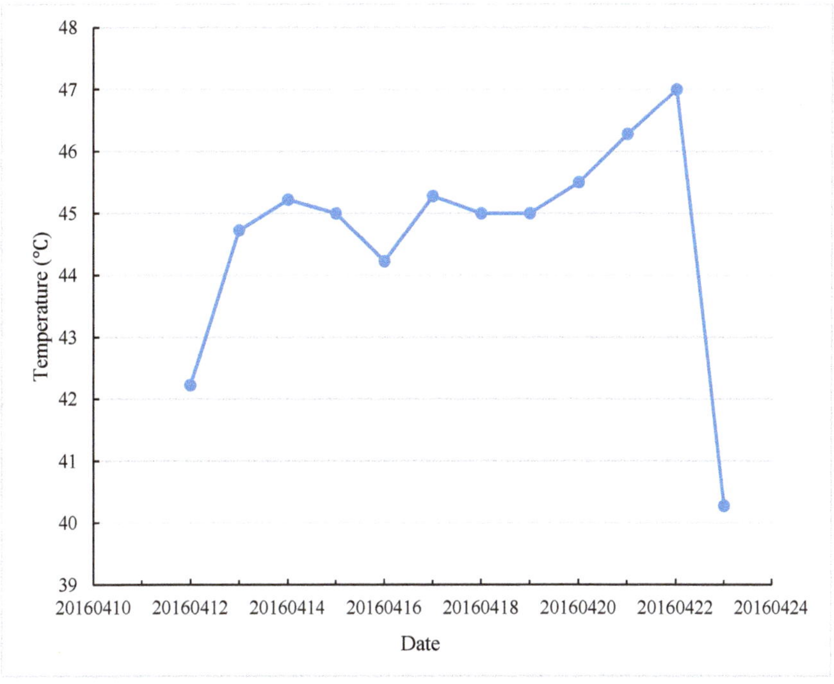

Figure 2. Average maximum temperature for 7 consecutive days over 30 years in certain African region (Data are from the reference [13]).

Calculation formula for design temperature of pavement is given in SHRP research [14], as shown by Equation (2). The temperature used to calculate PG high temperature grade is that of 2 cm below road surface.

$$T_{20mm} = T_{air} - 0.00618Lat^2 + 0.2289Lat + 42.2 \times 0.9545 - 17.78, \tag{2}$$

where:

$T_{20\,mm}$—the maximum temperature at 2 cm below road surface (°C);
T_{air} —average maximum air temperature for 7 consecutive days (°C);
Lat—local latitude (°).

According to Equation (2), the highest design temperature of local pavement is 67.9 °C. Therefore, evaluation of high temperature performance of asphalt and mixture at 60 °C does not reflect actual working condition of pavement in summer. In this paper, considering a long life design of the pavement, 70 °C is used as the temperature for evaluating ultimate temperature performance of mixture and 0.7 MPa is used as load for rutting test load. In order to meet the needs of pavement material design for application, the dynamic stability of porous asphalt is not less than 3000–5000 times/mm. Considering an application for light traffic case, this paper suggests that 3000 times/mm is quite reasonable. Calculated by Equation (1), the required dynamic viscosity of asphalt at 60 °C shall be no less than 260,124 Pa·s. Then 270,000 Pa·s is chosen as lower limit of dynamic viscosity at 60 °C, so as to ensure the performance of the mixture at locally high temperatures.

At the same time, as HVMA has greater viscosity than ordinary modified asphalt, construction workability must be considered in evaluating performance of modified asphalt. A large number of engineering practices show that Brookfield viscosity of high-viscosity modified asphalt at 170 °C shall not exceed 3 Pa·s [5].

2.4. Influence of HVA Content on Performance of Asphalt

Firstly, high-viscosity asphalt samples are prepared in the lab by using the four kinds of different base materials (as shown in Tables 1 and 2) and HVA with different contents. Then performances of high-viscosity asphalt are tested, including penetration, softening point, dynamic viscosity at 170 °C and dynamic viscosity at 60 °C. Variations of these parameters caused by different HVA contents are analyzed as well. Five different mixing contents of HVA are used, namely 8%, 10%, 12%, 14% and 16%. The relationship between penetration and softening point of high-viscosity asphalt prepared by different matrix asphalt and HVA content is shown in Figure 3. Dynamic viscosity at 60 °C for different HVA content is shown in Figure 4. For better illustration of the tendency of the dynamic viscosity changing with the HVA content, the dynamic viscosity at 60 °C is shown in logarithmic form.

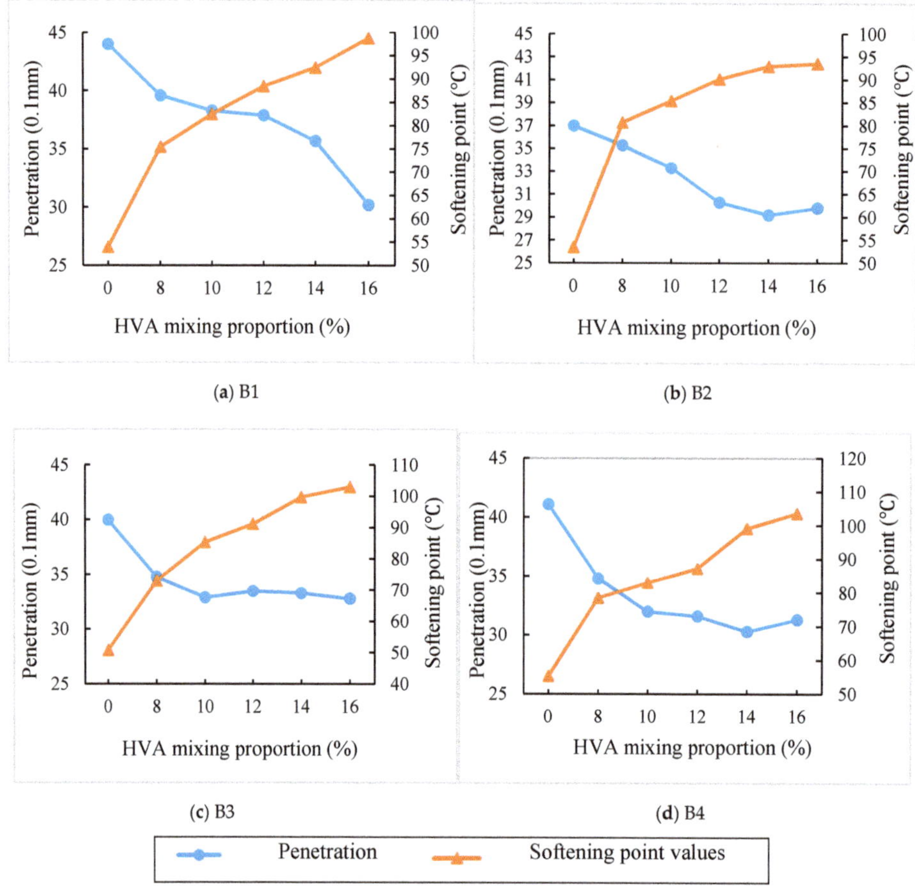

Figure 3. Penetration, softening point values of four HVMA with different HVA mixing proportion.

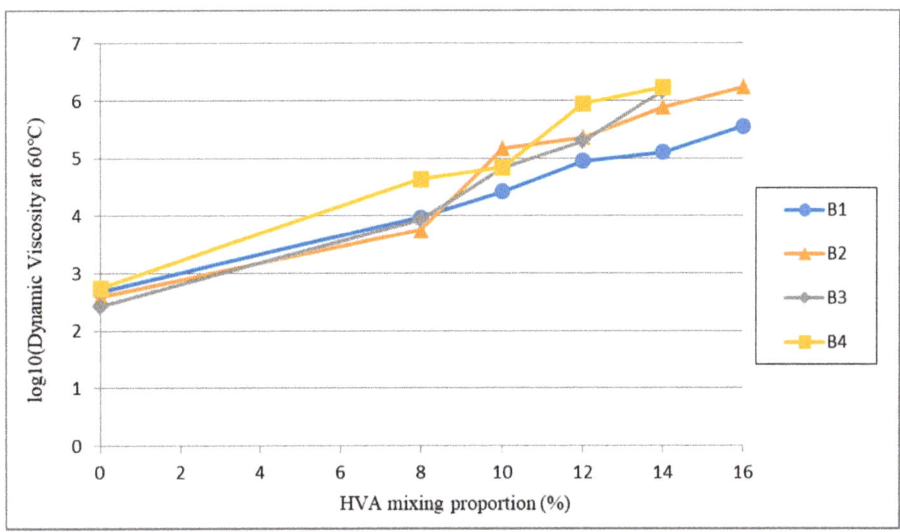

Figure 4. Dynamic viscosity at 60 °C of asphalt with different HVA mixing proportions.

(1) Penetration

In many countries, penetration is regarded as an index for asphalt grade, which reflects the asphalt consistency. It can be seen from Figure 3 that the penetration of the four matrix asphalt after modification is significantly reduced, which is mainly due to absorption of light components in asphalt by HVA addition. It makes the asphalt thicken and become harder. When HVA content increases from 0% to 14%, penetration values for B1 to B4 matrix asphalt reduces by 19%, 21%, 17% and 26%, respectively. In terms of reduction in penetration, B4 is more sensitive to HVA modification.

(2) Softening point

As shown in Figure 3, the softening point of the four modified asphalt increases with the HVA content but the increasing trends slow down after it rises to 14%. Compared with matrix asphalt, the softening points of the four asphalt are all above 90 °C when 14% HVA is added.

(3) Dynamic viscosity at 60 °C

In Figure 4, it shows that dynamic viscosities of the four modified asphalt increase exponentially with HVA content. For asphalt B4, when HVA content is 8%, its dynamic viscosity at 60 °C reaches 43,914 Pa·s and its dynamic viscosity is the most sensitive to mixing content of HVA. The variation on dynamic viscosity at 60 °C of asphalt B1 is less affected by content of HVA, which shows a poor compatibility between the matrix asphalt and the modifier at current contents.

2.5. Determination on Optimum Range of HVA Content

Based on test results, dynamic viscosity at 60 °C and Brookfield viscosity at 170 °C of different HVMA are correlated with HVA mixing proportion and regression curves are shown for expressing the relationships in Figure 5. From Figure 5, it can be seen that when expressed in logarithmic form, there are good linear relationships between the dynamic viscosity at 60 °C and HVA contents, as well as for Brookfield viscosity at 170 °C.

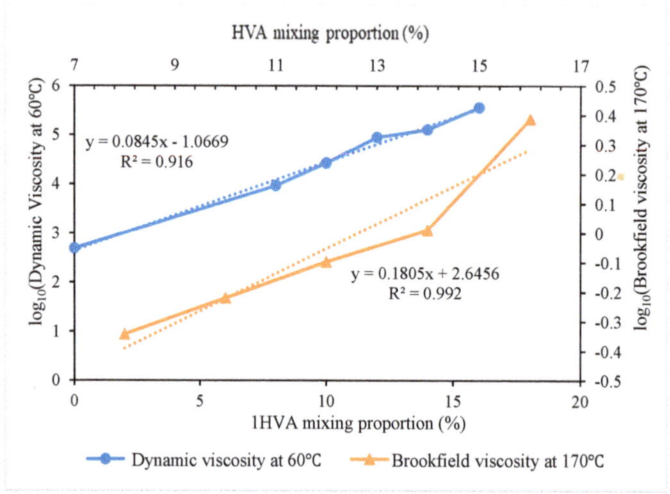

(**a**) Made by matrix asphalt B1

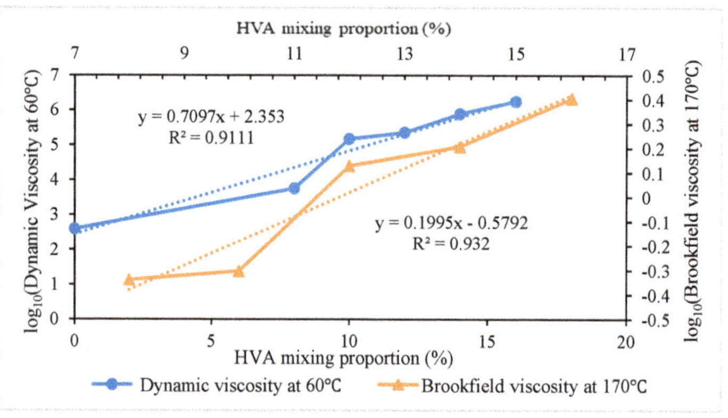

(**b**) Made by matrix asphalt B2

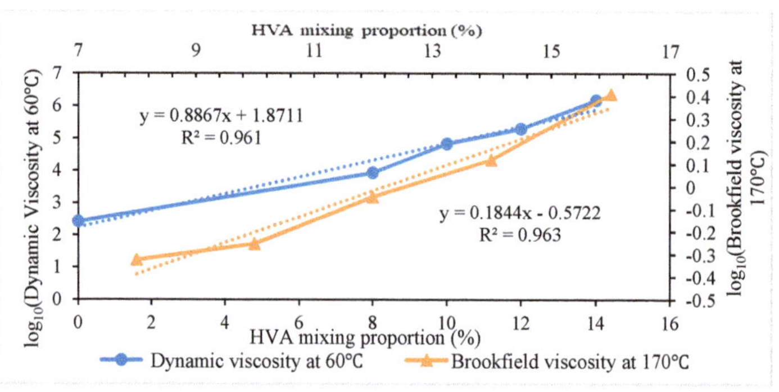

(**c**) Made by matrix asphalt B3

Figure 5. *Cont.*

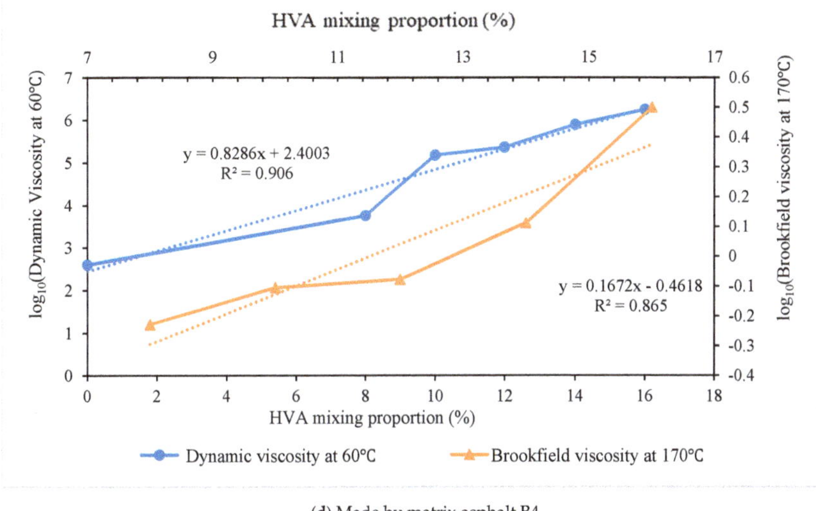

(**d**) Made by matrix asphalt B4

Figure 5. Dynamic viscosity and Brookfield viscosity for different HVA mixing proportion.

According to analysis in Section 2.3, Brookfield viscosity at 170 °C (not more than 3.0 Pa·s) is used as an index to control maximum content of HVA and dynamic viscosity at 60 °C (not less than 270,000 Pa·s) is used as index to determine minimum content of HVA. The proper ranges of HVA content corresponding to different matrix asphalt are calculated based on the regression relationship in Figure 5. The results for B1–B4 are respectively: 15.5%~18.2%, 12.5%~16.6%, 12.3%~17.3% and 11.0%~17.2%.

3. Experimental Study on Rheological Properties of High-Viscosity Modified Asphalt

In this section, rheological properties of HVMA made from various matrix asphalt are studied. In order to facilitate the investigation and considering the cost effective for using HVA in practical engineering, the HVA content 14% is used for the high-viscosity modified asphalt preparation, namely the ratio between the mass of asphalt and HVA is 86:14. For matrix asphalt B1, dynamic viscosity at 60 °C after mixing of 14% HVA cannot meet the requirement in Section 2.3, so it is not further studied in the experimental research on rheological properties in this paper.

3.1. High Temperature Performance Grade

High temperature performance grade (PG) is calculated from complex shear modulus $|G^*|$ and phase angle δ, which are measured by dynamic shear rheometer. In this section, dynamic shear rheometer (DSR) is used for the test. The experimental parameters are set to be 10 rad/s as angular frequency, 1 mm above and below of parallel plate, 12% of strain value for original asphalt and 10% for aged asphalt. In the Strategic Highway Research Program (SHRP) research program, rutting resistance is characterized by rutting factor $|G^*|/\sin\delta$. PG grading requires $|G^*|/\sin\delta$ of original asphalt to be greater than or equal to 1.0 kPa and $|G^*|/\sin\delta$ of short-term aging residue to be greater than or equal to 2.2 kPa. PG grading of asphalt sample is carried out and the test results are shown in Table 3.

It can be seen from Table 3 that: The PG grade for matrix asphalt B2, B3 and B4 are 70, 64 and 70 respectively. After modified by 14% HVA, there are increases by three grades for the three types of matrix asphalt, which means great improvement of high temperature performance. PG high temperature grade of B3 + 14% HVA can reach PG82 which is the highest temperature grade and that of the other two even exceeds the highest grade.

Table 3. Test results of high temperature performance grade.

Test Temperature (°C)	B2	B2 + 14%HVA	B3	B3 + 14%HVA	B4	B4 + 14%HVA
Original asphalt($|G^*|/\sin(\delta)$ (|KPa|))						
58	8.51	-	5.87	21.8	10.9	-
64	3.6	-	2.47	11.5	4.71	-
70	1.63	-	1.13	7.26	2.14	-
76	0.779	-	0.55	5.26	1.04	9.93
82	-	4.84	-	4.14	-	6.9
88	-	3.15	-	3.33	-	5.04
After short-term aging($|G^*|/\sin(\delta)$ (|KPa|))						
58	13.4	-	10.1	-	22.8	-
64	5.59	-	4.04	-	9.81	-
70	2.48	-	1.76	-	4.39	-
76	1.16	-	-	-	2.06	-
82	-	4	-	2.74	-	7.97
88	-	2.31	-	1.68	-	4.61

3.2. Asphalt Temperature Sweep Test

DSR measurement is used for temperature sweep. The temperature sensitivity of different asphalt is analyzed by measuring complex shear modulus $|G^*|$, phase angle δ and rutting factor $|G^*|/\sin\delta$ of asphalt. With a fixed loading frequency of 10 rad/s and a stress level of 0.1 kPa, temperature sweep test of the three kinds of matrix asphalt and corresponding high-viscosity modified asphalt with 14% HVA are carried out at test temperature from 30 °C to 80 °C. Parallel plate of 25 mm is used and the space between upper and lower parallel plates is fixed at 1 mm. For different kinds of asphalt, complex shear modulus $|G^*|$, phase angle δ and rutting factor $|G^*|/\sin\delta$ in accordance with different temperature are shown in Figure 6.

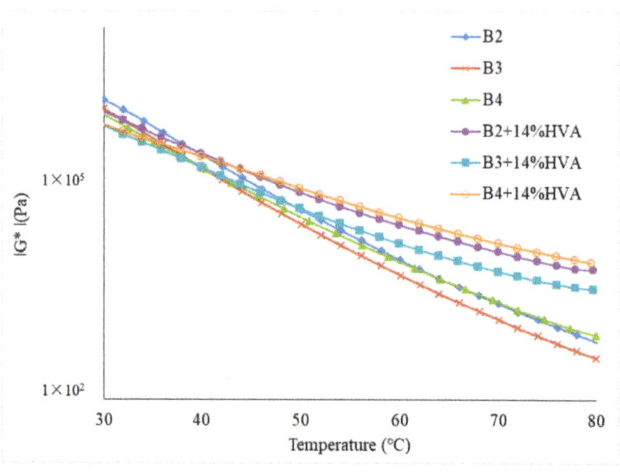

(**a**) Dynamic shear modulus

Figure 6. *Cont.*

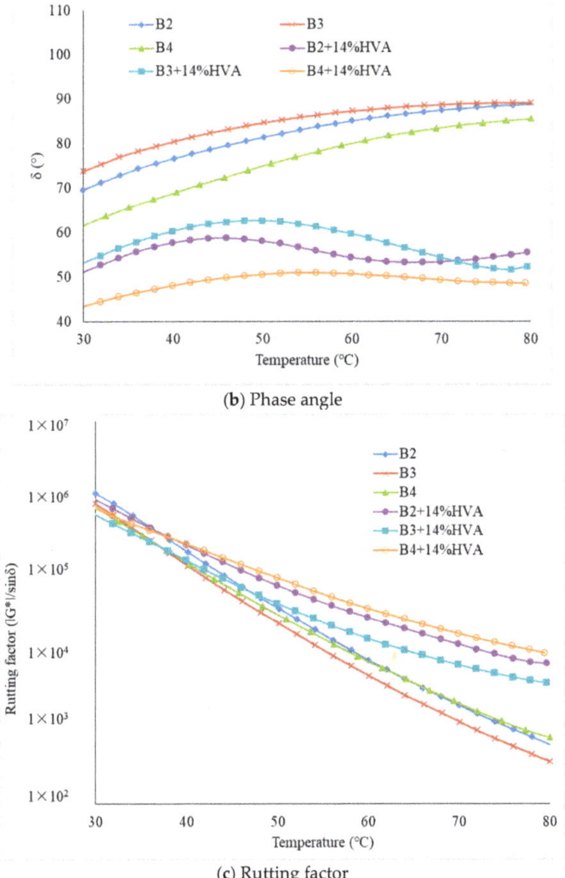

(**b**) Phase angle

(c) Rutting factor

Figure 6. Temperature sweep test results

As can be seen from Figure 6a, the dynamic shear modulus of all the asphalt samples have little difference from 30 °C to 45 °C. This is because of the low penetration and high consistency at low temperatures of matrix asphalt. When test temperature is higher than 45 °C, with the increase of temperature, the dynamic shear modulus of the three matrix asphalt decreases rapidly. For HVMA, with the increase of temperature, swelling effects between the HVA and light component in asphalt causes the increase of heavy component content in the asphalt, which in turn makes the asphalt viscous and hard macroscopically, meanwhile the flexibility is improved and higher value of shear modulus is obtained. The dynamic shear modulus of B4 + 14% HVA is the largest among the three types of high-viscosity modified asphalt.

In Figure 6b, it shows that phase angle of B4 is the smallest at the same test temperature among three matrix asphalt, which indicates that it has less viscous components and better resistance to permanent deformation. The phase angles of three matrix asphalt are greatly reduced after adding 14% of HVA, indicating that addition of HVA improves resistance ability to permanent deformation. Phase angles of three matrix asphalt gradually become larger as the test temperature increases, which shows more viscous properties. However, with the increase of temperature, phase angles of three high-viscosity asphalt firstly increase and then decrease, with small overall phase angle, showing greater elastic properties.

As can be seen from Figure 6c, with the increase of test temperature, rutting factor |G*|/sinδ of both matrix asphalt and high-viscosity asphalt added with HVA gradually decreases and the asphalt rutting factor and test temperature show an exponential decreasing trend. Compared with matrix asphalt, the three high-viscosity asphalt binders are less sensitive to temperature. As sweep temperature rises above 70 °C, the sequence of rutting factor value for different high-viscosity asphalt is: B4 + 14% HVA > B3 + 14% HVA > B2 + 14% HVA.

3.3. Multiple Stress Creep Recovery Test

The multiple stress creep recovery (MSCR) test method is incorporated in AASHTO TP70 [15], which is a new-generation test method for evaluating elasticity resuming performance of modified asphalt. MSCR specimen is characterized by less test loading times and easy calculation of parameters [16,17]. A dynamic shear rheometer set with 25 mm parallel plate and 1 mm gap is used in MSCR test. Loading stress at the first stage is 0.1 kPa and that at the second stage is 3.2 kPa. Under controlled-stress mode, different stress levels are used for repeated loading-unloading tests of asphalt, with 10 cycles per stage. For each cycle, asphalt is loaded for 1 s and unloaded for 9 s.

At the test temperature of 70 °C, the strains obtained at different loading times for the three matrix asphalt and high-viscosity modified asphalt after adding 14% HVA are shown in Figures 7 and 8 respectively.

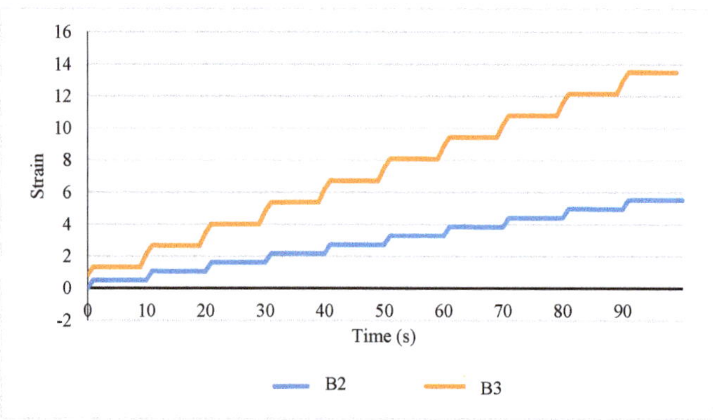

(**a**) Time-strain diagram of matrix asphalt under 0.1 kPa stress

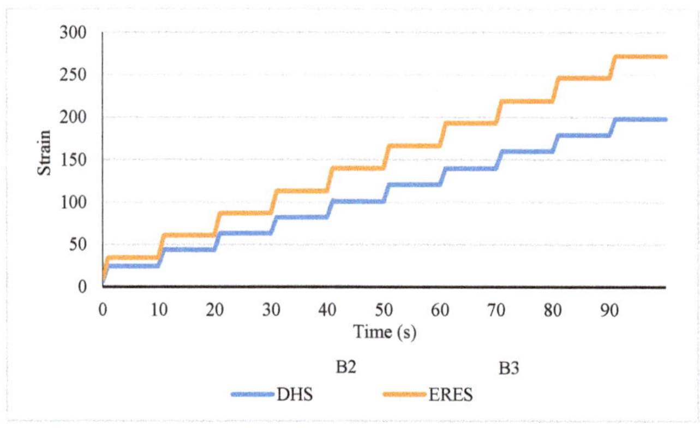

(**b**) Time-strain diagram of matrix asphalt under 3.2 kPa stress

Figure 7. MSCR test results of matrix asphalt.

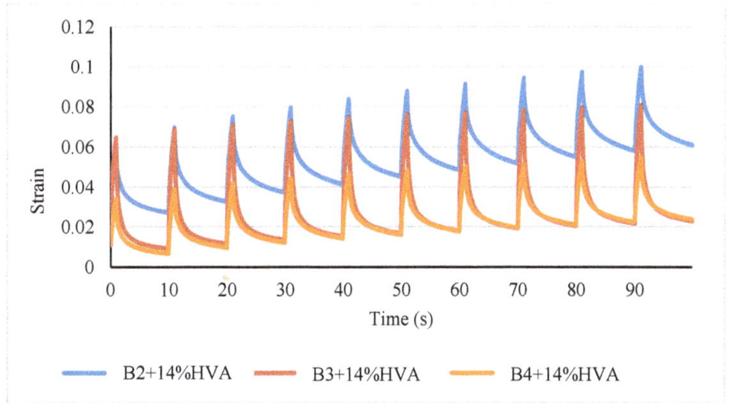

(**a**) Time-strain diagram of three high-viscosity asphalt under 0.1 kPa stress

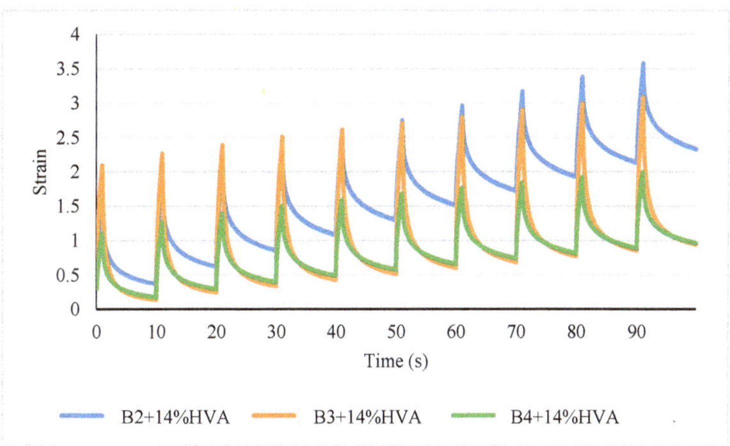

(**b**) Time-strain diagram of three high-viscosity asphalt under 3.2 kPa stress

Figure 8. Test results of high-viscosity modified asphalt.

As can be seen from Figures 7 and 8, with increasing loadings cycles, strain of asphalt is enlarged and creep recovery decreases. Strain of the same kind of asphalt increases significantly under a larger stress level. When placed under load, matrix asphalt has greater strain than high-viscosity asphalt. Besides, after loading stage, matrix asphalt has weak recovery ability, without strong resistance to deformation.

In AASHTO [15], deformation recovery capacity and high temperature rutting resistance of asphalt are evaluated by deformation recovery rate R and non-recoverable creep compliance J_{nr}. Essentially, J_{nr} represents the viscosity (non-recoverable) in creep compliance of the material. The two parameters are calculated as follows:

$$R = (r_p - r_{nr})/(r_p - r_0) \tag{3}$$

$$J_{nr} = (r_{nr} - r_0)/\tau \tag{4}$$

where r_p represents peak strain in a loading cycle, r_{nr} represents residual strain in a loading cycle, r_0 represents initial strain in a loading cycle and τ the shear stress.

R and J_{nr} of the first 10 creep cycles are calculated according to Equations (3) and (4) respectively. Then the average creep recovery rate ($R_{0.1}$) and average non-recoverable creep compliance ($J_{nr0.1}$) at stress level of 0.1 kPa are obtained. In a similar way, average creep recovery rate ($R_{3.2}$) and average non-recoverable creep compliance ($J_{nr3.2}$) at stress level of 3.2 kPa can be obtained as well. The results of $R_{0.1}$, $R_{3.2}$, $J_{nr0.1}$ and $J_{nr3.2}$ of three matrix asphalt and high-viscosity asphalt with addition of 14% HVA are shown in Table 4.

Table 4. Results of creep recovery rate and non-recoverable creep compliance.

Asphalt Type	Parameters			
	$R_{0.1}$ (%)	$J_{nr0.1}$ (kPa^{-1})	$R_{3.2}$ (%)	$J_{nr3.2}$ (kPa^{-1})
B2	0.0265	27.5500	0.0068	31.9725
B2 + 14%HVA	0.8926	0.7057	0.8426	0.6644
B3	−0.0199	40.1736	−0.0201	43.9400
B3 + 14%HVA	0.9570	0.3833	0.9427	0.6235
B4	0.0030	37.4957	−0.0422	47.7564
B4 + 14%HVA	0.9091	0.1650	0.8836	0.3931

From Table 4, it is known that at the same temperature, there is a dramatic increase of recovery rate and a significant reduction of non-recoverable creep compliance after modification by 14% HVA. The creep recovery rate of B3 is negative at both 0.1 kPa and 3.2 kPa, which may be due to poor thermal stability of matrix asphalt and the test temperature of 70 °C exceeds its softening point. As stress level increases from 0.1 kPa to 3.2 kPa, recovery rate of all asphalt is reduced, which proves that actual pavement is more likely to produce rutting when subjected to higher pressure. At stress levels of 0.1 kPa and 3.2 kPa, the B3 + 14% shows highest deformation recovery capability among the three, which is mainly due to the greater content of aromatic and resin of the matrix asphalt. As to the non-recoverable creep compliance, the B4 + 14% modified asphalt presents a lowest value at both stress levels of 0.1 kPa and 3.2 kPa. This is because of a higher content of asphaltene in the matrix asphalt, which makes it less sensitive to a loading impact at high temperature. J_{nr} indicates capability of asphalt to recover from deformation and its value has negative correlation with rutting resistance, that is, the higher is non-recoverable creep compliance, the poorer performance on deformation resistance of asphalt pavement.

3.4. Accelerated Fatigue Test

The asphalt accelerated fatigue test [18] can quickly evaluate and predict asphalt fatigue performance to determine fatigue resistance of asphalt qualitatively. In this paper, DSR linear amplitude sweep (LAS) is used to test the shear stress with different shear strain value. The test method is as follows: firstly, short-term aging of asphalt is carried out and then measurement is carried out under the condition of 10 Hz, 20 °C (lower than the working temperature of asphalt in this area), strain sweep range 0.1%~30%, sweep rate LAS-5. The stress-strain curves of the three types of matrix asphalt and high-viscosity modified asphalt with 14% HVA are shown in Figure 9.

From Figure 9, it can be seen that shear stress of matrix asphalt first appears a peak value with increase of strain, then it decreases continuously, which is mainly due to elastic and plastic deformation is generated in the process. The gradual decrease of shear stress after the peak value indicates that yield stress of asphalt occurs under corresponding strain, which is called yield strain. In general, the higher the yield strain, the better the elastic properties are.

The high stress levels are observed when the strain is between 5% and 10% and the decline is relatively rapid. After addition of 14% HVA, B3 asphalt can maintain high stress for a larger range of strain in comparison with the matrix asphalt B2. B2 + 14% HVA and B4 + 14% HVA asphalt do not show a decreasing trend of stress within 30% of the strain and does not reach the fatigue failure point within 30% strain range, which indicate a better fatigue resistance performances after modified by HVA.

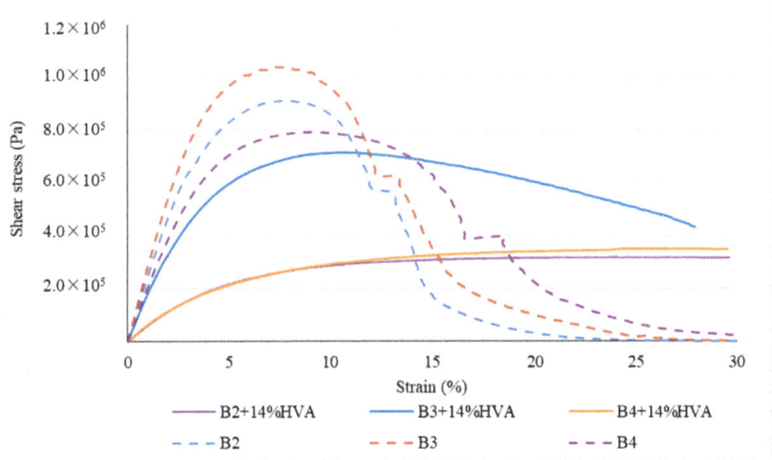

Figure 9. Stress-strain curves of three kinds of matrix asphalt and high-viscosity asphalt with 14% HVA.

4. Study on Viscosity-Temperature Characteristics of Low-Grade High-Viscosity Modified Asphalt

This section mainly aims at appropriate temperature range for the construction of porous asphalt with different high-viscosity modified asphalt. For porous asphalt mixture prepared by high-viscosity modified asphalt, the viscosity range of ordinary asphalt cannot be used for determining the construction temperature range. So a new method is needed to determine mixing and compaction temperatures of the mixtures.

The relationship between viscosity and temperature of high-viscosity modified asphalt can be expressed by the Saal formula:

$$\lg \lg \eta \times 10^3 = n - m \lg(T + 271.13) \tag{5}$$

where,

T—Temperature, °C:

η—Viscosity, Pa·s:

n—Regression constant, the higher the value, the greater the viscosity at the same temperature;

m—Regression constant, indicating asphalt temperature sensitivity. The larger the absolute value, the worse the temperature sensitivity.

Brookfield viscosity of three types of HVMA (Asphalt: HVA= 86:14) at 135 °C, 150 °C, 175 °C are measured by Brookfield viscometer and the curve of regression by Saal formula is shown in Figure 10.

According to the relation given by Saal formula and based on above data, the viscosity-temperature curves of three different asphalt with 14% HVA are as follows:

$$\begin{aligned}
&\text{B2} + 14\%\text{HVA: } y = -2.3404x + 6.5527 \quad (R^2 = 0.9942)\\
&\text{B3} + 14\%\text{HVA: } y = -3.4245x + 9.3507 \quad (R^2 = 0.9968)\\
&\text{B4} + 14\%\text{HVA: } y = -3.4914x + 9.5844 \quad (R^2 = 0.9964)
\end{aligned} \tag{6}$$

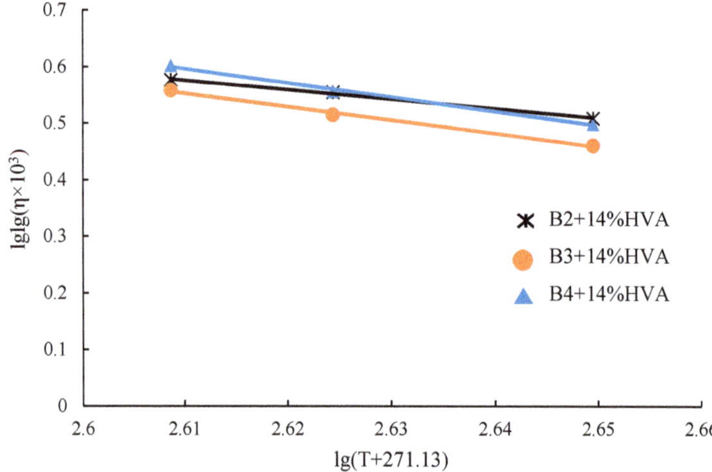

Figure 10. Viscosity-temperature curve of three asphalt with 14% HVA on double logarithmic coordinates.

From existing studies, the viscosity range for mixing and compaction temperatures of modified asphalt are considered [19]:

$$\text{Mixing temperature}: \ \eta = 0.275 \pm 0.03 Pa{\cdot}s$$

$$\text{Compaction temperature}: \ \eta = 0.550 \pm 0.04 Pa{\cdot}s$$

Based on the viscosity range above and the viscosity-temperature curve, the mixing and compaction temperatures of HVMA with 14% HVA are obtained and shown in Table 5. From the table, it can be seen that HVMA has a relatively high-viscosity at low temperature in use such as 60~70 °C but temperatures required for mixing and compaction are not too high. Besides, porous asphalt mixture generally has coarse aggregates content, it is favorable for the high-viscosity additive to be grinded in the mixing. For B3 + 14% HVA, considering the heating temperature and the workability of the mixing, it is also suggested the mixing temperature to be 160 °C. Then, from the study, the mixing temperature for HVMA from low grade asphalt is considered to be 160 °C and the compaction temperature is between 130 °C and 145 °C, depending on the matrix asphalt type.

Table 5. Mixing and compaction temperature for different HVMA.

Asphalt Type	Mixing Temperature (°C)	Compaction Temperature (°C)
B2 + 14%HVA	160	139
B3 + 14%HVA	143	130
B4 + 14%HVA	160	145

5. Conclusions and Recommendations

Performances of high-viscosity modified asphalt prepared by typical low grade asphalt in Africa were studied in this research. The conclusions drawn from the research are as follows:

(1) According to temperature characteristics of rainy environments in Africa, the lower limit of dynamic viscosity at 60 °C is considered to be 270,000 Pa·s for satisfying the rutting resistance requirement.

(2)   The high-viscosity modification process can greatly improve PG high temperature grade and high rutting resistance of the low grade asphalt, as well as reducing the sensitivity of asphalt to loading frequency.

(3)   For matrix asphalt with higher asphaltene content, a larger dynamic viscosity at 60 °C is achieved after high-viscosity modification and it also shows relatively good resistance to deformation caused by high temperature and shearing. For a matrix with lower asphaltene content but higher aromatic and resin, the most outstanding elastic recovery capability can be obtained after high-viscosity modification.

(4)   For application in porous asphalt in Africa, the suitable mixing temperature is considered to be 160 °C, while the compaction temperature is suggested to be from 130 °C to 145 °C for various types of high-viscosity modified asphalt binders.

The research can help guide the design of porous asphalt mixture based on low-grade and high-viscosity modified asphalt and provide a basis for implementation on the trial section of asphalt pavement in Africa. Studies on mixtures by using high-viscosity modified asphalt and observations of trial section will be further carried out in future work.

Author Contributions: Conceptualization, F.Z.; Data curation, M.L.; Formal analysis, R.X. and J.L.; Investigation, R.X. and J.L.; Methodology, M.L.; Project administration, F.Z.; Resources, D.C.; Supervision, F.Z.; Validation, D.C.; Visualization, J.L.; Writing—original draft, M.L. and R.X.; Writing—review & editing, D.C.

Funding: This work was supported by the National Key Research and Development Program of China (No. 2016YFE0108200), Central Public-interest Scientific Institution Basal Research Fund from Research Institute of Highway Ministry of Transport of China (No. 2017-9079 and No. 2019-0117).

Conflicts of Interest: The authors declare no conflict of interest.

References

1.   Moore, L.; Hicks, R.; Rogge, D. Design, construction, and maintenance guidelines for porous asphalt pavements. *Transp. Res. Rec. J. Transp. Res. Board* **2001**, *1778*, 91–99. [CrossRef]

2.   Nakanishi, H. *History of Porous Asphalt Pavement in Japan*; Taiyu Kensetsu Co., Ltd.: Nagoya, Japan, 2004.

3.   Takahashi, S. Comprehensive study on the porous asphalt effects on expressways in Japan: Based on field data analysis in the last decade. *Road Mater. Pavement Des.* **2013**, *14*, 239–255. [CrossRef]

4.   Moriyoshi, A.; Jin, T.; Nakai, T. Evaluation methods for porous asphalt pavement in service for fourteen years. *Constr. Build. Mater.* **2013**, *42*, 190–195. [CrossRef]

5.   Cao, D.; Liu, Q.; Tang, G. *Porous Asphalt Pavement*; China Communications Press: Beijing, China, 2009. (In Chinese)

6.   Liu, Z.; Wang, X.; Luo, S.; Yang, X.; Li, Q. Asphalt mixture design for porous ultra-thin overlay. *Constr. Build. Mater.* **2019**, *217*, 251–264. [CrossRef]

7.   Qin, X.; Zhu, S.; He, X.; Jiang, Y. High temperature properties of high-viscosity asphalt based on rheological methods. *Constr. Build. Mater.* **2018**, *186*, 476–483. [CrossRef]

8.   Yourong, T.; Tan, Y.; Zhang, H.; Cao, D.; Xia, L.; Du, R.; Shi, Z.; Wang, X. Study on cohesion and adhesion of high-viscosity modified asphalt. *Int. J. Transp. Sci. Technol.* **2019**, in press. [CrossRef]

9.   Cai, J.; Song, C.; Zhou, B.; Tian, Y.; Li, R.; Zhang, J.; Pei, J. Investigation on high-viscosity asphalt binder for permeable asphalt concrete with waste materials. *J. Clean. Prod.* **2019**, *228*, 40–51. [CrossRef]

10.   Xu, B.; Chen, J.; Li, M.; Cao, D.; Ping, S.; Zhang, Y.; Wang, W. Experimental investigation of preventive maintenance materials of porous asphalt mixture based on high-viscosity modified bitumen. *Constr. Build. Mater.* **2016**, *124*, 681–689. [CrossRef]

11.   Liu, M.; Huang, X.; Xue, G. Effects of double layer porous asphalt pavement of urban streets on noise reduction. *Int. J. Sustain. Built Environ.* **2016**, *5*, 183–196. [CrossRef]

12.   Luong, J.; Bueno, M.; Vázquez, V.F.; Paje, S.E. Ultrathin porous pavement made with high-viscosity asphalt rubber binder: A better acoustic absorption? *Appl. Acoust.* **2014**, *79*, 117–123. [CrossRef]

13.   Climate Data Online. Available online: https://www7.ncdc.noaa.gov/CDO/cdodateoutmod.cmd (accessed on 3 November 2018).

14. Cominsky, R.J.; Huber, G.A.; Kennedy, T.W.; Anderson, M. *the Super Pave Mix Design Manual for New Construction and Overlays*; Rep. No. SH RP -A -407. Strategic Highway Research Program; National Research Council: Washington, DC, USA, 1994.

15. American Association of State Highway and Transportation Officials (AASHTO). *Standard Method of Test for Multiple Stress Creep Recovery (MSCR) Test of Asphalt Binder Using a Dynamic Shear Rheometer (DSR)*; AASHTO TP 70-10; AASHTO: Washington, DC, USA, 2010.

16. D'angelo, J.A. The Relationship of the MSCR test to rutting road materials and pavement. *ICAM Road Mater. Pavement Des.* **2009**. [CrossRef]

17. Masad, E.; Huang, C.W.; D'Angelo, J.; Little, D. Characterization of asphalt binder resistance to permanent deformation based on nonlinear viscoelastic analysis of multiple stress creep recovery (mscr) test. *Asph. Paving Technol. Assoc. Asph. Paving Technol. Proc. Tech. Sess.* **2009**, *78*, 535–562.

18. American Association of State Highway and Transportation Officials (AASHTO). *Standard Method of Test for Estimating Fatigue Resistance of Asphalt Binders Using the Linear Amplitude Sweep*; AASHTO TP 101–2012; AASHTO: Washington, DC, USA, 2010.

19. Li, Y. Study on the Application of Domestic TPS in Drainage Asphalt Pavement. Master's Dissertation, Changan University, Xi'an, China, 2013. (In Chinese).

 *materials*

Article

Evaluation of Asphalt with Different Combinations of Fire Retardants

Guangji Xu [1], Xiao Chen [1,2], Shichao Zhu [3], Lingdi Kong [4], Xiaoming Huang [1], Jiewen Zhao [1] and Tao Ma [1,*]

[1]   School of Transportation, Southeast University, Jiulonghu, Nanjing 211189, China;
   guangji_xu@seu.edu.cn (G.X.); xche@seu.edu.cn (X.C.); huangxm@seu.edu.cn (X.H.);
   zhaojiewen@gmail.com (J.Z.)
[2]   State Engineering Laboratory of Highway Maintenance Technology,
   Changsha University of Science and Technology, Changsha 410114, China
[3]   Qilu Transportation Development Group, 1 Longaoxi, Jinan 200101, China; shichaozhu01@163.com
[4]   Shandong Guilu Expressway Construction Co. Ltd., 23 Changrun, Liaocheng 252000, China;
   mhxkld@163.com
*   Correspondence: matao@seu.edu.cn; Tel.: +86-1580-516-0021

Received: 4 March 2019; Accepted: 16 April 2019; Published: 18 April 2019

Abstract: When a fire takes place in a tunnel, the surface of the asphalt pavement will burn and release a large amount of smoke, which is toxic to human health. Thus, in order to prevent the combustion of the asphalt pavement under fire, it is necessary to propose some methods to retard its physical and chemical reaction under the high temperature. In this study, ten different combinations of fire retardants and a control case where no fire retardant was applied were prepared for evaluation. The thermogravimetric (TG)–mass spectrometry (MS) tests were used to evaluate their effect on the fire retardance from mass and energy perspectives and the Fire Dynamics Simulator (FDS) software was used to evaluate the fire retardance from temperature and smoke distribution perspectives. In experimental analysis, the TG (thermogravimetric) and DTG (differential thermogravimetric) curves were used to analyze the mass loss rate and residual mass of the asphalt and the activation energy was calculated and analyzed as well. In addition, decay rate of mass loss rate and increasing rate of activation energy were proposed to evaluate the ease of combustion of the asphalt with and without fire retardants. The results show that in laboratory experiments, the fire retardant combination which includes 48% aluminum hydroxide, 32% magnesium hydroxide, 5% expanded graphite, and 15% encapsulated red phosphorous would lead to an improved effect of fire retardance. In numerical modeling, the temperature and smoke height distribution over time were adopted to evaluate the fire retardance effect. The temperature distribution was found to be symmetrical on both sides of the combustion point and the same combination as proposed in experimental analysis was found to have the best effect on fire retardance due to the largest decrease in temperature. Additionally, because of the highest smoke height distribution, an improved effect on smoke suppression was also found when this combination was applied.

Keywords: asphalt; fire retardance; smoke suppression; thermogravimetry; differential thermogravimetry; activation energy; temperature distribution; smoke height distribution

---

1. Introduction

Asphalt binder is a kind of complex mixture composed by macromolecular hydrocarbons and non-hydrocarbons. It contains carbon, hydrogen, and a small number of other elements like sulfur, nitrogen and oxygen [1]. Asphalt mixture is a composite of three-phase materials consisting of aggregates, asphalt binder, and air voids. Under different conditions, the asphalt mixture would represent linear or nonlinear viscoelastic properties, leading to different road performances [2–4]. The modification of the asphalt was found to have an impact on the mechanical or rheological properties of the pavement. Karimi (2018) found that activated carbon modification was a robust binder-based conductive component for induced heating and healing of asphalt mixtures [5]. Jahanbakhsh (2016) evaluated the modified asphalt with commonly used modifiers in low temperature performance and found that 0.1 weight% of sulfur could improve thermal cracking resistance [6]. From the existing studies of the asphalt, it can be inferred that the modification could, to some extent, improve the asphalt property, making the asphalt pavement perform better. Normally, the asphalt pavement constructed by base asphalt is easy to combust under fire and will release CO, CO_2, SO_2, NO, NO_2, CH_4, and other gases, making the environment dangerous for people to survive in due to poisoning or suffocation [7]. Once a fire occurs in a tunnel, it is difficult for people to escape and the difficulty of rescue, evacuation, and fire protection is much higher than that of ordinary roads [8]. Therefore, it is necessary to research the potential modification of asphalt to improve its fire retardance and make it more stable under fire. The study of the fire retardance mechanism of asphalt can fundamentally reduce casualties caused by tunnel fires. Moreover, the impact of recycled asphalt and mixture gradation on the combustion needs further investigation [9–11].

Currently, there are many methods that can be used to analyze the combustion process, including experimental and numerical methods. Thermogravimetric (TG)–mass spectrometry (MS) detection is a typical experimental method. It has been applied to many thermal decomposition studies; the material mass change can be monitored and used for building material combustion dynamics models and for speculating the combustion process of the micro-reaction [12–14]. Xu et al. (2011) established a multi-order combustion model by using thermogravimetrics and a Fourier infrared spectrometer to study the combustion process [15]. Michelle et al. (2011) studied the thermal degradation parameters of three different asphalt samples and found that one asphalt had the highest activation energy according to the thermogravimetric curve obtained from TG–MS results [16].

For numerical analysis, the Fire Dynamics Simulator (FDS) is a computational fluid dynamics (CFD) model which describes the flow of the smoke [17]. It could be used to analyze the performance of the proposed combinations of fire retardants. Temperature and smoke distribution around the fire point could also be obtained to demonstrate the effectiveness of the fire retardants. It solves the Navier–Stokes equations using an explicit finite difference scheme [18]. FDS has been subjected to numerous validation and calibration studies; some cases can be found elsewhere [19]. Currently, FDS has been applied for the performance-based fire safety design, design of smoke control systems, sprinkler/detector activation studies, and so on. M.O. et al. (2013) used FDS to analyze the main feature of fire smoke on human body injury and discussed the impact of smoke and temperature change in the train personnel in emergency evacuation [20]. Kang et al. (2015) adopted FDS to make a risk management of main control board and estimate the time for main control board abandonment [21]. Additionally, Oliveira et al (2019) evaluated how the FDS correlated with experimental fire heights and temperature profiles around a small-scale pool fire and finally provided detailed temperature field around the height of the fire [22].

The aim of this study is to figure out whether the modifiers, i.e., fire retardants, could achieve satisfactory results of fire retardance and evaluate the effect of different modifiers through experimental and numerical analysis. In this study, the base asphalt with ten different fire-retardant combinations and a control case were used to evaluate their fire retardance in both laboratory experiments and numerical modelling. The effects of the different fire-retardant combinations on smoke suppression were also analyzed in numerical modeling.

2. Experimental Materials and Methods

2.1. Combinations of Fire Retardants and Asphalt

Aluminum hydroxide and magnesium hydroxide are two kinds of widely used fire retardants. Their fire retardance mechanisms are to reduce the combustion temperature by dehydration during the combustion process, thus limiting the combustion speed. Because they can catalyze and oxidize the gas, the amount and the escape rate of the smoke can be reduced [23,24]. Furthermore, the MgO and Al_2O_3 generated by aluminum hydroxide and magnesium hydroxide can absorb the smoke and precipitate it into ash [25]. Thus, the aluminum hydroxide and magnesium hydroxide can suppress the smoke during combustion as well. Encapsulated red phosphorous is also an effective fire retardant; it has a strong thermal stability and can reduce the oxygen content on the surface of the combustor, which can limit the occurrence of combustion. Expanded graphite is another effective fire retardant; it can expand and rapidly increase its surface area to form an insulating layer when heated, thus reducing the rate of combustion [26]. Therefore, these four components were selected as basic fire retardants. From the smoke suppression perspective, aluminum hydroxide and magnesium hydroxide were chosen to be the main fire-retardant components. The asphalt used in this study is 70# base asphalt and the appearance of the fire retardants are shown in Figure 1. The related characteristics of fire retardants used in this study are shown in Table 1.

Figure 1. Four fire retardants. (**a**) Aluminum hydroxide; (**b**) magnesium hydroxide; (**c**) expanded graphite; (**d**) encapsulated red phosphorous.

Table 1. Fire retardant characteristics.

	Aluminum Hydroxide	Magnesium Hydroxide	Expanded Graphite	Encapsulated Red Phosphorous
Moisture Content	0.15%	0.3%	0.1%	0.65%
Mean Size	20 nm	1.2 μm	7.5 μm	12.5 μm
Density	0.15 g/cm^3	2.36 g/cm^3	2.1 g/cm^3	2.2 g/cm^3
Purity	99.9%	96.0%	99.0%	70%
Specific Surface Area	5 m^2/g	200 m^2/g	150 m^2/g	-

2.2. TG–MS Test

The combustion experiments were carried out with a TG–MS system, including a NETZSCH STA 409 thermogravimetric analyzer (TGA) (Bavaria, Germany) and a NETZSCH QMS 403C mass spectrometer (Bavaria, Germany). To start an experiment, a sample of about 10 mg was tiled at the bottom of an Al$_2$O$_3$ crucible and the internal atmosphere of the TGA was set to simulate the air atmosphere. The N$_2$ was set to be 40 mL/min and O$_2$ was 10 mL/min. During the test, the temperature remained unchanged at 220 °C. The scanning mode of the mass spectrometry analyzer was ion scanning, which was performed every 105 s. The heating rate was 15 K/min and at least two parallel tests were conducted. The results include TG (thermogravimetric) curves and differential thermogravimetric (DTG) curves, which represent the residual mass of the asphalt and the mass loss rate, respectively.

2.3. Manufacturing Technique and Sample Preparation

First of all, the asphalt was heated to about 180 °C, then different combinations of fire retardants were added into the asphalt equally four times. The mixture was mixed manually until the fire retardants were evenly distributed in the asphalt and no powdery floaters could be seen. The high-speed shear emulsifier was used to ensure the fire retardants were fully distributed in the asphalt. The asphalt was kept to over 100 °C to ensure its rheological property and the initial shear speed was set to be 1000 r/min for 5 min, then increased to 3000 r/min for 10 min, and finally decreased to 1000 r/min for 5 min. After shearing, the modified asphalt was stirred manually for 2 min and was used after natural cooling.

2.4. Evaluation Index in Laboratory Experiments

2.4.1. Thermogravimetric Analysis

Thermogravimetric analysis (TGA) is a method of thermal analysis in which changes in physical and chemical properties of materials are measured as a function of increasing temperature (with constant heating rate) or as a function of time (with constant temperature and/or constant mass loss). It provides a rapid quantitative method to examine the overall combustion process, especially under non-isothermal conditions, and enables one to estimate the effective kinetic parameters for the overall decomposition reactions. The TG and DTG curves obtained from the TG–MS tests were two main methods in TGA. This technique has been widely used in recent years for investigation of combustion or structural characteristics of fossil fuels. The TG curve represents the variation of the residual mass of the asphalt as the temperature increases and a higher residual mass indicates a better fire retardance. DTG is the difference in thermogravimetry ratio of measurement of Dm (weight loss or weight increase) at heating/cooling/isotherm, with interpretation by Dm over temperature or time. In other words, it indicates the variation of mass loss as the temperature increases and the mass loss rate can also be obtained. The mass loss rate represents the efficiency of the mass loss during the combustion process. A higher mass loss rate means that the asphalt is easier to burn. In addition, to compare the difficulty of the asphalt to combust before and after adding fire retardants, decay rate of the maximum mass loss rate was proposed. A higher decay rate of mass loss rate corresponds to a better effect on fire retardance. The definition of the decay rate of mass loss rate is shown in Equation (1).

$$\eta = \frac{DTG_O - DTG_A}{DTG_O} \times 100\%,\tag{1}$$

where DTG_O = maximum mass loss rate of asphalt during combustion without fire retardant (%/min); DTG_A = maximum mass loss rate of asphalt during combustion or pyrolysis with fire retardant (%/min); η = decay rate of mass loss rate (%).

2.4.2. Growth Rate of Activation Energy

Activation energy is an internal factor that determines the rate of the chemical reaction; it can be used to evaluate the ease of combustion [27]. The higher the activation energy is, the more energy is needed for the combustion of the material and vice versa. In this study, the Coats–Redfern integral method was adopted to calculate the activation energy, which has been applied elsewhere [28]. In addition, the growth rate of activation energy was proposed in this study to evaluate the ease of combustion before and after adding fire retardants. The definition of activation energy is shown in Equation (2).

$$\lambda = \frac{AE_A - AE_O}{AE_O} \times 100\%,\tag{2}$$

where AE_O = activation energy of asphalt during combustion without fire retardant (KJ/mol); AE_A = activation energy of asphalt during combustion with fire retardant (KJ/mol); λ = growth rate of activation energy.

2.5. Thermodynamic Properties of Four Basic Fire Retardants

The thermodynamic properties of basic fire-retardant components may be different, therefore, it is necessary to analyze their thermodynamic behaviors accurately to determine the appropriate ratio of each component. The thermodynamic characteristics of the four basic fire retardants were studied in order to evaluate their fire retardance effects and provide a theoretical basis for the design of fire retardant combinations. Figure 2 shows the TG and DTG curves of the four basic fire retardants.

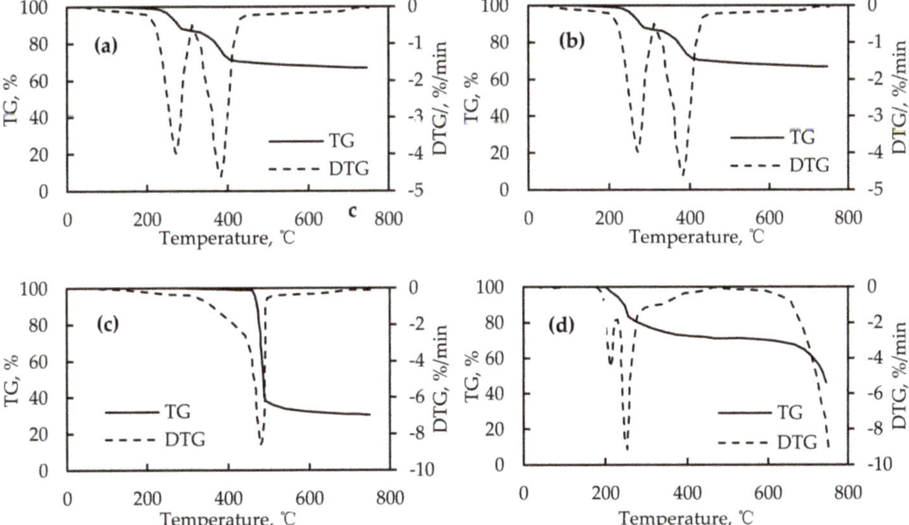

Figure 2. Thermogravimetric–differential thermogravimetric (TG–DTG) curves of four fire retardants. (**a**) Aluminum hydroxide; (**b**) magnesium hydroxide; (**c**) encapsulated red phosphorous; (**d**) expanded graphite.

As shown in Figure 2, the aluminum hydroxide had two peak temperatures: 269 °C and 383 °C, respectively. Its peak thermal weight loss rate was only −4% per minute, while its duration was long and the operating temperature ranged from 200 to 480 °C. It can be inferred that aluminum hydroxide is a relatively long-lasting fire retardant, but its fire retardance efficiency is relatively low. Thus, its proportion in the fire-retardant combination should be relatively high in order to improve the fire retardance efficiency. Magnesium hydroxide had a peak temperature of 285 °C and a peak thermal weight loss efficiency of −8% per minute. However, its duration was short and its working temperature ranged from 200 to 350 °C. It can be inferred that magnesium hydroxide is not a very lasting fire retardant, but its fire retardance efficiency is high. Thus, its content should be relatively low in the fire-retardant combination. With regard to the encapsulated red phosphorous, it had a peak temperature of 495 °C and its peak thermal weight loss efficiency was −8% per minute. However, its duration was very short and its working temperature ranged from 450 to 500 °C. It can be inferred that the fire retardance effect of encapsulated red phosphorus only works when the asphalt component exhibits a relatively high peak temperature. Additionally, there were two peak temperatures of expanded graphite at an early stage of the combustion, which were 212 °C and 254 °C, respectively. Its peak thermal weight loss efficiency was −8% per minute. Because the expanded graphite is primarily used for heat insulation, its chemical reaction should precede that of other fire retardants to effectively form a heat insulation layer. Therefore, its dosage should be strictly controlled to avoid affecting the road performance because it would cause great expansion when heated.

2.6. Experiment Design

Given that the expanded graphite would expand greatly after heating, some preliminary experiments were conducted to determine the appropriate ratio of basic fire retardants. According to the test results, the dosage of the expanded graphite was limited to 5% of the total mass of the fire retardants and the best ratio of the encapsulated red phosphorous to the expanded graphite was decided to be 3:1. Table 2 shows the different combinations of fire retardants.

Table 2. Fire-retardant combinations.

Combination Number	Aluminum Hydroxide (%)	Magnesium Hydroxide (%)	Expanded Graphite (%)	Encapsulated Red Phosphorous (%)
1	0	100	0	0
2	20	80	0	0
3	40	60	0	0
4	60	40	0	0
5	80	20	0	0
6	100	0	0	0
7	16	64	5	15
8	32	48	5	15
9	48	32	5	15
10	64	16	5	15
11	0	0	0	0

Note: The mass ratio of the fire retardants is determined to be 10% of the asphalt.

3. Numerical Analysis Model and Methods

3.1. Model Establishment

In this study, FDS was adopted to simultaneously calculate the temperature and smoke field in a fire. The size of the tunnel model in this study was 100 m long, 10 m wide, and 7.2 m high. 20-cm thick concrete was built in the tunnel and the pavement at the bottom of the tunnel was divided into three layers. Two asphalt surface layers were, respectively, 4 cm and 6 cm from top to bottom and the third concrete layer was set to be 10-cm thick. Figure 3 shows the numerical model and its grid division.

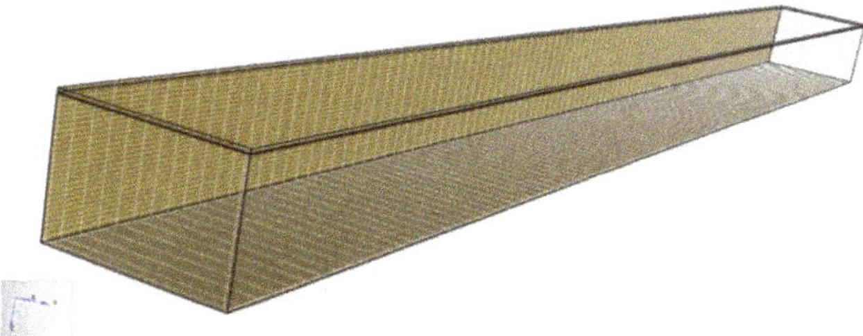

Figure 3. Numerical model and grid generation of the tunnel.

According to the model size, the grid was divided into 100 grids (1 m for each grid) in the longitudinal direction of the tunnel, 40 grids (0.25 m for each grid) in the transverse section, and 28 grids (0.26 m for each grid) in the vertical direction. Two kinds of materials, which were, respectively, cement concrete and asphalt concrete, were created in the model. The surface parameters of the fire source, asphalt concrete, and cement concrete were established in the "SURFACE" module. A tunnel model was established by the "OBSTRUCTURE" module and a fire source (2 m × 2 m on the road surface at the center of the tunnel), two asphalt pavement layers, a cement concrete layer, and vents on both sides of the tunnel were set in the "VENT" module in FDS. Observation points were created at different horizontal and vertical distances from the fire source to record the distribution of the temperature and smoke during the combustion process. Finally, the running time of FDS was selected to be 100 s in this study.

The temperature and smoke height distributions over time were recorded. The monitor points of the temperature were along the transverse and longitudinal directions of the tunnel; 9 of them were along the transverse direction and 6 of them were along the longitudinal direction. For the smoke height, a total of 6 monitor points was set up along the longitudinal direction of the tunnel.

3.2. Evaluation Index in Numerical Analysis

In the numerical modeling, the distribution of the temperature field after combustion can directly reflect the effect of fire retardance. The rate of increase in temperature would be reduced due to the fire retardants, which would in turn affect the distribution of the temperature field. In addition, the height distribution of the smoke can also be obtained in simulation; it is an important index for evaluating the survival probability of disaster victims [29,30] and is also an important parameter for analyzing the smoke suppression effect of fire retardants. As the smoke height decreased, its density correspondingly decreases, making the smoke flow upward. Therefore, the higher the smoke distributes, the better the smoke is suppressed. Furthermore, the cloud charts in FDS could be used to show the smoke height distribution over time [31]. Therefore, the temperature and smoke height distribution were used to evaluate the fire retardance and smoke suppression in the numerical analysis.

4. Results

4.1. Experimental Analysis of Fire Retardance Effects Based on the TG Curve

Figure 4 shows a typical TG curve of the asphalt with the combination of fire retardants. From Figure 4, it can be noted that the fire retardants do have an effect on retarding combustion due to the reduced mass loss. Compared to the control case, the temperature for the asphalt with fire retardants increased more than 300 °C before dehydration, which can ensure the asphalt a stable status under the high temperature. Thus, the fire retardants can help the asphalt maintain the mass. The residual mass percentage is listed in Table 3. From the results of combination 1 and combination 6, it can be found that the aluminum hydroxide solely had a better fire retardance than magnesium hydroxide because of its higher residual mass percentage. Furthermore, adjusting the basic fire-retardant content in the asphalt only leads to a little variation in fire retardance and no remarkable trend or relationship is found. Among all the tests, combination 9, which included 48% aluminum hydroxide, 32% magnesium hydroxide, 5% expanded graphite, and 15% encapsulated red phosphorous, results in the best fire retardance and combination 5, which included 80% aluminum hydroxide and 20% magnesium hydroxide leads to the least remarkable effect.

Table 3. Residual mass results of the asphalt.

Combination Number	Aluminum Hydroxide (%)	Magnesium Hydroxide (%)	Expanded Graphite (%)	Encapsulated Red Phosphorous (%)	Residual Mass Percentage (%)
1	0	100	0	0	14
2	20	80	0	0	16
3	40	60	0	0	15
4	60	40	0	0	20
5	80	20	0	0	12
6	100	0	0	0	17
7	16	64	5	15	17
8	32	48	5	15	19
9	48	32	5	15	23
10	64	16	5	15	19
11	0	0	0	0	0

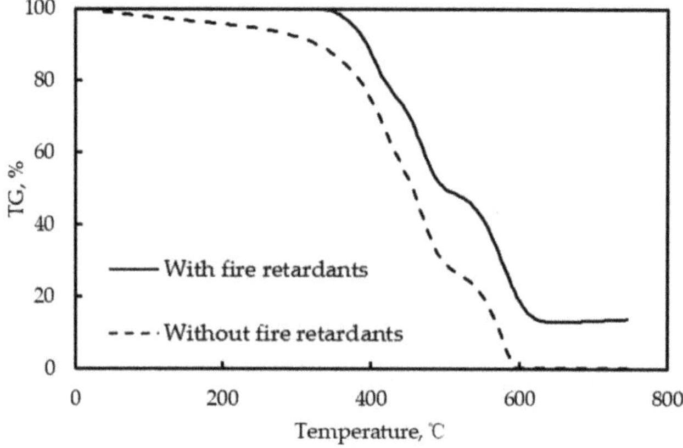

Figure 4. Typical TG curve of the asphalt with or without fire retardant.

4.2. Experimental Analysis of Fire Retardance Effects Based On the DTG Curve

Figure 5 shows the typical DTG (differential thermogravimetric) curve of the asphalt with the combination of fire retardants. As seen, the largest mass loss rate of the asphalt without the fire retardants was larger than that with fire retardants, indicating that the fire retardants can retard the asphalt combustion and in turn retard the mass loss. Table 4 shows the maximum mass loss rate and the decay rate of mass loss rate of different cases. It can be found that the variation in the basic fire-retardant content only leads to a little change in the two indexes and the aluminum hydroxide indicated a better fire retardance than magnesium hydroxide due to its lower maximum mass loss rate and higher decay rate of mass loss rate, which are similar to the results in TG curve analysis. Among all the cases, combination 9 shows the lowest mass loss rate and the highest decay rate of mass loss rate, indicating the best fire retardance effect. This result is the same as the analysis based on the TG curves. However, the combination which leads to the least remarkable fire retardance differs from that in the analysis based on the TG curves.

Table 4. Decay rate of mass loss rate of the asphalt.

Combination Number	Maximum Mass Loss Rate (%/min.)	Decay Rate Of Mass Loss Rate (%)
1	−7.9	13.4
2	−7.9	13.4
3	−8.0	12.3
4	−7.4	18.9
5	−7.9	13.4
6	−7.8	14.5
7	−7.8	14.5
8	−7.5	17.8
9	−7.2	21.1
10	−7.8	14.5
11	−9.2	-

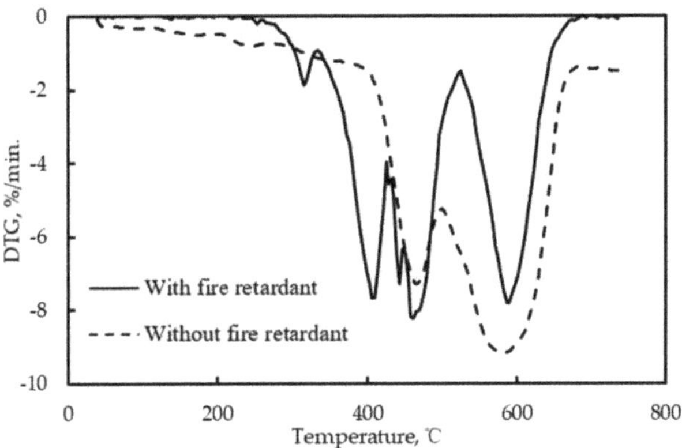

Figure 5. Typical DTG Curve of the Asphalt.

4.3. Experimental Analysis of Fire Retardance Effects Based on Activation Energy

Table 5 shows the activation energy and the corresponding increasing rate of each combination of fire retardants. As seen, the aluminum hydroxide had a higher activation energy and increasing rate than magnesium hydroxide, indicating a better fire retardance effect. Combination 9 shows the highest activation energy and increasing rate, which means the asphalt is the most difficult to burn when these fire retardants are added. This result correlates well with the analysis based on the TG and DTG curves. However, the lowest activation energy and increasing rate were found in combination 1, so this combination may have the least remarkable effect on fire retardance, which is different from neither TG analysis nor DTG analysis. Thus, further research is needed.

Table 5. Activation energy and its increasing rate of the asphalt.

Combination Number	Activation Energy (kJ/mol)	Increasing Rate of Activation Energy (%)
1	117.12	5.17
2	124.8	12.07
3	126.72	13.79
4	128.64	15.52
5	120.96	8.62
6	126.72	13.79
7	124.8	12.07
8	126.72	13.79
9	132.48	18.97
10	120.96	8.62
11	103.62	3.21

4.4. Numerical Analysis of Fire Retardance Effects Based on Temperature Distribution over Time

Figure 6 shows the distribution of the temperature along the transverse direction at the combustion point (0 m, 1 m, 2 m, 3 m, 4 m in the transverse direction of the combustion point) when no fire retardant was added and Figure 6b shows a typical distribution of the temperature along the transverse direction at the combustion point when fire retardants were added. It can be found that the temperature approximately presented a symmetrical distribution from the center of the combustion point. In the control case where no fire retardant was applied, the maximum temperature at each time point increased as the temperature increased and the highest temperature reached nearly 900 °C. However, after the addition of fire retardants, the temperature at all points decreased to different degrees and the decrease in the maximum temperature of the asphalt with fire-retardant combination 9 was only 759 °C, which fully demonstrated that the addition of fire retardants effectively inhibited the increase of temperature. Also, the asphalt with fire-retardant combination 6 exhibited a lower temperature at all points from the combustion point than combination 1, indicating that the aluminum hydroxide had a better effect on fire retardance than magnesium hydroxide.

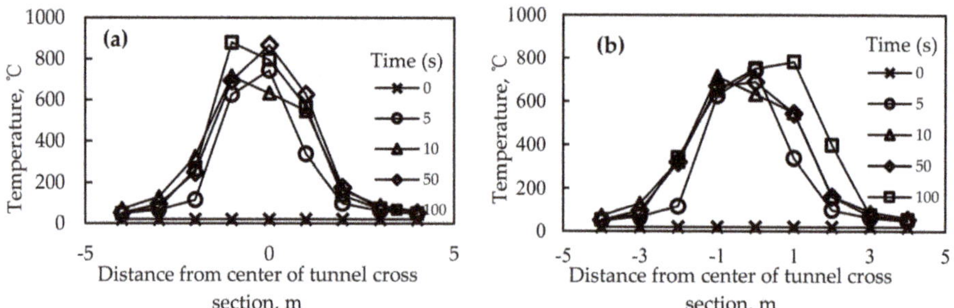

Figure 6. Distribution of temperature along the transverse direction; (**a**) without fire retardant; (**b**) with fire retardant.

Figure 7 shows a typical distribution of the temperature at the combustion point (0 m, 5 m, 10 m, 15 m, 20 m, 25 m in the longitudinal direction of the combustion point). It can be found that the temperature increased with time and the temperatures were all lower than 50 °C when it was 5 m away from the combustion point, indicating that the affected distance of the combustion along the longitudinal direction was less than 5 m.

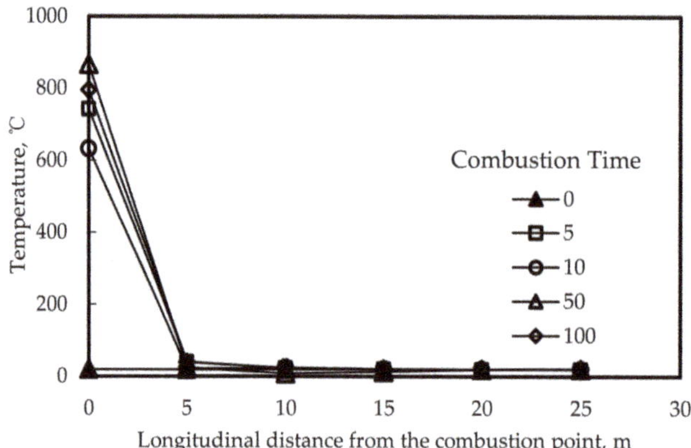

Figure 7. Distribution of the temperature along the longitudinal direction.

4.5. Numerical Analysis of Fire Retardance Effects Based on the Smoke Height Distribution over Time

Figure 8 shows the smoke height distribution over time when no fire retardant was added. Due to the restriction of the software function in the current version, only the smoke cloud chart without a legend could be obtained and many other researchers had used it to achieve satisfactory results [31,32]. Its distribution was analyzed to evaluate the effect on smoke suppression.

It can be seen that after the combustion, the smoke gradually expanded to the exits on both sides of the tunne and the height of the smoke in the tunnel decreased continuously. Due to the limitation of paper space, the smoke height distribution in the following analysis was reflected by smoke height distribution instead of cloud charts. Figure 9 shows the smoke height distribution along the longitudinal direction at the combustion point when no fire retardant was added. It shows that the smoke height was 2 m when it was 5 m away from the combustion point. Given that the height of people is normally under 2 m, thus, people in this range would be poisoned by the smoke in the tunnel.

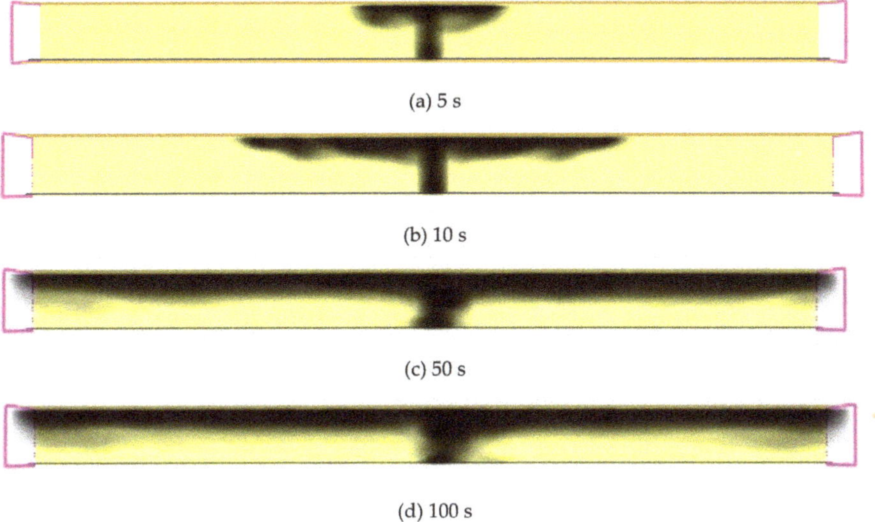

(a) 5 s

(b) 10 s

(c) 50 s

(d) 100 s

Figure 8. Smoke height distribution over time without fire retardant additive.

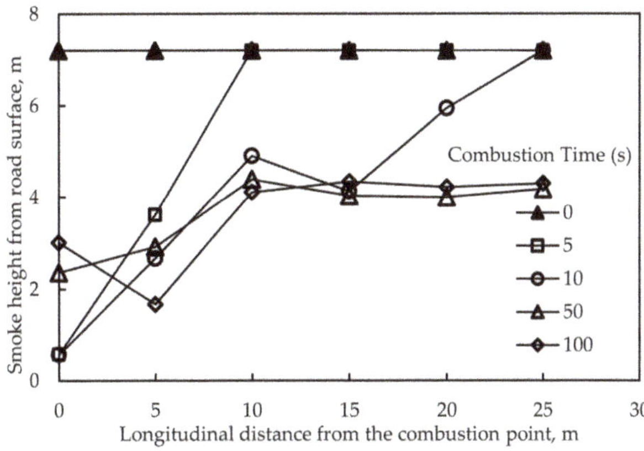

Figure 9. Smoke height distribution along the longitudinal direction (without fire retardant).

Figure 10 is a typical smoke height distribution along the longitudinal direction. Among them, the smoke height increased to different degrees when it was 5 m away from the combustion point and combinations 1 and 9 whose smoke height were the highest led to the best effect on smoke suppression. Taking the temperature distribution into consideration, the combination of fire retardants which would lead to optimal fire retardance and smoke suppression was combination 9. It is also consistent with the test results in laboratory experiments.

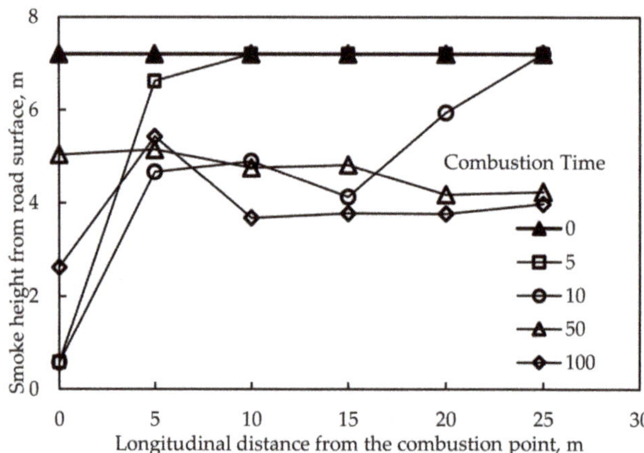

Figure 10. Smoke height distribution along the longitudinal direction (with fire retardants).

5. Conclusions

In this study, in order to research the effect of different modifiers on fire retardance, various combinations of fire retardants were added into the base asphalt and analyzed by laboratory experiments and numerical modeling. The following conclusions were drawn:

1. Among all the combinations of fire retardants used in this study, the combination of 48% aluminum hydroxide, 32% magnesium hydroxide, 5% expanded graphite, and 15% encapsulated red phosphorous leads to the best effect on fire retardance in experimental analysis. In numerical modeling, that combination also leads to an improved smoke suppression effect, while further research is needed to evaluate it in the real scenario.
2. The aluminum hydroxide indicated a better effect on fire retardance and smoke suppression than the magnesium hydroxide in both experimental and numerical analysis.
3. The temperature distribution on both sides of the combustion point is basically symmetrical. When the fire retardants were added, the temperature at each time and space point decreased to different degrees. Among all the combinations adopted in this study, it led to the largest decrease when 48% aluminum hydroxide, 32% magnesium hydroxide, 5% expanded graphite, and 15% encapsulated red phosphorous were added into the asphalt.
4. When no fire retardant was added, the smoke height at 5 m away from the combustion point was about 2 m. Among all the combinations adopted in this study, when fire retardants were added, the smoke height increased to different degrees and the height distribution became the highest when 48% aluminum hydroxide, 32% magnesium hydroxide, 5% expanded graphite, and 15% encapsulated red phosphorous were added.

Author Contributions: Conceptualization, G.X. and X.H.; methodology, T.M.; investigation, G.X. and J.Z.; writing—original draft preparation, J.Z. and X.C.; data curation, S.Z. and L.K.; writing—review and editing, G.X. and X.C.; supervision, T.M. and X.H.

Funding: This research was funded by the National Natural Science Foundation of China (grant number 51808116), Natural Science Foundation of Jiangsu Province (No. BK20180404), Fundamental Research Funds for the Central Universities with grant number 2242019K40204, Open Fund of State Engineering Laboratory of Highway Maintenance Technology (Changsha University of Science & Technology, No. kfj160104), and Science and Technology Projects of Shandong Transportation Department (No. 2017B60).

Conflicts of Interest: The authors declare no conflict of interest.

References

1. Petersen, J.C. Chapter 14 Chemical Composition of Asphalt as Related to Asphalt Durability. *Dev. Pet. Sci.* **2000**, *40*, 363–399.
2. Bazzaz, M.; Darabi, M.K.; Little, D.N.; Garg, N. A Straightforward Procedure to Characterize Nonlinear Viscoelastic Response of Asphalt Concrete at High Temperatures. *Transp. Rec.* **2018**, *2672*, 481–492. [CrossRef]
3. Bazzaz, M. Experimental and Analytical Procedures to Characterize Mechanical Properties of Asphalt Concrete Materials for Airfield Pavement Applications. Ph.D. Thesis, University of Kansas, Lawrence, KS, USA, August 2018; p. 247.
4. Liu, K.; Lu, L.; Wang, F.; Liang, W. Theoretical and experimental study on multi-phase model of thermal conductivity for fiber reinforced concrete. *Constr. Build. Mater.* **2017**, *148*, 465–475. [CrossRef]
5. Karimi, M.M.; Jahanbakhsh, H.; Jahangiri, B.; Nejad, F.M. Induced heating-healing characterization of activated carbon modified asphalt concrete under microwave radiation. *Constr. Build. Mater.* **2018**, *178*, 254–271. [CrossRef]
6. Jahanbakhsh, H.; Karimi, M.M.; Nejad, F.M.; Jahangiri, B. Viscoelastic-based approach to evaluate low temperature performance of asphalt binders. *Constr. Build. Mater.* **2016**, *128*, 384–398. [CrossRef]
7. Xu, T.; Wang, H.; Huang, X.M.; Guo, F.L. Inhibitory action of fire retardant on the dynamic evolution of asphalt pyrolysis volatiles. *Fuel* **2013**, *105*, 757–763. [CrossRef]
8. Qiu, J.; Yang, T.; Wang, X.; Wang, L.; Zhang, G. Review of the flame retardancy on highway tunnel asphalt pavement. *Constr. Build. Mater.* **2019**, *195*, 468–482. [CrossRef]
9. Ding, X.; Chen, L.; Ma, T.; Ma, H.; Gu, L.; Chen, T.; Ma, Y. Laboratory investigation of the recycled asphalt concrete with stable crumb rubber asphalt binder. *Constr. Build. Mater.* **2019**, *203*, 552–557. [CrossRef]
10. Li, J.; Zhang, J.; Qian, G.; Zheng, J.; Zhang, Y. Three-Dimensional Simulation of Aggregate and Asphalt Mixture Using Parameterized Shape and Size Gradation. *J. Mater. Civ. Eng.* **2019**, *31*, 04019004. [CrossRef]
11. Chen, J.; Li, J.; Wang, H.; Huang, W.; Xie, P.; Xu, T. Preparation and Effectiveness of Composite Phase Change Material for Performance Improvement of Open Graded Friction Course. *J. Clean. Prod.* **2019**, *214*, 259–269. [CrossRef]
12. Huang, J.; Liu, J.; Chen, J.; Xie, W.; Kuo, J.; Lu, X.; Chang, K.; Wen, S.; Sun, G.; Cai, H.; et al. Combustion behaviors of spent mushroom substrate using TG-MS and TG-FTIR: Thermal conversion, kinetic, thermodynamic and emission analyses. *Bioresour. Technol.* **2018**, *266*, 389–397. [CrossRef] [PubMed]
13. Kmita, A.; Fischer, C.; Hodor, K.; Holtzer, M.; Roczniak, A. Thermal decomposition of foundry resins: A determination of organic products by thermogravimetry–smoke chromatography–mass spectrometry (TG–GC–MS). *Arab. J. Chem.* **2018**, *11*, 380–387. [CrossRef]
14. Jayaraman, K.; Kok, M.V.; Gokalp, I. Thermogravimetric and mass spectrometric (TG-MS) analysis and kinetics of coal-biomass blends. *Renew. Energy* **2017**, *101*, 293–300. [CrossRef]
15. Xu, T.; Huang, X.M. Combustion properties and multistage kinetics models of asphalt binder filled with fire retardant. *Combust. Sci. Technol.* **2011**, *183*, 1027–1038. [CrossRef]
16. Michelle, G.M.; Leni, F.M.L.; Cheila, G.M. Kinetic parameters of different asphalt binders by thermal analysis. *J. Therm. Anal. Calorim.* **2011**, *106*, 279–284.
17. Shen, T.-S.; Huang, Y.-H.; Chien, S.-W. Using fire dynamic simulation (FDS) to reconstruct an arson fire scene. *Build. Environ.* **2008**, *43*, 1036–1045. [CrossRef]
18. Sellami, I.; Manescau, B.; Chetehouna, K.; de Izarra, C.; Nait-Said, R.; Zidani, F. BLEVE fireball modeling using Fire Dynamics Simulator (FDS) in an Algerian smoke industry. *J. Loss Prev. Process Ind.* **2018**, *54*, 69–84. [CrossRef]
19. Wahlqvist, J.; Van Hees, P. Validation of FDS for large-scale well-confined mechanically ventilated fire scenarios with emphasis on predicting ventilation system behavior. *Fire Saf. J.* **2013**, *62*, 102–114. [CrossRef]
20. Mo, S.J.; Li, Z.R.; Liang, D.; Li, J.X.; Zhou, N.J. Analysis of Smoke Hazard in Train Compartment Fire Accidents Base on FDS. *Procedia Eng.* **2013**, *52*, 284–289. [CrossRef]

21. Kang, D.I.; Kim, K.; Jang, S.-C.; Yoo, S.Y. Risk assessment of main control board fire using fire dynamics simulator. *Nucl. Eng. Des.* **2015**, *289*, 195–207. [CrossRef]
22. Oliveira, R.L.F.; Doubek, G.; Vianna, S.S. On the behaviour of the temperature field around pool fires in controlled experiment and numerical modelling. *Saf. Environ. Prot.* **2019**, *123*, 358–369. [CrossRef]
23. Li, M.; Pang, L.; Chen, M.; Xie, J.; Liu, Q. Effects of Aluminum Hydroxide and Layered Double Hydroxide on Asphalt Fire Resistance. *Materials* **2018**, *11*, 1939. [CrossRef] [PubMed]
24. Chai, H.; Duan, Q.; Jiang, L.; Sun, J. Effect of inorganic additive flame retardant on fire hazard of polyurethane exterior insulation material. *J. Therm. Anal. Calorim.* **2018**, *135*, 2857–2868. [CrossRef]
25. Zhao, H.; Li, H.-P.; Liao, K.-J. Study on Properties of Flame Retardant Asphalt for Tunnel. *Pet. Sci. Technol.* **2010**, *28*, 1096–1107. [CrossRef]
26. Shen, M.Y.; Chen, W.J.; Tsai, K.C.; Kuan, C.F.; Kuan, H.C.; Chou, H.W.; Chiang, C.L. Preparation of expandable graphite and its flame retardant properties in HDPE composites. *Polym. Compos.* **2017**, *38*, 2378–2386. [CrossRef]
27. Ahn, J.; Eastwood, C.; Sitzki, L.; Ronney, P.D. Gas-phase and catalytic combustion in heat-recirculating burners. *Proc. Combust. Inst.* **2005**, *30*, 2463–2472. [CrossRef]
28. Naqvi, S.R.; Tariq, R.; Hameed, Z.; Ali, I.; Naqvi, M.; Chen, W.-H.; Ceylan, S.; Rashid, H.; Ahmad, J.; Taqvi, S.A.; et al. Pyrolysis of high ash sewage sludge: Kinetics and thermodynamic analysis using Coats-Redfern method. *Renew. Energy* **2019**, *131*, 854–860. [CrossRef]
29. Xu, P.; Jiang, S.; Xing, R.; Tan, J. Full-scale immersed tunnel fire experimental research on smoke flow patterns. *Tunn. Undergr. Space Technol.* **2018**, *81*, 494–505. [CrossRef]
30. Gao, Z.; Ji, J.; Fan, C.; Li, L.; Sun, J. Determination of smoke layer interface height of medium scale tunnel fire scenarios. *Tunn. Undergr. Space Technol.* **2016**, *56*, 118–124. [CrossRef]
31. Valasek, L.; Glasa, J. Simulation of the course of evacuation in tunnel fire conditions by FDS+ Evac. In Proceedings of the 2013 International Conference on Applied Mathematics and Computational Methods in Engineering, Almería, Spain, 24–27 June 2013; pp. 288–295.
32. Li, Q.; Li, S.H.; Wang, Z.-H. Research on Smoke Exhaust Effect at Different Installation Height of Mechanical Exhaust Port in Ring Corridor of High-Rise Building. *Ind. Eng. Manag.* **2016**, *5*, 327–335. [CrossRef]

Article

Experimental Study on the Phase Transition Characteristics of Asphalt Mixture for Stress Absorbing Membrane Interlayer

Guang Yang [1], Xudong Wang [1,2,*], Xingye Zhou [2] and Yanzhu Wang [1]

[1]   School of Transportation Science and Engineering, Harbin Institute of Technology, Harbin 150090, China; guang.yang@rioh.cn (G.Y.); wangyzh0102@163.com (Y.W.)
[2]   Research Institute of Highway Ministry of Transport, Beijing 100088, China; zhouxingye1982@163.com
*   Correspondence: xd.wang@rioh.cn

Received: 29 November 2019; Accepted: 16 January 2020; Published: 19 January 2020

Abstract: Asphalt mixtures used in stress absorbing membrane interlayers (SAMIs) play a significant role in improving the performance of asphalt pavement. To investigate the rheological properties and phase transition characteristics of asphalt mixtures used in SAMI with temperature changes, twenty-seven candidate mixtures with different binders, gradation types and binder contents were selected in this research. During the study, dynamic mechanical analysis method was employed to evaluate their temperature-dependent properties and a series of wide-range temperature sweep tests were conducted under a sinusoidal loading. Some critical points and key indexes from the testing curves such as glass transition temperature (Tg) can be obtained. Test results show that phase transition characteristics can better reflect the rheological properties of asphalt mixtures at different temperatures. Crumb rubber modified asphalt mixtures (AR) provide a better performance at both high and low temperatures. Additionally, the range of AR asphalt mixtures' effective functioning temperature ΔT is wider, and the slope K value is greater than the others, which indicates that AR asphalt mixtures are less sensitive to temperature changes. Additionally, gradation type and asphalt content also influence the properties: finer gradation and more asphalt content have a good effect on the low-temperature performance of the asphalt mixtures; while mixtures with a coarser gradation and less asphalt content perform better at high temperature and they are less sensitive to temperature changes. Finally, AR asphalt mixture is more suitable to be applied in the SAMI due to its phase transition characteristics from this method.

Keywords: stress absorbing membrane interlayer; dynamic mechanical analysis; temperature sweep test; phase transition; crumb rubber modified asphalt mixture

---

1. Introduction

Stress Absorbing Membrane Interlayer (SAMI) is a thin and soft layer composed of asphalt layers [1], which is widely used as a functional layer between semi-rigid base and asphalt layer or between old asphalt pavement and overlay. Usually, it is used to prevent the reflection cracking and to prevent water entering the base course. Molenaar et al. [2] thought application of a SAMI prevented reflection of cracks through an overlay. Some studies have indicated that the SAMI can delay reflective cracking and plays a role in isolating the overlay from relative deflection of the cracked underlying layer due to traffic loading [3]. For its function of absorbing stress and diffusing deformation, researchers have mainly focused on the mechanical properties of SAMI, such as their shear resistance, flexural-tensile and anti-fatigue characteristics [4,5]. However, their properties with temperature changes have rarely been studied, and extreme weather conditions may lead them to fail quickly.

Asphalt mixtures are typical temperature-sensitive materials and exhibit totally different characteristics at different temperatures. They are usually stiff at low temperature and deform very little under stress, so thermal cracking is induced more easily when a larger thermal stress exceeds their tensile limit; meanwhile, at high temperature, they become more like a fluid, and rutting may occur [6]. With a gradual increase in temperature, the phase of asphalt mixture changes from a glassy state to a highly elastic state, until it finally reaches a viscous state. In engineering applications, asphalt binders are usually modified in order to improve their performance. There are several ways to modify asphalt binders: sulphur, fatty acid amides, polyphosphoric acid, waxes and polymers [7,8]. Furthermore, among polymers, styrene-butadiene-styrene copolymer (SBS), styrene-butadiene rubber (SBR), polyethylene (PE) and ethylene vinyl acetate (EVA) are the most widely used [9,10]. However, the application of modified asphalt binders influences the phase transition process of mixtures, and their rheological characterization requires further study [11]. Phase transition and rheological properties are closely related to pavement distress, so it is very important to evaluate the performance of asphalt mixtures by studying them.

Dynamic Mechanical Analysis (DMA) is an efficient method for studying the rheological properties of viscoelastic materials. Relevant research has demonstrated that the viscoelastic behavior of asphalt binder or cement and asphalt mortar can be studied by this method [12,13]. DMA can be simply described as applying an oscillating force to a sample and analyzing the material's response to that force. As a typical viscoelastic material, asphalt mixture can be investigated by this method at different temperatures.

From the literature review it can be concluded that recent studies mainly focused on the mechanical properties of asphalt mixtures used in SAMI, such as shear resistance and anti-fatigue performance. However, the effects of environmental factors on the characterization of mixtures have rarely been considered. Investigating the phase transition characteristics of asphalt mixtures with changes in temperature is a more efficient approach to evaluating the high- and low-temperature performance and temperature sensitivity of asphalt mixtures. In this paper, different types of asphalt mixtures are selected to determine which is more suitable for the material in SAMI.

The objective of the present study was to develop a new test method for exploring the rheological and visco-elastic properties of asphalt mixtures used in SAMI with broad temperature changes, and this method can be applied to selecting the optimum asphalt mixture for the SAMI layer. To achieve these objectives, a dynamic mechanical analysis was considered, and temperature sweep tests were conducted on multiple asphalt mixtures with a variety of binder types, gradations and binder contents. High- and low-temperature performance and temperature sensitivity can be compared and evaluated in order to select the optimum mixtures in the SAMI.

2. Materials and Methods

2.1. Asphalt Binder

Three typical kinds of asphalt were used to study the effect of them on properties of mixtures: neat asphalt binder A70 (70 means the penetration of needle in a binder at 25 °C ranges from 60 to 80 according to the Chinese specification, JTG F40-2004 [14]), SBS modified asphalt, and crumb rubber modified asphalt (AR). They are widely used in the actual production of asphalt mixtures but have obvious differences in properties with temperature and load changing. Table 1 lists the technical index of the three binders.

Table 1. Technical index of asphalt binder.

Technical Index	A70	SBS	AR
Penetration/0.1 mm (25 °C)	71.5	63.4	37.8
Softening point/°C	48.4	72.7	70.0
Ductility/cm (5 cm/min, 5 °C)	-	28.5	6.7

2.2. Asphalt Mixture

2.2.1. Gradation

The gradations of asphalt mixtures in this paper were classified by smooth function curves that were fitted by two critical control points. For AC-5 asphalt mixtures (the maximum nominal particle size is less than 5 mm), the passing rate of nominal maximum size and 0.075 mm sieve size were marked as two critical points: their coordinates were (d_{max}/d_{max}, passing rate at 4.75 mm) and ($d_{0.075mm}/d_{max}$, passing rate at 0.075 mm), respectively [15]. Based on the function of asphalt mixtures used for SAMI, the passing rate of 0.075 mm sieve was set as 10%, and that of the nominal maximum size was 100%. As a result, two points (1, 100) and (0.075/4.75 = 0.01579, 10) were calculated. Then, three different functions (logarithmic, power and exponential respectively) were employed to fit the gradation curves, as shown in Figure 1.

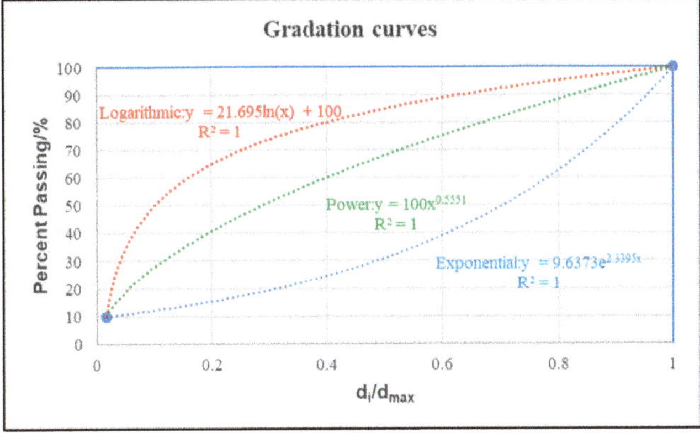

Figure 1. Fitting of gradation curves.

Finally, three different gradations were determined, marked as M, Z and D. As can be seen from Figure 2, the mixture with gradation D has more fine aggregate, and that with gradation Z is coarser than the other two.

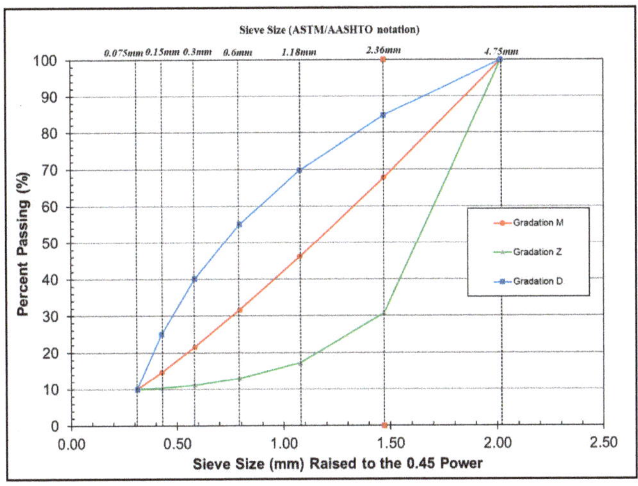

Figure 2. Gradation of AC-5 mixtures.

2.2.2. Asphalt Content

Not only the asphalt type and gradation, but also asphalt content directly affects the properties of asphalt mixtures. To make a better comparison of three asphalt mixtures, three asphalt-aggregate ratios were selected by weight for each mixture (5.5%, 6.5% and 7.5%).

To sum up, 27 asphalt mixtures were designed in this test with three asphalt types, three gradations and three asphalt contents, shown in Table 2 and Figure 3.

Table 2. Asphalt mixture types.

Asphalt Type	Gradation	Asphalt Content
A70, SBS and AR	M, Z and D	5.5%, 6.5% and 7.5%

Figure 3. 27 AC-5 asphalt mixtures.

2.3. Experimental Program

2.3.1. Specimen Preparation

The above mixtures were compacted by gyratory compaction method with a diameter of 150 mm. Then both ends were cut to eliminate the uneven distribution of air void. The remaining parts were cut into slices with a high-precision cutting machine (Presi, France). The slices were 60 mm in length, 15 mm in width and 3 mm in thickness. There are six parallel samples for each type of asphalt mixture to ensure the reliability of results.

2.3.2. Experimental Device

Dynamic Mechanical Analysis (DMA for short) was a practical method to study the rheology properties of viscoelastic materials [16,17]. A Dynamic Mechanical Analyzer called DMA Q800 (TA Instruments, New Castle, DE, USA) was used in this test, shown in Figure 4. It can achieve a wider range of temperature and frequency than other devices for asphalt mixtures and simulate the real environment condition of asphalt pavement at a short time. A dual cantilever clamp was selected to measure the samples under flexural load, shown in Figure 5.

Figure 4. DMA Q800 device.

Figure 5. Sample with dual cantilever clamp.

2.3.3. Test Method

The temperature sweep test is an efficient method for studying the materials' rheology properties in a wide range of temperatures [18–20]. The materials were tested under a sinusoidal load with the temperature continuously increasing or decreasing. According to a series of trials, experiments with 1 Hz frequency were chosen as the final parameters. The temperature range is from −40 °C to 80 °C, and the heating rate was set as 2 °C/min, which is more applicable to asphalt mixture.

Before testing, the linear elastic range should be determined for each specimen to ensure elastic deformation within a certain strain level. Finally, 25 $\mu\varepsilon$ was used as the strain level of temperature sweep test.

2.3.4. Parameter Acquisition

From the DMA test, a series of parameters can be obtained by measuring the specimen variation under the sine wave load: Complex modulus E*, storage modulus E′, loss modulus E″ and tangent of phase angle tanδ (see Figure 6). The complex modulus E * is composed of real and imaginary components, referred to as the storage modulus E′ and loss modulus E″. Their correlations can be expressed by the following Equations (1)–(4). The phase transition process from glassy state to highly elastic state and then to viscous state can be obtained by measuring the change in these parameters.

$$E' = \sigma_0 \cos\delta / \varepsilon_0, \tag{1}$$

$$E'' = \sigma_0 \sin\delta / \varepsilon_0, \tag{2}$$

$$E^* = E' + iE'', \tag{3}$$

$$\tan\delta = E'' / E', \tag{4}$$

where: σ_0 is stress amplitude, MPa; ε_0 is the strain amplitude; δ is the phase angle; E′ is storage modulus, MPa; E″ is loss modulus, MPa; E* is complex modulus, MPa; i = $\sqrt{-1}$.

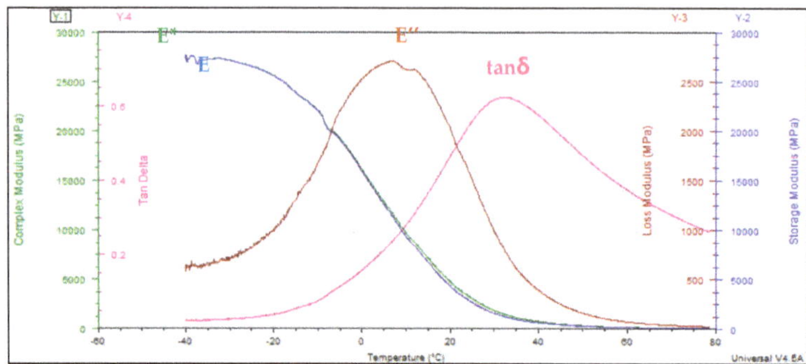

Figure 6. Parameters from temperature sweep test.

3. Results

As can be seen from Figure 7, complex modulus-temperature is a smooth curve, and there exist two platforms at high and low temperatures. With increasing temperature, the complex modulus E* starts to decrease gradually, finally reaching a steady state at high temperature. According to the characteristics of the curve, it can be fitted by the Bolzmann function [21], as shown in Equation (5).

$$y = \frac{A_1 - A_2}{1 + e^{\frac{x-x_0}{dx}}} + A_2 \tag{5}$$

where A_1 is the maximum modulus; A_2 is the minimum modulus; x_0 and dx are two parameters that describe the shape of the curve.

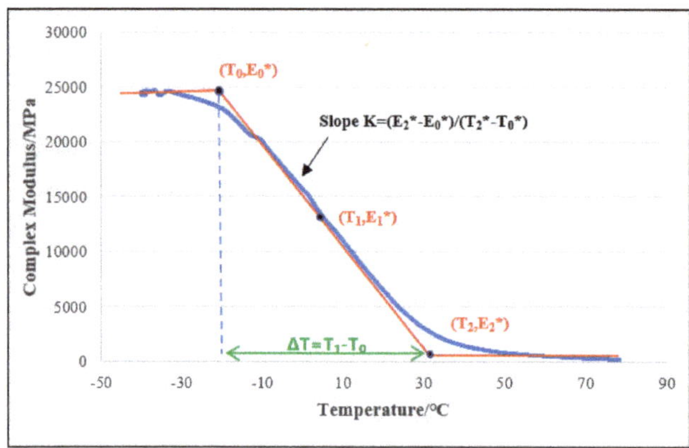

Figure 7. Complex modulus curve.

Additionally, there was an obvious peak in loss modulus-temperature curve (Figure 6) and the Gauss function can be used to fit the curve and determine the value of the peak, shown in Equation (6).

$$y = y_0 + \frac{A}{\omega \sqrt{\pi/2}} e^{-2\frac{(x-x_c)^2}{\omega^2}} \tag{6}$$

where x_c is the temperature where the maximum of loss modulus occurs; y_0 is the minimum of the loss modulus; A and ω are two parameters that describe the shape of the curve.

Likewise, there was also a peak in the tangent of phase angle-temperature curve. The GaussAmp function can be used to fit this curve better and estimate the value of the peak, as shown in Equation (7).

$$y = y_0 + Ae^{-\frac{(x-x_c)^2}{2\omega^2}}$$

(7)

where y_0 is the minimum of tanδ; x_c is the temperature where the maximum of tanδ occurs; A and ω are two parameters that describe the shape of the curve.

Based on the above functional fitting, critical points of different curves can be obtained and rheological properties of asphalt mixtures with temperature change can be represented by these points.

3.1. Glass Transition Temperature Tg

Glass transition is the reversible transition from a glassy state into a viscous or rubbery state with increasing temperature. Some researchers have stated that Tg is closely related to the low-temperature performance of asphalt mixture [22]. There are several test methods that can obtain Tg, such as DSC, NMR and DMA. Research indicates that the DMA method can measure the asphalt Tg of asphalt-filler mastics accurately by applying a sine wave load on the sample [23]. Lower Tg temperature corresponds to higher fracture energy, which means better low temperature performance of asphalt pavement [24]. Asphalt mixture with lower Tg means it transfers from the high-elastic state to the glassy state at a lower temperature, which indicates that it has a better performance in resisting low-temperature impact. From the DMA method, three critical points can be used as the Tg: onset temperature of E′ curve, temperature at peak of loss modulus and temperature at peak of tanδ.

The objective of study in this paper was asphalt mixture, which was a mix of asphalt, aggregate and mineral powder. The corresponding temperature at peak of tanδ was usually above 40 °C, and it was in the high-temperature region for asphalt pavement. At the same time, the onset temperature of E′ was not easy to determine, because of the curve shape. Ultimately, the temperature at the peak of the loss modulus was selected as the Tg. The results are shown in Figure 8.

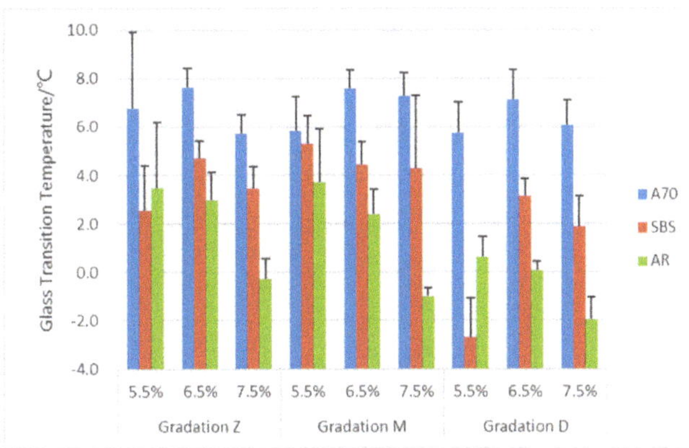

Figure 8. Tg of asphalt mixtures.

Test results indicate that AR asphalt mixtures have a lower Tg (except SBS mixture with gradation D and 5.5% asphalt content) no matter what the gradation or asphalt content is. It is indicated that AR asphalt has a better low-temperature performance than the other two mixtures. Compared with the other two mixtures, those with neat asphalt A70 have a higher Tg, which means that polymer modifier and crumb rubber have a positive effect on the asphalt mixture to improve their low-temperature performance.

Additionally, gradation type also affected the value of Tg, but the results differed with different asphalt types. For SBS and AR asphalt mixtures, there is a decreasing trend of Tg with the increasing content of fine aggregates from Z to D. In particular, Tg of the finest gradation D decreased obviously than the other two gradations. However, there is little change for Tg of A70 asphalt mixtures with three gradations. It can be concluded from the result that the increasing of fine aggregate content would benefit the low-temperature performance of asphalt mixtures, but it has little effect on the mixtures with neat asphalt.

Moreover, the influence of asphalt content on the Tg was different for the three binder types. For AR asphalt mixtures, Tg decreased a lot with increasing asphalt content, especially at 7.5%. The sufficient asphalt rubber can help mixture release the thermal stress with decreasing temperature. For A70 asphalt mixtures, there exists a peak at 6.5% with a higher Tg, and there was no clear law for SBS asphalt mixtures.

In summary, asphalt type in the mixtures plays a more significant role to the low-temperature performance, followed by gradation type and asphalt content. AR asphalt mixtures have a better low-temperature performance than the other two mixtures.

3.2. Stiffness at Extreme Low Temperature $E_0{}^*$

The stiffness of asphalt mixtures at low temperature is also an indication of its performance. Some studies have demonstrated that lower stiffness has a better low-temperature performance [18,22]. It can be seen from Figure 7, the complex modulus approaches a constant when the temperature is extreme low for asphalt pavement (usually below −30 °C). From the fitting curve of the complex modulus, the critical point $(T_0, E_0{}^*)$ (Figure 7) can be obtained and $E_0{}^*$ can be used to represent the stiffness of mixture at extreme low temperature, shown in Figure 9.

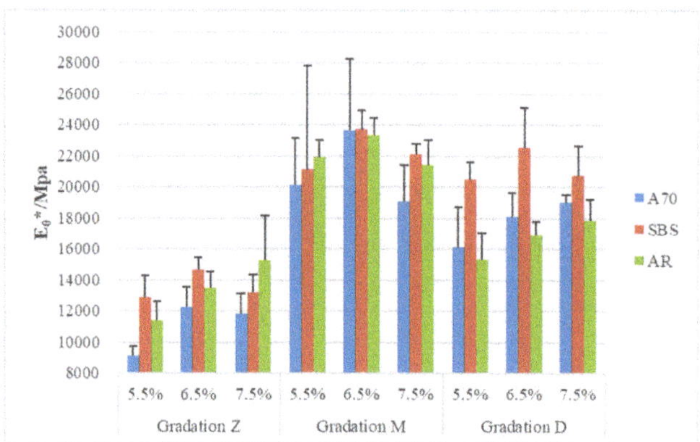

Figure 9. $E_0{}^*$ of asphalt mixtures (MPa).

Based on these results, asphalt mixtures with different asphalt types do not have obvious trends in stiffness at extreme low temperature. For gradation M, the stiffness of asphalt mixtures was almost the same among three asphalt types. Similarly, values of mixtures with different asphalt contents fluctuated within a narrow range. This is due to the condition that the mixtures are in the glassy state.

On the contrary, gradation type played a key role on the stiffness. Mixtures with gradation Z had a lower stiffness mainly due to their high air voids. A large number of holes can be seen on the sample of A70 asphalt mixtures with gradation Z. As a result, asphalt mixtures with gradation Z may not have advantage in resistance to water damage. The stiffness reached a peak at gradation M with increasing compactibility, but higher stiffness means that they cannot make a larger deformation under the same stress. From the gradation M to D, the content of fine aggregates increased further, but the stiffness decreased instead, which is an optimal condition for low-temperature performance of asphalt mixtures.

Factors that influence the stiffness were more complex than those on the Tg. For example, increasing air void may decrease the stiffness of asphalt mixtures such as gradation Z of A70 asphalt, but this was adverse to the low-temperature performance and waterproofness. Lower stiffness with less air void, like gradation D, is more beneficial with respect to resistance to thermal cracking.

3.3. High-Temperature Properties T_{tan}

As mentioned previously, Tg can be determined by the peak of tangent of phase angle. However, the value of peak was usually above 40 °C for the asphalt mixture, which is in the region of high temperature for asphalt pavement. A higher T_{tan} means it transfers from a highly elastic state to a viscous state at a higher temperature, which is beneficial to high-temperature performance. From the test, it can be observed that after the peak point of tanδ, the sample started to yield, and the stiffness of the asphalt remained at a lower level. Therefore, it can be used to evaluate the high-temperature performance of asphalt mixtures and it has a clear physical meaning. The results of T_{tan} are shown in Figure 10.

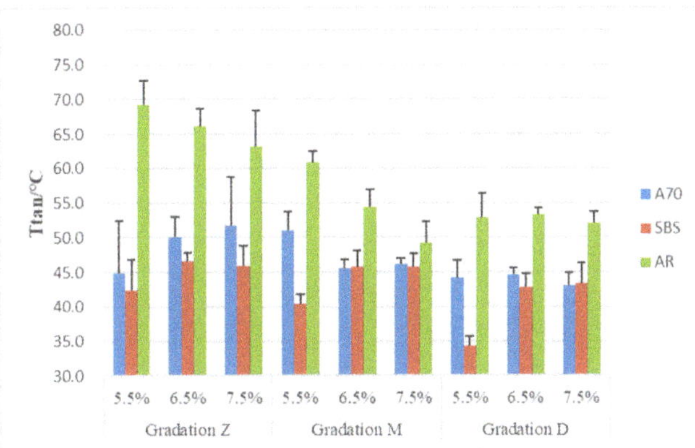

Figure 10. T_{tan} of asphalt mixtures.

The results show that AR asphalt mixtures have an obviously higher T_{tan} than the other two mixtures. It can be found that they performed better at higher temperature. Additionally, gradation Z with coarser aggregates have a higher T_{tan}. This is mainly because coarser aggregates interlock with each other and form a skeleton structure to resisting permanent deformation.

3.4. Range of Effective Function Temperature ΔT

Based on the E* graph, the curve starts to turn with increasing temperature. T_0 can be used to describe the turning temperature of phase transition. With a gradual increase in temperature, it reaches a platform, and T_2 can also be marked as the turning temperature for another phase transition. As a result, $\Delta T = T_2 - T_0$ shows the range of temperature that can work well for this asphalt mixture, and material in this range was neither too glassy nor too soft. A larger ΔT means it is highly adaptable to temperature change and has a better performance. The results of ΔT are shown in Figure 11.

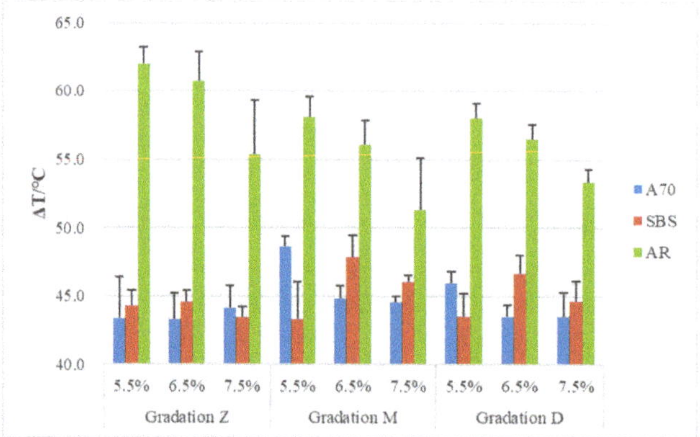

Figure 11. ΔT of asphalt mixtures.

Results indicate that AR asphalt mixtures have a wider effective functional zone than the other two mixtures. For A70 and SBS asphalt mixtures, the effects of gradation and asphalt content on ΔT are not obvious. For AR asphalt mixtures, there are no obvious differences between different gradation types. However, with increasing asphalt content, ΔT becomes smaller, which means more asphalt makes its range narrower and mixtures are more sensitive to temperature change.

3.5. Temperature Sensitivity K Value

From the E* curve, the value of modulus decreases from 10^4 to 10^2 MPa with the temperature increasing, which means a significant change to asphalt pavement. The slope K of the curve can be used to represent the sensitivity of materials under the effect of temperature. K value can be calculated by Equation (8). The results are shown in Figure 12:

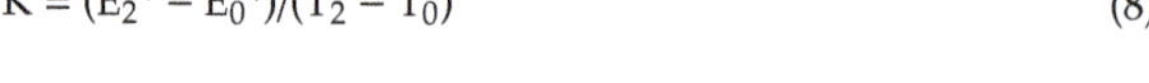

$$K = (E_2{}^* - E_0{}^*)/(T_2 - T_0) \tag{8}$$

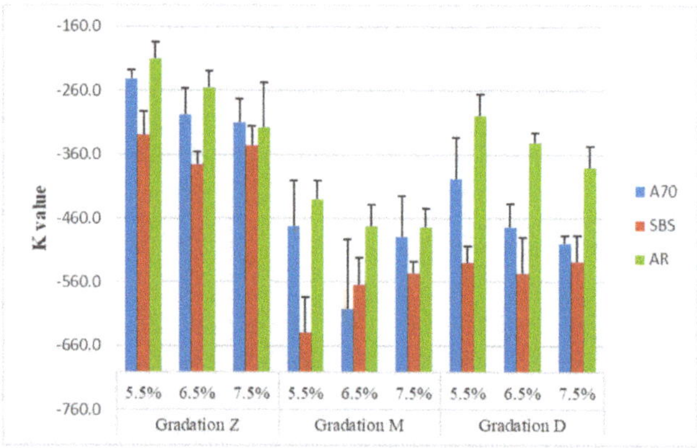

Figure 12. K value of asphalt mixtures.

The results indicate that asphalt mixtures with gradation Z have a higher K, which means their complex moduli have the minimum change with increasing temperature. One reason is that their lower moduli at low temperature cause lower ΔE, on the other hand, coarser gradation has a better ability to release stress than the other two mixtures.

As for asphalt type, AR asphalt mixtures have a higher K value than the others, which indicates that they have a smaller change of moduli. The K values of A70 mixtures are higher than those of SBS mixtures. The results indicate that SBS asphalt mixtures are most sensitive to temperature changes.

As for binder content, the results show that AR asphalt mixtures containing more asphalt were more sensitive to temperature changes. A70 asphalt mixtures with gradation Z and D have a similar rule. There are no obvious rules for SBS asphalt mixtures.

In terms of gradation type, mixtures of gradation M have a lower K, followed by D and Z. If air void is considered, gradation D is more applicable in SAMI.

In conclusion, the effects of asphalt type, gradation type and binder content on the different indicators can be concluded in Table 3.

Table 3. Summary of the results.

Index	Binder Type	Gradation Type	Binder Content
Tg	* AR < SBS < A70	D < M ≈ Z	7.5 < 6.5 < 5.5 (AR)
E_0^*	Not clear	* Z < D < M	Not clear
T_{tan}	* AR > A70 ≈ SBS	Z > M > D	Not clear
ΔT	* AR > SBS ≈ A70	Not clear	5.5 > 6.5 > 7.5 (AR)
K value	AR > A70 > SBS	* Z > D > M	5.5 > 6.5 > 7.5 (AR)

* means the key impact factor among three variables.

The above five indicators can be used to describe phase transition characteristics of asphalt mixtures with temperature changes, and then a comprehensive evaluation of their performance can be made. Tg and E* can characterize the low-temperature properties of mixtures. T_{tan} reflects their stability at higher temperature. ΔT and K value show their temperature sensitivity and effective function range. The test results indicate that the three variables have different effects on the five indicators, and the key impact factor changes with each indicator.

4. Conclusions

Based on the testing and analysis, there are some conclusions that can be summarized as follows:

(1) By measuring the glass transition temperature Tg, it can be concluded that crumb rubber modified asphalt mixtures AR with finer gradation D and higher asphalt content have a better low-temperature performance. Asphalt type has the largest impact on the Tg, which shows that asphalt plays a key role in low-temperature properties.

(2) For stiffness at extreme low temperature, values of E_0^* among different asphalt types are similar. The gradation type plays a major role for E_0^* and mixtures with gradation M are larger than the others. Stiffness of mixtures with gradation Z is very low due to its higher air void and it is detrimental to waterproofness. AR asphalt mixtures with gradation D has a relative lower stiffness and it is beneficial to the low-temperature performance.

(3) According to phase transition characterization, T_{tan} (temperature at peak of tanδ) can be used to evaluate the high-temperature performance of asphalt mixtures. Coarser gradation has a good effect on the high-temperature performance. Mixtures with AR asphalt have an obvious higher T_{tan} and they have a better high-temperature stability.

(4) Range of effective function temperature ΔT can be used to describe the range of temperatures where asphalt mixtures can stay in a high-elastic state and work well. Slope K value can be used to evaluate the sensitivity of stiffness to temperature changes. Results show that AR asphalt mixtures have a wider ΔT and a lower K, which indicates that they are less sensitive to various temperatures. Besides, mixtures with more asphalt content and finer gradation are more sensitive to temperature changes.

(5) Considering the comprehensive evaluation of various indicators, AR asphalt mixtures are more suitable materials to be applied in SAMI. An optimum gradation type and asphalt content can be determined from the actual environmental condition and application requirement.

Author Contributions: Conceptualization, G.Y. and X.W.; Data curation, G.Y. and Y.W.; Formal analysis, G.Y. and Y.W.; Funding acquisition, X.W.; Investigation, X.Z.; Methodology, G.Y. and Y.W.; Project administration, X.W.; Writing—original draft, G.Y.; Writing—review & editing, G.Y. and X.Z. All authors have read and agreed to the published version of the manuscript.

Funding: This research was supported by National Key R&D Program of China (2017YFC0840200).

Acknowledgments: This study is supported by Fundamental Research Innovative Center from Research Institute of Highway, Ministry of Transport. The efforts of group members are highly appreciated. Also sincere thanks to Pavement Materials Research Group from Harbin Institute of Technology and the China Scholarship Council (award to Guang Yang for 18 months' study abroad at University of California, Davis).

Conflicts of Interest: The authors declare no conflict of interest.

References

1. Barksdale, R.D. Fabrics in asphalt overlay and pavement maintenance. In *National Cooperative Highway Research Program Synthesis of Highway Practice 171*; Transportation Research Board: Washington, DC, USA, 1991; pp. 7–9.
2. Molenaar, A.A.A.; Heerkens, J.C.P.; Verhoeven, J.H.M. *Effects of Stress Absorbing Membrane Interlayers*; Association of Asphalt Paving Technologists: St. Paul, MN, USA, 1986; Volume 55, pp. 453–481.
3. Ogundipe, O.; Thom, N.; Collop, A. Evaluation of performance of stress-absorbing membrane interlayer (SAMI) using accelerated pavement testing. *Int. J. Pavement Eng.* **2013**, *14*, 569–578. [CrossRef]
4. Tang, W.; Sheng, X.; Sun, L. Research on Asphalt Mixture Performance of Stress Absorbing Membrane Interlayer (SAMI). *J. Build. Mater.* **2009**, *12*, 173–176.
5. Wu, K. Study on Design and Performance of Stress Absorbing Membrane Interlayer (SAMI) Mixture. *J. Highw. Transp. Res. Dev.* **2006**, *23*, 18–20.
6. Kim, H.; Wagoner, M.P.; Buttlar, W.G. Micromechanical fracture modeling of asphalt concrete using a single-edge notched beam test. *Mater. Struct.* **2008**, *42*, 677–689. [CrossRef]
7. Bahia, H.U.; Hislop, W.P.; Zhai, H.; Rangel, A. Classification of asphalt binders into simple and complex binders. *J. Assoc. Asphalt Paving Technol.* **1998**, *67*, 1–41.
8. Salas, M.Á.; Pérez-Acebo, H.; Calderó, V.; Gonzalo-Orden, H. Bitumen modified with recycled polyurethane foam for employment in hot mix asphalt. *Ing. Investig.* **2018**, *38*, 60–66. [CrossRef]
9. García-Morales, M.; Partal, P.; Navarro, F.J.; Martínez-Boza, F.; Gallegos, C.; González, N.; González, O.; Munoz, M. Viscous properties and microstructure of recycled eva modified bitumen. *Fuel* **2004**, *83*, 31–38. [CrossRef]
10. Wang, K.; Yuan, Y.; Han, S.; Yang, Y. Application of FTIR spectroscopy with solvent-cast film and PLS regression for the quantification of SBS content in modified asphalt. *Int. J. Pavement Eng.* **2019**, *20*, 1336–1341. [CrossRef]
11. Bahia, H.U.; Hanson, D.I.; Zeng, M.; Zhai, H.; Khatri, M.A.; Anderson, R.M. *Characterization of Modified Asphalt Binders in Superpave Mix Design*; Transportation Research Board: Washington, DC, USA, 2001; pp. 32–37.
12. Fu, Q.; Xie, Y.; Long, G.; Niu, D.; Song, H. Dynamic Mechanical Thermo-analysis of Cement and Asphalt Mortar. *Powder Technol.* **2017**, *313*, 36–43. [CrossRef]
13. Yuan, Q.; Liu, W.; Pan, Y.; Deng, D.; Liu, Z. Characterization of Cement Asphalt Mortar for Slab Track by Dynamic Mechanical Thermoanalysis. *J. Mater. Civ. Eng.* **2016**, *28*, 04015154. [CrossRef]
14. Ministry of Transport of the People's Republic of China. *Technical Specification for Construction of Highway Asphalt Pavement JTG F40-2004*; China Communication Press: Beijing, China, 2004; pp. 12–14.
15. Wang, X. *Multi-Performance Design of Hot Mix Asphalt Based on Stone Interlock Structure Theory*; China Communication Press: Beijing, China, 2014; pp. 338–353.
16. Hevin, P.M. *Dynamic Mechanical Analysis—A Practical Introduction*, 2nd ed.; CRC Press: Boca Raton, FL, USA, 2008; pp. 1–4.

17. Hesami, E.; Ghafar, A.N.; Birgisson, B.; Kringos, N. *Multi-scale Characterization of Asphalt Mastic Rheology*; Springer Science and Business Media LLC: Berlin, Geramny, 2013; pp. 45–61.

18. Soliman, H.; Shalaby, A. Characterizing the Low-Temperature Performance of Hot-Pour Bituminous Sealants Using Glass Transition Temperature and Dynamic Stiffness Modulus. *J. Mater. Civ. Eng.* **2009**, *21*, 688–693. [CrossRef]

19. Kang, Y.; Song, M.; Pu, L.; Liu, T. Rheological behaviors of epoxy asphalt binder in comparison of base asphalt binder and SBS modified asphalt binder. *Constr. Build. Mater.* **2015**, *76*, 343–350. [CrossRef]

20. Fang, L.; Yuan, Q.; Deng, D.; Pan, Y.; Wang, Y. Effect of Mix Parameters on the Dynamic Mechanical Properties of Cement Asphalt Mortar. *J. Mater. Civ. Eng.* **2017**, *29*, 04017080. [CrossRef]

21. Cercignani, C. The boltzmann equation. In *The Boltzmann Equation and Its Applications*; Springer: New York, NY, USA, 1988; pp. 40–103.

22. Lei, Z.; Bahia, H.; Yi-Qiu, T. Effect of bio-based and refined waste oil modifiers on low temperature performance of asphalt binders. *Constr. Build. Mater.* **2015**, *86*, 95–100. [CrossRef]

23. Guo, M.; Tan, Y.Q.; Zhang, L. Effect of Filler on Glass Transition of Asphalt Mastics. *Adv. Eng. Forum* **2012**, *5*, 376–381. [CrossRef]

24. Lei, Z.; Yi-Qiu, T.; Bahia, H. Relationship between glass transition temperature and low temperature properties of oil modified binders. *Constr. Build. Mater.* **2016**, *104*, 92–98. [CrossRef]

Article

Performance Characterization of Semi-Flexible Composite Mixture

Weiguang Zhang [1,*], Shihui Shen [2], Ryan Douglas Goodwin [3], Dalin Wang [1] and Jingtao Zhong [1]

[1] School of Transportation Engineering, Southeast University, Nanjing 211189, China; 230199144@seu.edu.cn (D.W.); zhongjingtao@seu.edu.cn (J.Z.)

[2] Department of Engineering, Pennsylvania State University, Altoona, PA 16601, USA; szs20@psu.edu

[3] Department of Civil and Environmental Engineering, Washington State University, Pullman, WA 99164, USA; ryan.goodwin@wsu.edu

* Correspondence: wgzhang@seu.edu.cn

Received: 2 December 2019; Accepted: 6 January 2020; Published: 11 January 2020

Abstract: Semi-flexible composite mixture (SFCM) is developed based on a unique material design concept of pouring cement mortar into the voids formed by open graded asphalt mixture. It combines the flexibility of asphalt concrete and the stiffness of Portland cement concrete and has many advantages comparing to conventional roadway paving materials. The main objective of this paper was to evaluate the engineering properties of SFCM and assess the constructability of the SFCM. A slab SFCM sample was fabricated in the laboratory to simulate the filling of cement mortar in the field. Performance testing was carried out by indirect tensile (IDT) test because it was found to be able to correlate with the field performance of asphalt mixtures at low, intermediate, and high temperatures. They were used in this study to evaluate the thermal cracking, fatigue, rutting, as well as moisture resistance of SFCM. A control hot mix asphalt (HMA) mixture was used to compare with the results of SFCM. Based on the testing results, it was found that the designed SFCM showed good filling capability of cement mortar. SFCM had higher dynamic modulus than the control HMA. It had good resistance to rutting and moisture damage. Based on fracture work, SFCM showed better resistance to thermal cracking while lower resistance to fatigue cracking.

Keywords: semi-flexible composite mixture; cement mortar; asphalt mixture; indirect tensile test

---

1. Introduction

High-performance cement pastes and pure cement paste (s) are respectively grouted into matrix asphalt mixtures to service as semi-flexible pavement materials [1], The resultant composite structure is referred as "resin modified pavement (RMP)" [2–4], combines the flexibility of asphalt pavement and the stiffness close to concrete pavement. It has traditionally been used as a special pavement surfacing due to its excellent rutting resistance and fuel spillage [5].

The advantages of the SFCM have been found which included but not limited to: (a) in contrast to Portland cement concrete (PCC), the SFCM does not require joints that are used to accommodate thermally induced movements [4,5]; (b) it has strong resistance to moisture damage due to the very impermeable structure [6]; (c) it provides a tough and durable pavement surface that can better resist rutting caused by heavy channelized traffic loads and traffic abrasion, and surface deterioration due to fuel spillage [2], rutting resistance is also good at early curing stage [7]; and (d) it can be quickly open to traffic, usually within 24 h after pouring cementious material [8].

Given the advantages of the SFCM, it has the potential of being used at heavy duty roads such as airport pavements and bridge deck surfaces [2]. The SFCM can be precast as slab and transported to the site to save construction time. The initial cost of a full depth of SFCM design using hot mix asphalt is about 50 to 80 percent higher than a dense-graded asphalt concrete while 30 to 60 percent less than a comparable PCC pavement design, whereas its cost can be dramatically reduced by using cold mix to substitute hot mix asphalt as core asphalt structure [4,8].

Limited research on SFCM has been carried out in both laboratory and fields. Using the specific design reported by Anderton [4], the SFCM was found to have about the same indirect tensile strength as AC (one type of asphalt mixture) at low testing temperatures, whereas two to three times higher strength than AC at moderate to high testing temperatures (i.e., 50 °C or higher). Its flexural strength and compressive strength were approximately 40–60% and 10–25% of a typical PCC mix, respectively. The thermal coefficients of a SFCM mix was in the same general range as PCC, and about two to three times lower than that of hot mix asphalt (HMA) mixture. Oliveira et al. [9,10] evaluated the fatigue performance of the laboratory prepared SFCM with rest periods considered. It was found that the fatigue performance was improved by using SFCM. A shift factor was also suggested by Oliveira et al. [11] to convert laboratory fatigue testing results of SFCM to field fatigue life. Studies by Hao and Ling [12,13] indicated that semi-flexible pavement have good low temperature cracking resistance and excellent rutting resistance. Huang et al. [14] used six emulsified asphalt-to-cement (A/C) ratios and found 0.3 was the optimum one that can balance performance of SFCM between elastic modulus and fatigue life [14]. Al-Qadi et al. [2,6] studied SFCM as a possible alternative for bridge deck overlays considering its improved engineering performance compared to a typical AC wearing surface. The study concluded that SFCM was two to three times more resistant to moisture and chloride intrusion than PCC due to its low air void content. Skid resistance of SFCM pavement was evaluated using a self-watering Mark V Mu-Meter method on airfield pavements to be compared with AC and PCC pavements of the same age [4]. This method is standard for the US Air Force and the Federal Aviation Administration. The wet surface coefficient of friction was measured at two speeds, 65 km/h (low-speed test) and 96 km/h (high-speed test). It was found that the SFCM had about the same coefficient of friction as control AC taxiways at low speed, and better coefficients of friction than that of the PCC taxiway tested at high speed. A long-term performance survey of SFCM over a five-year period indicates that there was no rutting observed throughout the duration of the project. For comparison, the control HMA section used PG82-22 binder was paved at the same time and in the same project with SFCM section. Rutting survey result by the end of the fourth year shows there is an average rutting depth of 0.25 inches for the control section. The skid resistance of SFCM was noted as a problem shortly after paving but it was improved within a few months, and kept at a good level by the end of the fourth year. The pavement condition rating (PCR) value of SFCM, which is a parameter to quantify pavement's overall performance based on measurements of roughness, surface distress, skid resistance and deflection, were acceptable by the end of the survey [14].

As seen, the current state-of-practice for SFCM is empirically developed based upon limited experiences in European countries and some airfield and military projects in the United States. Early performance data from test sections have shown great promise for expanded future usage of this technology as an alternative sustainable application for modern transportation engineering [4]. However, the lack of research methodology and comprehensive understanding about its engineering properties could hinder the quick and appropriate deployment of this technology. Moreover, the pouring capability, as one of the most crucial properties, is usually evaluated based on standard specimen size (i.e., 150 mm in height and 100 mm in diameter), which actually cannot simulate the field flow ability of the materials especially considering the side effects. Furthermore, most previous studies assumed that the materials are elastic whereas its very likely the materials would behave as visco-elastic due to the existance of asphalt; or most studies simply applied strength and moisture tests, whereas neglected advanced test methods such as fracture energy and dynamic modulus which could better capture the properties and failure mode of the SFCM materials, in contrast to the conventional HMA mixtures. Therefore, the objective of this paper is to assess the constructability of the SFCM mixture and compare the mechanical properties (fatigue, thermal, rutting, moisture damage, and dynamic modulus) of the SFCM with a typical hot mix asphalt (HMA) mixture. A higher PG grade binder (PG70-22) was used in the control HMA mix to evaluate how the SFCM compares with a higher-grade mix. A SFCM slab was made in laboratory and cored into cylindrical specimens (100 mm in diameter) for further testing. All the tests were performed under the indirect tensile mode which has been found to be able to characterize the fatigue, rutting, and thermal cracking properties of asphalt mixtures and reasonably correlate with field performance [15,16].

2. Mix Design and Sample Preparation of SFCM

The mix design of SFCM includes two components, asphalt mixture and cement mortar. The strength and other mechanical properties of the SFCM were found to be greatly affected by the proportion of the two components [17]. The main purpose of asphalt mixture design procedure was to obtain optimum asphalt binder content, as well as target air voids. In contrast, the design of cement mortar was targeted to balance cementitious material with high strength and good fluidity.

2.1. Aggregate Gradation

Aggregate gradation was designed to form a mixture skeleton to allow adequate room for asphalt binder and cement mortar. According to the recommendation by Battey and Whittington [17], gradation used in this study is coarse graded to realize high air voids in asphalt mixtures. Figure 1 shows the gradation used in this study.

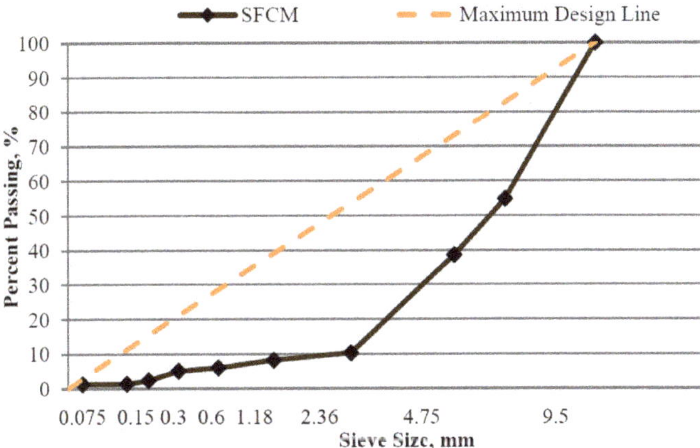

Figure 1. Aggregate gradation of asphalt mixture.

2.2. Optimum Asphalt Content

This study used PG64-22 binder with 0.1% of anti-stripping agent (by weight of asphalt binder). Optimum asphalt content (OAC) was calculated based on an empirical equation suggested by Roffee [18], as shown in Equation (1),

$$Optimum\ asphalt\ content = 3.25aS^{0.2} \tag{1}$$

where,

$a = 2.65/SG$
SG = apparent specific gravity of the combined aggregates
S = conventional specific surface area = $0.21G + 5.4S + 7.2s + 135f$
G = percentage of material retained on 4.75 mm (No.4) sieve
S = percentage of material passing 4.75 mm (No.4) and retained on 600 μm (No.30) sieve
s = percentage of material passing 600 μm (No.30) and retained on 75 μm (No.200) sieve
f = percentage of material passing 75 μm (No.200) sieve

The OAC was then determined as 3.6% by mass of aggregates. The OAC value is less than typical results mainly due to the relative high ratio of coarse aggregates and high design air voids.

2.3. Mixing and Compaction of Asphalt Mixture

Asphalt mixture was mixed at a high temperature of 156 °C with mixing duration of 30 min. As soon as the mixing was completed, they were loaded into a square mold with 530 mm in length and width by 100 mm in depth. Mixtures were compacted slightly using a small vibration compactor to get target sample height and an even surface. The mold was made of plywood and was well sealed to avoid outflow. Preparation of slab sample allows simulating the field paving condition and evaluating the mortar filling capability. Additional asphalt mixtures were also prepared to check the theoretical maximum specific gravity (G_{mm}) according to AASHTO T209. The averaged G_{mm} value was calculated as 1.848.

2.4. Mix Design of Cement Mortar

Materials and proportions (by weight) used to prepare cement mortar included water (25%, 13.15 kg in weight), type I-II cement (50%, 26.3 kg in weight), fly ash (23%, 12.1 kg in weight) and superplasticizer (2%, 1.05 kg in weight), with the total weight of the specimen of 52.6 kg. Study indicates that fly ash can enhance porosity, permeability and shrinkage characteristics of SFCM [3]. The dosage of superplasticizer was used as recommended by the literature [19] to reduce the water to cement ratio at the same time maintain a high workability and fluidity of the mortar to assist pouring process.

2.5. Viscosity Check of Cement Mortar

The viscosity of cement mortar was checked to determine its fluidity using the Marsh funnel method (ASTM D6910). The Marsh funnel viscosity is defined as the time required for 964 mL of a slurry to flow into a graduated cup from a funnel with specific dimensions. For 964 mL of water at the temperature of 21 ± 3 °C, the flow duration is 26 ± 0.5 s. Following the method described in the specification, cement mortar was prepared and tested in lab using the same proportions shown above. The room temperature (25 °C) and the temperature of cement mortar (20 °C) were also recorded for information purpose. Three replicates were performed and flow time were measured as 144, 146, and 148 s, which is very flowable compared with the testing results of fresh cement mortar (240–300 s) without using superplasticizer [20].

2.6. Cement Mortar Pouring

The asphalt sample was waited for approximately 24 h to cool down before starting the pouring. The cement mortar was mixed for several minutes to ensure the homogeneity. During pouring, high workability and infiltration rate were observed and the total filling procedure finished in less than 20 min. After pouring was complete, the slab sample was cured for 14 days at room temperature. Water was sprayed on sample frequently and the sample was sealed securely to maintain a humid curing environment.

Figure 2a,b shows the asphalt mixture sample before and after filling, and Figure 2c presents cement mortar during filling. Sample mold was removed after 14 days of curing and sample bottom was well-filled with cement mortar as shown in Figure 2d.

Figure 2. Sample preparation before and after cement mortar pouring. (**a**) Sample before filling cement mortar, (**b**) sample after filling cement mortar, (**c**) cement mortar pouring, (**d**) sample bottom.

2.7. Samples Coring and Sawing

The demolded sample was cored into 9 specimens with 100 mm in diameter and 100 mm in height. Figure 3a,b present figures during and after sample coring. Among the 9 specimens, four of them were sawed into 63.5 mm in height to perform indirect tensile (IDT) strength testing and the rest were cut into 38 mm in height to carry out other IDT testing. Figure 3c,d shows a comparison before and after sample sawing. As seen, asphalt mixture was filled up with cement mortar pretty well.

Figure 3. *Cont.*

(c)   (d)

Figure 3. Sample before and after sawing. (**a**) Sample coring, (**b**) slab sample after coring, (**c**) sample before sawing, (**d**) sample after sawing.

3. Performance Testing

In this study, five types of IDT tests were carried out to evaluate the performance of SFCM specimens, including IDT dynamic modulus testing, IDT fatigue and thermal testing, IDT tensile strength, and IDT high temperature strength testing. IDT testing results of a control PG70-22 HMA mixture was also presented to conduct a comparison with SFCM specimens. A higher PG grade binder than the PG binder of the SFCM core asphalt structure (PG64-22) was used to show how SFCM could compare with higher PG grade HMA.

This paper was majorly focusing on comparing the material properties of the SFCM specimens with the conventional hot mix asphalt mixtures, whereas the performance comparison between the HMA mixtures and other mixtures with modified binders (i.e., warm mix asphalt, polymer, rubber, RAS, etc.) can be easily found within published articles. Actually, the performance comparison between the HMA and the SFCM provides useful information in determining in which scenario the SFCM can be best adopted.

3.1. IDT Dynamic Modulus Testing

IDT dynamic modulus testing was performed using three replicate specimens. Six temperatures (−20, −10, 0, 10, 20, and 30 °C) and five frequencies (0.1, 1, 5, 10, and 20 Hz) were applied to conduct the testing and 20 °C was selected as the reference temperature when plotting master curve. Control HMA specimens were also tested using the same specimen geometry and testing conditions. The control HMA specimens were compacted in the laboratory using loose mixtures taken from the plant. Other mixture design properties include: PG70-22 binder, nominal maximal size of 19 mm, target air voids of 4%, and 100 design numbers of gyrations. It should be noted that a higher PG grade (PG70-22 vs. PG64-22) binder was used for control mixture to compare the SFCM mixture with a higher grade HMA mixture.

Figure 4 shows the dynamic modulus master curve in a log-log scale for SFCM and the control HMA specimens using reference temperature of 20 °C. As shown, dynamic moduli of SFCM specimens are much higher than that of HMA specimens at all frequencies (temperatures). In addition, the dynamic modulus of SFCM varies with the change of frequency (temperature), indicating SFCM is also a viscoelastic material in nature. However, the SFCM seems to be less sensitive to frequency (temperature) change in contrast to the HMA mixture which indicates that SFCM can be used at an extensive environmental condition. The extra high dynamic moduli of SFCM specimens can be attributed to: (a) very low air voids of the mix; (b) high stiffness cementitious filling material; and (c) additional confining effect of cement mortar to asphalt structure especially at high temperatures. It is also interesting to note that the variability of the SFCM specimens at lower frequency was higher, which could be caused by the method that was utilized to generate the master curve: the current method was developed majorly based on asphalt mixtures, although the SFCM specimens are essentially visco-elastic whereas its properties are different from conventional from the conventional HMA samples. Thus, the current method may need to be adjusted to better fit the master curve generation procedure of the SFCM specimens.

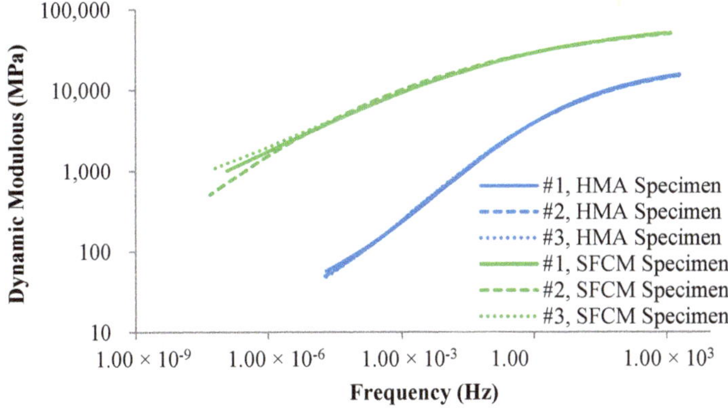

Figure 4. Comparison of dynamic modulus between semi-flexible composite mixture (SFCM) and hot mix asphalt (HMA) specimens.

3.2. IDT Tensile Strength Testing

IDT tensile strength ratio for wet and dry specimens, as recommended by AASHTO T283 test [21], is used to evaluate the moisture susceptibility of SFCM and the control HMA mixture. All the samples were sawed into 100 mm in diameter and 63.5 ± 2.5 mm in thickness. Two specimens were tested for each set, dry and wet.

For dry specimens, they were wrapped with plastic and placed in a 25 °C water bath for 2 h before testing. The other subset was firstly conditioned at a constant temperature of −16 °C for 24 h, and was then placed in a 60 °C water bath for another 24 h. It should be noted that after 24 h conditioning in the 60 °C water bath, a small quantity of asphalt binder was squeezed out from SFCM specimens. This could be caused by the very low air voids of specimens and a continuous exposure to high temperature environment. Specimens were then removed from the 60 °C water bath, wrapped and put into a 25 °C water bath for 2 h before testing. Chamber of MTS equipment which was used for IDT test was also pre-conditioned at a temperature of 25 °C so that the specimens can keep the same temperature during testing. Since SFCM specimens were assumed to have extremely low air voids, they were not applied vacuum pressure, nor was the degree of saturation determined. However, vacuum pressure was carried out on HMA specimens before conditioning. All the four dry and wet SFCM specimens were tested on the same day to avoid effects of curing. Tensile strength and tensile strength ratio (TSR) are calculated based on Equations (2) and (3).

$$S_t = \frac{2000P}{\pi t D} \tag{2}$$

where:

S_t = tensile strength, kPa
P = maximum load, N
t = specimen thickness, mm
D = specimen diameter, mm

$$TSR = \frac{S_2}{S_1} \tag{3}$$

where:

S_1 = average tensile strength of the dry subset, kPa (psi); and
S_2 = average tensile strength of the conditioned subset, kPa (psi).

The testing results are summarized in Table 1. As shown, the calculated TSR value of SFCM specimen is high, indicating that the SFCM has good resistance to moisture damage. When the SFCM specimens were broken after strength test, no obvious stripping was observed, showing a good bonding among asphalt binder, aggregates, and cement mortar. In contrast to the HMA specimens, the higher TSR of SFCM specimens could be attributed to its extreme low air voids: the moisture takes longer to penetrate into the specimens using the current moisture conditioning for HMA at lower air voids, furthermore, the penetration of moisture into the surface between asphalt and cement was even more difficult.

Table 1. Summary of indirect tensile (IDT) strength testing results (25 °C).

Sample	Sample Thickness, mm	Sample Diameter, mm	Maximum Load, N	S_t, MPa	Average S_t, MPa	TSR
Dry #1, SFCM	62.1	100	22,900	2.31	2.32	
Dry #2, SFCM	64.6	100	24,000	2.33		
Conditioned #1, SFCM	64.9	100	23,100	2.23	2.24	0.97
Conditioned #2, SFCM	64.4	100	23,200	2.26		
Dry #1, HMA	63.2	100	20,300	2.04	2.04	
Dry #2, HMA	64.1	100	20,500	2.04		
Conditioned #1, HMA	63.1	100	18,050	1.82	1.81	0.89
Conditioned #2, HMA	62.2	100	17,600	1.80		

3.3. IDT Fatigue Testing

By determining the fracture behavior of SFCM at intermediate temperature, the IDT test can be used to evaluate the fatigue cracking resistance of SFCM which takes into account both the strength and the ductility of the mixture [15]. In this study, fatigue test was performed at a temperature of 20 °C and a strain rate of 50.8 mm/min. Three replicate specimens were tested and stress-strain relationship is shown in Figure 5 for both the SFCM specimens and the control HMA mixtures. The figure indicates that generally the SFCM specimens experienced higher maximum stress than the control HMA specimens, whereas the stress value of SFCM specimens decreases faster than that of HMA specimens after the peak value, showing a more brittle behavior. Table 2 compares the maximum stress and fracture work between the SFCM specimens and the HMA specimens. The maximum stress corresponds to the stress value at which failure strain occurs, and fracture work is calculated as the area under the stress-strain curve, which can be adopted to relate to the field fatigue performance of asphalt mixture. The results in the table plus that in Figure 5 indicate that the fatigue resistance of the SFCM specimens is not as strong as that of the control HMA specimens.

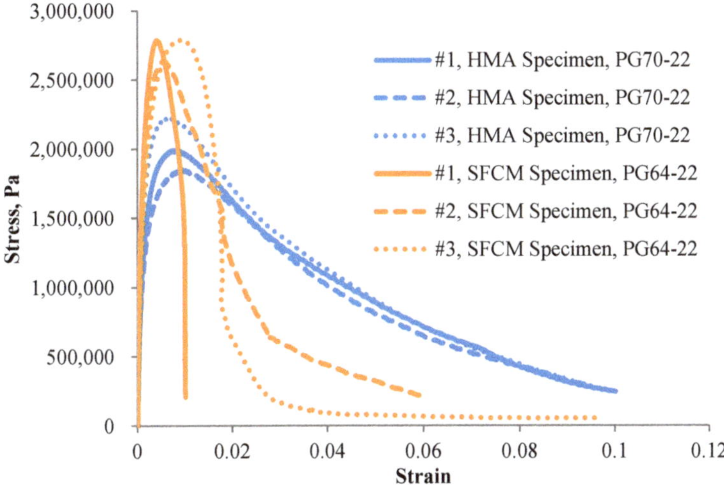

Figure 5. Stress-strain relationship of fatigue testing.

Table 2. Summary of fracture work and maximum stress (20 °C).

HMA Specimens			SFCM Specimens		
Specimen No.	Fracture Work, N·mm	Maximum Stress MPa	Specimen No.	Fracture Work, N·mm	Maximum Stress MPa
#1	38,337.23	1.99	#1	23,601.33	2.63
#2	40,518.21	2.23	#2	25,366.62	2.79
#3	45,387.78	1.85	#3	21,465.77	2.79
Average	41,414.41	2.02	Average	23,477.91	2.74
CV, %	8.7	9.4	CV, %	8.3	3.3

3.4. IDT Thermal Cracking Testing

IDT thermal cracking test was found to be able to correlate with the thermal cracking resistance of visco-elastic materials in the field [16]. By measuring load-deformation relationship at low temperature, stress and strain values can be calculated and fracture work can be calculated accordingly as well. In this study, thermal cracking was conducted at a temperature of −10 °C with a strain rate of 2.54 mm/min. Three replicate specimens were tested.

A summary of the testing results as well as the stress-strain relationship for both the SFCM specimens and the HMA specimens are shown in Figure 6 and Table 3. As presented, the average maximum stress of SFCM (3.62 MPa) is slightly lower than that of HMA mixtures (3.96 MPa), whereas the fracture work of SFCM (174,305 N·mm) is higher than that of the HMA mixtures (167,396 N·mm), indicating an overall better resistance to thermal cracking of SFCM. The asphalt binder become brittle at low temperatures which induces fast decrease of stress value with constant strain rate; in contrast, the asphalt was somehow protected by the cement mortar in the SFCM and thus its stress release ability still works well to endure load after failure strain. Additionally, it was observed that the variability of the SFCM specimen was higher than that of the control HMA specimens, which could be caused by the relative coarse aggregate gradation of the SFCM and will be validated in the future study.

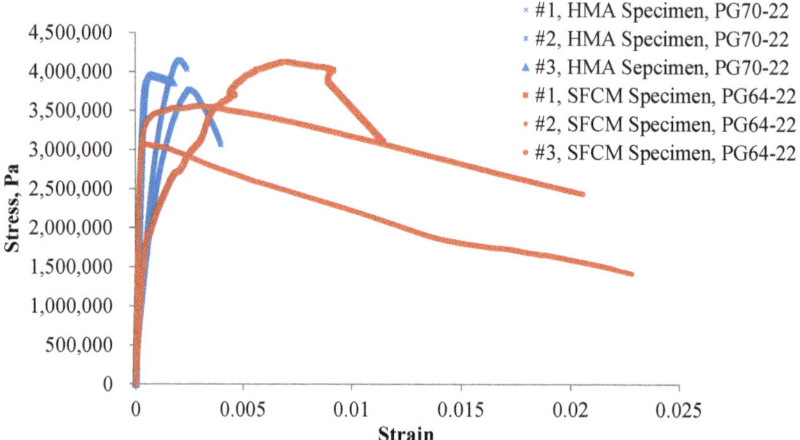

Figure 6. Stress-strain relationship of thermal cracking testing.

Table 3. Summary of fracture work and maximum stress (−10 °C).

HMA Specimens			SFCM Specimens		
Specimen No.	Fracture Work, N·mm	Maximum Stress, MPa	Specimen, No.	Fracture Work, N·mm	Maximum Stress, MPa
#1	166,267.90	4.15	#1	185,328.88	4.13
#2	163,591.74	3.77	#2	162,810.69	3.17
#3	172,328.52	3.95	#3	174,776.83	3.56
Average	167,396.05	3.96	Average	174,305.47	3.62
CV, %	2.7	4.8	CV, %	6.5	13.3

3.5. IDT High Temperature Strength

High temperature indirect tensile (HT-IDT) strength is a test that can be used to evaluate the rutting resistance of HMA mixtures [22–29]. The test was conducted at a temperature of 42 °C with a loading rate of 50 mm/min. The test temperature was selected as 10 °C lower than the 50% reliability, seven-day average maximum pavement temperature determined based on LTPPBind program Version 3.1. For both the SFCM mixture and the HMA mixture, three replicate specimens were tested and averaged.

Testing results are shown in Table 4 and the results indicate that the indirect tensile strength of the SFCM specimens is 12% higher than that of the control HMA specimens, indicating better rutting resistance. NCHRP report 673 [22] recommended minimum HT-IDT strength requirements based on ESALs. Specifically, for ESALs ranged from 3 to 10 million, 10 to 30 million and ESALs larger than 30 million, the minimum requirements are 270, 380, and 500 kPa, respectively. As seen from Table 4, the tensile strengths of all the six SFCM and HMA specimens are greater than the minimum requirement for 30 million ESALs (500 kPa), and the average values of SFCM (746.6 kPa) and HMA (664.6 kPa) are 49.3% and 32.9% more than the minimum requirement, respectively.

Table 4. High temperature indirect tensile strength testing results (42 °C).

Sample No.	Maximum Load, N	Sample Thickness, mm	Sample Diameter, mm	Indirect Tensile Strength, kPa
#1, SFCM	4140.0	38.1	100.0	691.8
#2, SFCM	5350.0	38.9	100.0	875.6
#3, SFCM	3950.0	37.4	100.0	672.4
Average,SFCM	4480.0	38.1	100.0	746.6
#1, HMA	4212.1	39.4	100.0	680.6
#2, HMA	4119.3	40.8	100.0	642.8
#3, HMA	3854.6	36.6	100.0	670.5
Average,HMA	4062.0	38.9	100.0	664.6

4. Conclusions and Recommendations

A slab SFCM sample was fabricated in the laboratory to simulate the mortar filling in the field, with 2% of superplasticizer was included to achieve good filling characteristics. Cylindrical specimens were cored from the slab to perform the IDT testing. By comparing the IDT testing results of SFCM mixture (with PG64-22 binder) and a control PG70-22 HMA, the following findings are obtained. It should be noted that the core asphalt structure was made out of a PG64-22 binder which is a lower grade than the control HMA for the purpose of comparing SFCM with higher grade HMA mixture. In addition, very limited compactive effort was applied to the SFCM slab sample while gyratory compaction with N_{design} of 100 was used for the PG70-22 HMA mixture.

1. SFCM has good performance in rutting resistance and moisture susceptibility which was mainly attributed to the addition of cemetious materials, its higher dynamic moduli in contrast to the conventional HMA specimens, as well as the less sensitive to the temperature (frequency) change of the SFCM materials.
2. Dynamic modulus testing results also show that SFCM is viscoelastic in nature. In addition, the IDT fatigue testing shows that SFCM is less fatigue resistant compared with the control PG70-22 HMA mixture, based on fracture work property. IDT thermal cracking test results show that the SFCM has better resistance to thermal cracking than control HMA at the low temperature, based on fracture work property.
3. Based on its strong rutting resistance and thermal cracking resistance, as well as relative low fatigue resistance, it is recommended to apply the SFCM materials on the top surface layer of a pavement structure, since the surface layer endures more compressive stress and thermal stress, and experiences very limited magnitude of tensile stress.
4. It should be noted that although the fatigue resistance of SFCM may be low due to the brittleness of the material. However, the overall fatigue performance of the SFCM should be evaluated comprehensively based on the pavement structure. The high strength high modulus properties of SFCM can significantly reduce the tensile strain responses in the pavement layer, therefore, compensate for its fatigue performance.

The good filling characteristics were concluded majorly based on eye observations during filling process, as well as after the cores were drilled. It is suggested to apply more advanced method to evaluate filling properties with deeper analysis. Additionally, the current method may need to be adjusted to better fit the master curve generation procedure of the SFCM specimens.

Additionally, in contrast to the conventional sample preparation procedure which was fabricated for mix design purpose, the sample preparation process proposed in this article can better simulate the field construction procedure. However, more research work was necessary to evaluate how the materials flow during paving, as well as how they are performing under real trafficking and climatic conditions.

Author Contributions: Conceptualization, W.Z.; methodology, S.S.; software, D.W.; validation, D.W. and J.Z.; investigation, W.Z.; data curation, R.D.G.; formal analysis, R.D.G.; writing—original draft preparation, W.Z.; writing—review and editing, W.Z. and S.S.; visualization, J.Z.; supervision, S.S.; All authors have read and agreed to the published version of the manuscript.

Funding: This research received no external funding.

Conflicts of Interest: The authors declare no conflict of interest.

References

1. Cai, J.; Pei, J.; Luo, Q.; Zhang, J.; Li, R.; Chen, X. Comprehensive service properties evaluation of composite grouting materials with high-performance cement paste for semi-flexible pavement. *Constr. Build. Mater.* **2017**, *153*, 544–556. [CrossRef]
2. Al-Qadi, I.L.; Gouru, H.; Weyers, R.E. Asphalt portland cement concrete composite: Laboratory evaluation. *J. Transp. Eng.* **1994**, *120*, 94–108. [CrossRef]
3. Anderton, G.L. *User's Guide: Resin Modified Pavement*; U.S. Army Corps of Engineers Waterways Experiment Station: Vicksburg, MS, USA, 1996.
4. Anderton, G.L. Engineering Properties of Resin Modified Pavement (RMP) for Mechanistic Design. In *Final Report ERDC/GLTR-00-2*; U.S. Army Engineer Research and Development Center: Vicksburg, MS, USA, 2000.
5. Oliveira, J.; Thom, N.H.; Zoorob, S. Fracture and Fatigue Strength of Mortared Macadams. In Proceedings of the 10th International Conference on Asphalt Pavements 2006, Quebec City, QC, Canada, 12–17 August 2006.
6. Al-Qadi, I.L.; Prowell, D.B.; Weyers, R.E.; Dutta, T.; Gouru, H.; Berke, N. *Concrete Bridge Protection and Rehabilitation: Chemical and Physical Techniques*; SHRP-S-666; Strategic Highway Research Program, National Research Council: Washington, DC, USA, 1993.
7. Hassan, K.E.; Setyawan, A.; Zoorob, S.E. Effect of Cemenetious Mortars on the Properties of Semi-flexible Bituminous Pavements, Performance of Bituminous and Hydraulic Materials in Pavements. In Proceedings of the Fourth European Symposium on Performance of Bituminous and Hydraulic Materials in Pavments, Nottingham, UK, 11–12 April 2002.
8. Zoorob, S.E.; Hassan, K.E.; Setyawan, A. Cold mix, Cold Laid Semi-flexible Mortared Macadams, Mix Design and Properties, Performance of Bituminous and Hydraulic Materials in Pavements. In Proceedings of the Fourth European Symposium, Nottingham, UK, 11–12 April 2002.
9. Oliveira, J.; Zoorob, S.E.; Thoml, N.H.; Pereira, P.A.A. A Simple Approach to the Design of Pavements Incorporating Mortared Macadams. In Proceedings of the 4th International Conference Bituminous Mixtures and Pavements, Thesaloniki, Greece, 19–20 April 2007.
10. Oliveira, J.; Thom, N.H.; Zoorob, S.E. Design of Pavements Incorporating Mortared Macadams. *J. Transp. Eng.* **2008**, *134*, 7–14. [CrossRef]
11. Oliveira, J.; Sangiorgi, C.; Fattorini, G.; Zoorob, S.E. Investigating the fatigue performance of grouted macadams. *J. Proc. Inst. Civ. Eng. Transp.* **2009**, *162*, 115–123. [CrossRef]
12. Hao, P.W.; Cheng, L.; Lin, L. Pavement Performance of Semi-flexible Pavement in Laboratory. *J. Xi'an Highw. Univ.* **2003**, *23*, 1–6.
13. Ling, T.; Zhao, Z.; Xiong, C.; Dong, Y.; Liu, Y.; Dong, Q. The application of semi-flexible pavement on heavy traffic roads. *Int. J. Pavement Res. Technol.* **2009**, *2*, 211–217.
14. Huang, C.; Liu, J.; Hong, J.; Liu, Z. Improvement on the crack resistance property of semi-flexible pavement by cement-emulsified asphalt mortar. In *Key Engineering Materials*; Trans Tech Publications Ltd.: Zurich, Switzerland, 2012; Volume 509, pp. 26–32.
15. Wen, H.; Bhusal, S. A Laboratory Study to Predict the Rutting and Fatigue Behavior of Asphalt Concrete Using the Indirect Tensile Test. *J. Test. Eval.* **2013**, *41*, 299–304. [CrossRef]
16. Zborowski, A. *Development of a Modified Superpave Thermal Cracking Model for Asphalt Concrete Mixtures Based on the Fracture Energy Approach*; Arizona State University: Tempe, Arizona, 2007.
17. Battey, R.L.; Jordan, S.W. *Construction, Testing and Performance Report on the Resin Modified Pavement Demonstration Project*; Transportation Research Board: Washington, DC, USA, 2007.
18. Roffee, J.C. *Salviacim-Introducing the Pavment*; Jean Lafebvre Enterprise: Paris, France, 1989.
19. Chandra, S.; Björnström, J. Influence of Superplasticizer Type and Dosage on the Slump Loss of Portland Cement Mortars-Part II. *Cem. Concr. Res.* **2002**, *32*, 1613–1619. [CrossRef]

20. Svermova, L.; Sonebi, M.; Bartos, P. Influence of mix proportions on rheology of cement mortars containing limestone powder. *Cem. Concr. Comp.* **2003**, *25*, 737–749. [CrossRef]

21. Azari, H. *Precision Estimates of AASHTO T283: Resistance of Compacted Hot Mix Asphalt (HMA) to Moisture-Induced Damage*; Transportation Research Board: Washington, DC, USA, 2010; p. 166.

22. Christensen, D.W. *A Manual for Design of Hot Mix Asphalt with Commentary*; Transportation Research Board: Washington, DC, USA, 2011; p. 673.

23. Christensen, D.W.; Bonaquist, R.; Anderson, D.A.; Gokhale, S. *Indirect Tension Strength as a Simple Performance Test. New Simple Performance Tests for Asphalt Mixes, Report E-C068*; Transportation Research Board: Washington, DC, USA, 2004; pp. 44–57.

24. Ding, X.; Ma, T.; Gu, L.; Zhang, Y. Investigation of Surface Micro-Crack Growth Behaviour of Asphalt Mortar Based on the Designed Innovative Mesoscopic Test. *Mater. Des.* **2020**, *185*, 108238. [CrossRef]

25. Ma, T.; Zhang, D.; Zhang, Y.; Wang, S.; Huang, X. Simulation of Wheel Tracking Test for Asphalt Mixture Using Discrete Element Modelling. *Road Mater. Pavement Des.* **2018**, *19*, 367–384. [CrossRef]

26. Zhang, Y.; Ma, T.; Meng, L.; De, Z.; Xiao, H. Prediction of Dynamic Shear Modulus of Asphalt Mastics by Using the Discretized Element Simulation and Reinforcement Mechanisms. *J. Mater. Civ. Eng.* **2019**, *31*, 04019163. [CrossRef]

27. Ma, T.; Zhang, D.; Zhang, Y.; Huang, X. Effect of Air Voids on the High-temperature Creep Behavior of Asphalt Mixture based on Three-dimensional Discrete Element Modeling. *Mater. Des.* **2015**, *89*, 304–313. [CrossRef]

28. Ma, T.; Wang, H.; He, L.; Zhao, Y.; Huang, X.; Chen, J. Property Characterization of Asphalt Binders and Mixtures Modified by Different Crumb Rubbers. *J. Mater. Civ. Eng.* **2017**, *29*, 04017036. [CrossRef]

29. Ding, X.; Chen, L.; Ma, T.; Ma, H.; Gu, L.; Chen, T.; Ma, Y. Laboratory investigation of the recycled asphalt concrete with stable crumb rubber asphalt binder. *Constr. Build. Mater.* **2019**, *203*, 552–557. [CrossRef]

Article

Effects of Aggregate Mesostructure on Permanent Deformation of Asphalt Mixture Using Three-Dimensional Discrete Element Modeling

Deyu Zhang [1,2], Linhao Gu [1] and Junqing Zhu [1,*]

[1] School of Transportation, Southeast University, Nanjing 210096, China; zhangdy@njit.edu.cn (D.Z.); gulinhao@seu.edu.cn (L.G.)
[2] School of Architecture Engineering, Nanjing Institute of Technology, Nanjing 211167, China
* Correspondence: zhujunqing@seu.edu.cn; Tel.: +86-151-5065-9102

Received: 26 September 2019; Accepted: 31 October 2019; Published: 2 November 2019

Abstract: This paper investigated the effects of aggregate mesostructures on permanent deformation behavior of an asphalt mixture using the three-dimensional (3D) discrete element method (DEM). A 3D discrete element (DE) model of an asphalt mixture composed of coarse aggregates, asphalt mastic, and air voids was developed. Mesomechanical models representing the interactions among the components of asphalt mixture were assigned. Based on the mesomechanical modeling, the uniaxial static load creep tests were simulated using the prepared models, and effects of aggregate angularity, orientation, surface texture, and distribution on the permanent deformation behavior of the asphalt mixtures were analyzed. It was proven that good aggregate angularity had a positive effect on the permanent deformation performance of the asphalt mixtures, especially when approximate cubic aggregates were used. Aggregate packing was more stable when the aggregate orientations tended to be horizontal, which improved the permanent deformation performance of the asphalt mixture. The influence of orientations of 4.75 mm size aggregates on the permanent deformation behavior of the asphalt mixture was significant. Use of aggregates with good surface texture benefitted the permanent deformation performance of the asphalt mixture. Additionally, the non-uniform distribution of aggregates had a negative impact on the permanent deformation performance of the asphalt mixtures, especially when aggregates were distributed non-uniformly in the vertical direction.

Keywords: asphalt mixture; aggregate mesostructure; permanent deformation; discrete element method

1. Introduction

With the increase of traffic and heavy vehicles, rutting has become one of the main distresses of asphalt pavement, which seriously affects driving safety and comfort. The permanent deformation behavior of asphalt mixtures has become an important issue for road researchers. Within asphalt mixtures, aggregates account for more than 90% by weight and 80% by volume. Aggregates, as the main component of asphalt mixtures, have a significant impact on the permanent deformation behavior of asphalt mixtures. Previous studies on the effects of aggregates on the permanent deformation behavior of asphalt mixtures have been mainly related to aggregate gradations. It is difficult to investigate the effects of the mesostructural characteristics of aggregates on permanent deformation behavior of asphalt mixtures, due to the limitations of available laboratory tests.

In previous studies, X-ray computed tomography (CT) techniques and image processing technology have been used for aggregate mesostructural analysis. Masad proposed computer-automated image analysis procedures to quantify the internal structure of asphalt concrete in terms of aggregate orientation, aggregate contacts, and air void distribution, and developed a finite-element model of asphalt concrete mesostructure to study the influence of localized strain distribution on the mechanical response of asphalt concrete [1–3]. Kose investigated strain distribution within a binder using digitized images analyzed using finite-element procedures [4]. Wang developed a method to quantify the local volume fractions of voids and their spatial gradients using X-ray tomography imaging and image analysis [5,6]. You predicted the dynamic modulus of asphalt mixtures using both the two-dimensional and three-dimensional discrete element method, generated using X-ray computed tomography [7]. However, pre-compacted specimens should have been prepared in the laboratory for control purposes in the studies mentioned above. Variability of the components within specimens have inevitably occurred, and the distribution, morphological characteristics of aggregates, and air voids are not the same even within two mixture specimens of the same kind. Therefore, there are many factors that might disrupt investigation of the effect of a single mesostructure on the behavior of asphalt mixtures. Moreover, the mesostructures of the components cannot be desirably controlled in the laboratory, and the results to date have been consequently unconvincing. Therefore, it is necessary to control the internal structures of asphalt mixtures in order to accurately analyze the effects of a single mesostructure.

At present, commonly used numerical simulation methods at the meso-scale include the finite element method (FEM) [8], the discrete element method (DEM) [9], and the combined finite–discrete element method (FDEM) [10]. In recent years, DEM, as a numerical simulation method used to solve the problem of discontinuity media, has been used to analyze the mechanical properties of asphalt mixtures. Chang developed a model called ASBAL (TRUBAL for ASphalt) based on the discrete element method by modifying the TRUBAL program to simulate hot-mix asphalt mixtures [9]. Buttlar presented a microfabric discrete-element modeling approach (MDEM) for modeling asphalt concrete mesostructures using image analysis techniques [11]. You predicted the asphalt mixture complex modulus in extension/compression across a range of test temperatures and load frequencies using the MDEM approach, and simulated and analyzed the creep responses of an asphalt mixture with a 3D-mesostructure-based DE viscoelastic model [12,13]. Abbas presented a methodology for analyzing the viscoelastic response of asphalt mixtures using the DEM [14]. Liu developed a viscoelastic model of asphalt mixtures using the discrete element method, where the viscoelastic behaviors of asphalt mastics are represented by a Burger's model [15]. Chen built a mesomechanical model to investigate the stiffness anisotropy of asphalt concrete using the DEM [16,17]. Ma conducted simulated wheel-tracking tests on asphalt mixtures, predicted the rutting deformation of asphalt mixtures, and analyzed the mesomechanical response of aggregate skeletons within the asphalt mixtures during the tests [18,19]. Ma also investigated the effects of different parameters related to air voids on the creep behavior of asphalt mixtures based on a 3D DEM [20]. Ding proposed a new modeling method to reconstruct hollow shapes of aggregate particles using the DEM to accurately characterize the mesostructures of aggregates and efficiently predict the mechanical properties of aggregate skeletons [21]. It has been proven that the DEM can accurately capture the internal structure of asphalt mixtures, and desirably control the mesostructural characteristics of aggregates, asphalt binder, and air voids within the mixtures [22–25]. However, few researchers have focused on the permanent deformation behavior of asphalt mixtures, especially the effects of aggregate mesostructure on the permanent deformation behavior of asphalt mixtures using the DEM.

The objective of this study was to investigate the effects of aggregate mesostructures on the permanent deformation behavior of asphalt mixtures using the DEM. A 3D DE model of asphalt mixture including aggregates, asphalt mastics, and air voids was developed using the DEM software Particle Flow Code in three dimensions (PFC3D). Three-dimensional discrete element simulations of uniaxial static creep tests of asphalt mixtures were carried out. By accurately controlling the mesostructural characteristics of the aggregates, such as angularity, orientation, surface texture, and distribution, the effects of aggregate mesostructures on the permanent deformation behavior of the asphalt mixtures were carefully analyzed.

2. Materials and Methods

2.1. Materials

According to the Marshall mix design method based on the Chinese specifications [26], a dense-graded asphalt mixture, AC-20, was prepared in the laboratory. Aggregate gradation with a nominal maximum aggregate size of 19 mm is shown in Figure 1. Based on the Marshall mix design method, the volumetric properties of the AC-20 asphalt mixture were determined as shown in Table 1. Aggregates smaller than 2.36 mm, mineral filler, and asphalt binder were mixed as asphalt mastic with an asphalt content of 11.5%. The air void content of the asphalt mastic was supposed to be zero to maximize its flowability.

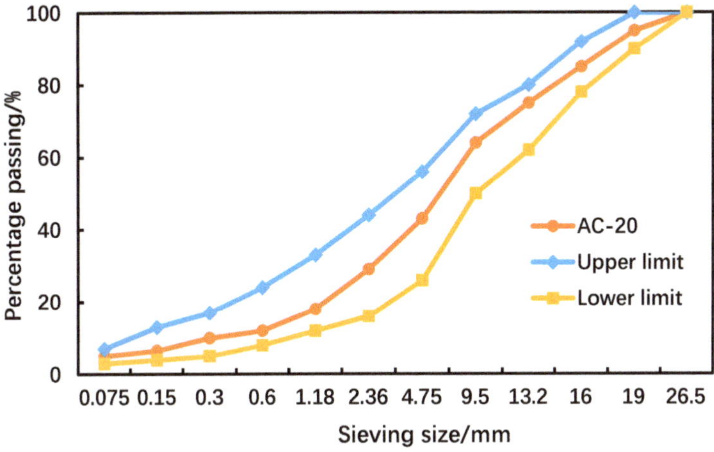

Figure 1. Gradation of the AC-20 asphalt mixture.

Table 1. Volumetric properties of the AC-20 asphalt mixture.

Asphalt Content (%)	Bulk Density (g/cm^3)	VV (%)	VMA (%)	VFA (%)
4.3	2.432	4.0	13.8	72.2

2.2. Laboratory Tests

The uniaxial static creep test is one of the simplest and most practical test methods, and it is an effective way to investigate the permanent deformation behaviors of asphalt mixtures and mastics at high temperatures. In this test, an instantaneous load is applied to a cylinder specimen in the axial direction and the load is kept constant, and the creep curve of the asphalt mixtures or mastics can be obtained. Uniaxial static creep tests were conducted in this paper to evaluate the permanent deformation behaviors of asphalt mixtures and mastics at the temperature of 60 °C using a universal test machine (UTM, IPC Global, Melbourne, Australia) [27]. Cylindrical specimens of asphalt mixtures with diameter of 100 mm and height of 150 mm were prepared using a gyratory compactor for the uniaxial static creep test. Cylindrical asphalt mastic specimens with diameter and height of 100 mm were prepared by vibration. The applied axial stresses for the asphalt mixture and asphalt mastic were set at 0.7 MPa and 0.07 MPa, respectively, in this study. The asphalt mastic specimens and the testing process are shown in Figure 2. Lab test results were used for parameters of DE modeling and to be compared with simulation test results, and are presented in the following sections. The experimental program for the asphalt mixture and mastics is summarized in Table 2.

(a)   (b)

Figure 2. (a) The asphalt mastic specimens; (b) the uniaxial static creep test testing process.

Table 2. Experimental program for the asphalt mixture and mastics.

Materials	Specimen Size	Test	Load/MPa
Asphalt mixture	Φ 100 mm × H 150 mm (cylinder) L 100 mm × L 150 mm (cubic)	Uniaxial static creep test	0.7
Asphalt mastic	Φ 100 mm × H 100 mm (cylinder)	Uniaxial static creep test	0.07

3. Discrete Element Modeling of Asphalt Mixtures

3.1. Discrete Element Modeling

An asphalt mixture, as a multiphase composite, is composed of aggregates, asphalt binder, and air voids. If the particle size of the fine aggregates and mineral filler is small, it will lead to a large increase of discrete elements in the DE model, which is bound to significantly reduce the computational efficiency if the fine aggregate and mineral powder are considered fully in DE models of asphalt mixtures. Therefore, the asphalt mixture was simplified into coarse aggregates (bigger than 2.36 mm), asphalt mastics (asphalt binder, fine aggregates smaller than 2.36 mm and mineral filler), and air voids to improve the calculation efficiency. The asphalt mastic was considered as a homogeneous material within the DE model.

During modeling, the spatial range of the mixture model was first constructed by "wall". The quantity of coarse aggregates in each grade was calculated according to aggregate gradation, asphalt content and air void content. The coarse aggregate balls were then delivered into the spatial range of the mixture model constructed by "wall". Force was generated between overlapped balls, as there were overlaps between the coarse aggregate balls during delivery. This force was eliminated by the "cycle" command. The coarse aggregate balls within the spatial range of the mixture model are shown in Figure 3a.

A regular array of discrete elements was then filled into the mixture space as the base of coarse aggregates and asphalt mastic, as shown in Figure 3b. The radius of the discrete element was set to 1 mm, considering the calculation efficiency.

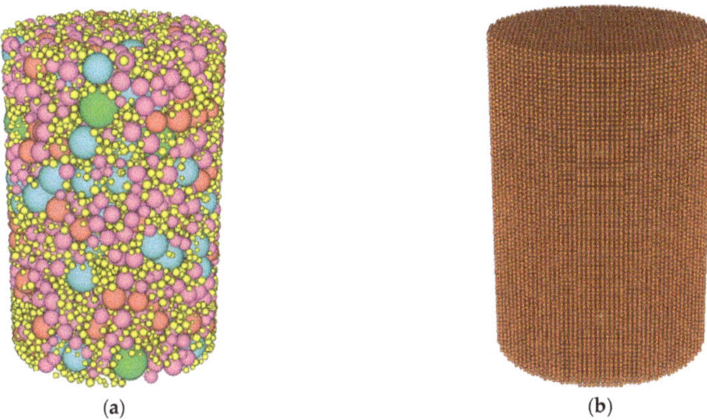

(a)   (b)

Figure 3. (**a**) Graded coarse aggregates; (**b**) uniform-sized packed discrete elements.

It has been proven that the geometry of coarse aggregates has a significant effect on the permanent deformation behavior of asphalt mixtures [28,29]. Therefore, the coarse aggregate geometry will directly determine the accuracy of the simulation results. In this paper, coarse aggregate was simplified as an irregular polyhedron. Random planes were used to cut a cube or a sphere to generate irregular polyhedral aggregates using a user-defined program. By traversing the regularly packed discrete elements, the positional relationship between the regular packing discrete elements and the irregular polyhedral aggregates could be evaluated. Discrete elements belonging to irregular polyhedral aggregate were regarded as aggregate elements, and were set as a clump, as seen in Figure 4. The original coarse aggregate balls were then deleted. Discrete elements outside the irregular polyhedron aggregate were considered as asphalt mastic. The initially developed 3D DE model of asphalt mixture is shown in Figure 5a.

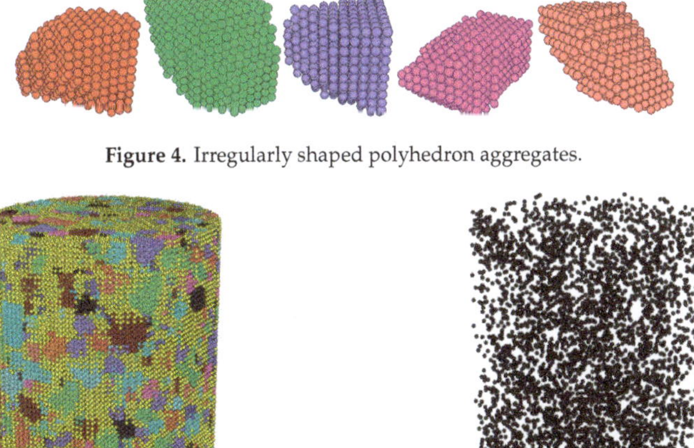

Figure 4. Irregularly shaped polyhedron aggregates.

(a)   (b)

Figure 5. (**a**) Discrete element (DE) model of asphalt mixture; (**b**) air voids within the DE model.

Considering the complexity of air void distribution within an asphalt mixture specimen, the air voids were assumed to be randomly distributed [30]. Asphalt mastic elements in the model were traversed and randomly deleted and regarded as air voids. The distribution of air voids is shown in Figure 5b.

3.2. Mesomechanical Models and Parameters

There are four types of contact within asphalt mixtures, including contacts between aggregate elements, contacts between adjacent aggregates, contacts between asphalt mastic elements, and contact between asphalt mastic and aggregates, as seen in Figure 6. In PFC3D, mesomechanical models are used to describe the contact behaviors between different components within asphalt mixtures. In this study, the mesomechanical models used in the model of asphalt mixtures included the stiffness model, slipping model, bonding model, and Burger's model.

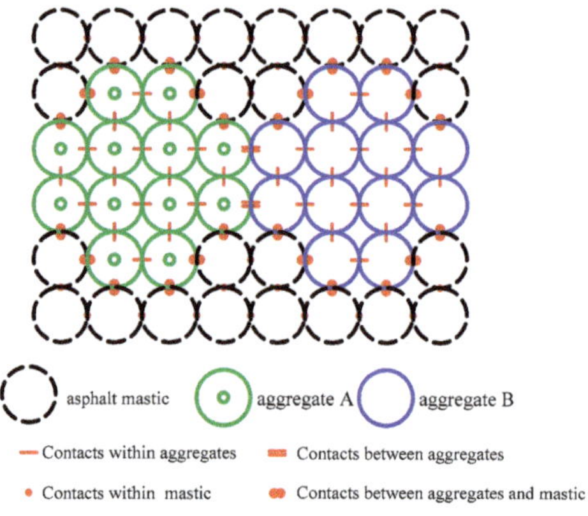

Figure 6. Interactions among the components within the asphalt mixture.

Due to the high stiffness of an aggregate, it can be approximately regarded as an elastic material. In this study, the stiffness model and slipping model were used to characterize the mesomechanical behavior between adjacent aggregates. As coarse aggregates within the model of asphalt mixtures were set as clumps, it was unnecessary to assign the mesomechanical model within aggregates. The mesomechanical parameters of the stiffness model could be obtained from the macro properties of the aggregates, as shown in Equations (1) and (2) [15,31,32]. The macro parameters for the aggregates are shown in Table 3 [9,12,33].

$$E = \frac{k_n}{4R}, \; k_s = \frac{k_n}{2(1 + v\prime)} \tag{1}$$

$$\mu_c = \mu_a \tag{2}$$

where E is the apparent Young's modulus of the aggregates, k_n and k_s are the stiffness in the normal and shear direction, respectively, R is the discrete element radius, $v\prime$ is the aggregate Poisson's ratio, μ_c is the friction coefficient between aggregates, and μ_a is the friction coefficient of aggregates.

Table 3. Macro parameters for asphalt mastic and aggregates.

E_1 (MPa)	η_1 (MPa·s)	E_2 (MPa)	η_2 (MPa·s)	v	E (GPa)	μ_a	$v\prime$
0.568	973.163	0.396	27.895	0.5	55.5	0.5	0.35

Asphalt mixtures show a macro viscoelastic behavior due to the viscoelastic characteristic of the asphalt mastic. Therefore, the mesomechanical model of the asphalt mastic directly affects the macro properties of asphalt mixtures. The meso Burger's model in PFC3D was well able to describe the mechanical properties of viscoelastic materials and was used to characterize the viscoelastic properties of the asphalt mastic in this study, as shown in Figure 7. It has been proven that there is a conversion between the parameters of the meso Burger's model and the macro Burger's model [15,31], as shown in Equations (3) and (4). Parameters of meso Burger's model could then be obtained from the macro Burger's model parameters. The macro Burger's model parameters of asphalt mastic were obtained by uniaxial static creep test, as shown in Table 3.

$$K_{mn} = E_1 L, \ C_{mn} = \eta_1 L, \ K_{kn} = E_2 L, \ C_{kn} = \eta_2 L \tag{3}$$

$$K_{ms} = \frac{E_1 L}{2(1+v)}, \ C_{ms} = \frac{\eta_1 L}{2(1+v)}, \ K_{ks} = \frac{E_2 L}{2(1+v)}, \ C_{ks} = \frac{\eta_2 L}{2(1+v)} \tag{4}$$

where E_1, η_1, E_2, and η_2, are parameters of the macro Burger's model; K_{mn}, C_{mn}, K_{kn}, and C_{kn} are parameters of the meso Burger's model in the normal direction; K_{ms}, C_{ms}, K_{ks}, and C_{ks} are parameters of the meso Burger's model in the shear direction; L is the distance between adjacent discrete elements; and v is Poisson's ratio of the asphalt mastic.

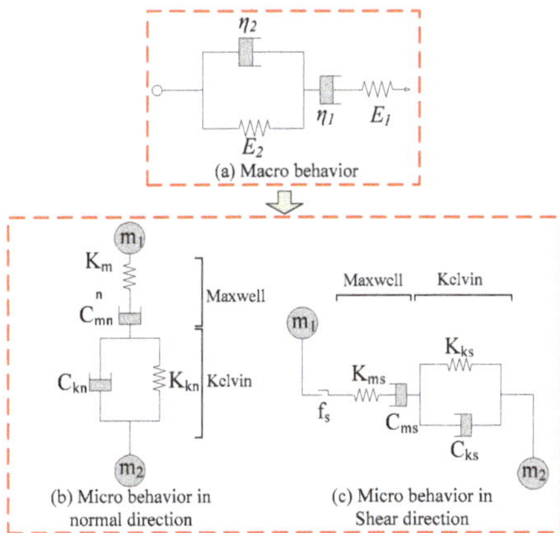

Figure 7. Burger's models. (**a**) Macro behavior; (**b**) micro behavior in the normal direction; (**c**) micro behavior in the shear direction.

The contact behavior between aggregates and the asphalt mastic can be characterized by the equivalent meso Burger's model, as shown in Figure 8. The equivalent meso model parameters can be obtained from the macro parameters of the aggregates and asphalt mastic, as expressed in Equations (5) and (6) [15,31]:

$$K'_{mn} = \frac{2EE_1}{E+E_1} L, \ C'_{mn} = 2\eta_1 L, \ K'_{kn} = 2E_2 L, \ C'_{kn} = 2\eta_2 L \tag{5}$$

$$K'_{ms} = \frac{2EE_1}{E(1+v)+2E_1(1+v)} L, \ C'_{ms} = \frac{\eta_1 L}{(1+v)}, \ K'_{ks} = \frac{E_2 L}{(1+v)}, \ C'_{ks} = \frac{\eta_2 L}{(1+v)} \tag{6}$$

where K'_{mn}, C'_{mn}, K'_{kn}, and C'_{kn} are the parameters of the equivalent meso Burger's model between the aggregates and mastic in the normal direction and K'_{ms}, C'_{ms}, K'_{ks}, and C'_{ks} are the parameters of the equivalent meso Burger's model between the aggregates and the mastic in the shear direction.

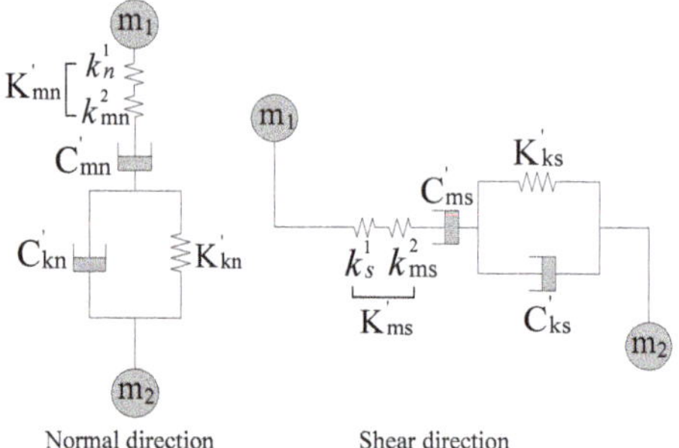

Figure 8. Equivalent meso Burger's model for contacts between aggregates and asphalt mastic.

3.3. Simulation of Uniaxial Creep Test

Based on the above-mentioned DE modeling procedures for asphalt mixtures, the simulated uniaxial creep test was carried out using PFC3D. The results were compared with laboratory test results under the same conditions. Figure 9 shows the axial strain curve obtained from the simulation and lab test, and it can be seen that the simulation curve was very close to the curve of the laboratory test. This indicates that the simulation of uniaxial creep test conducted using DEM could precisely estimate the permanent deformation behavior of asphalt mixtures.

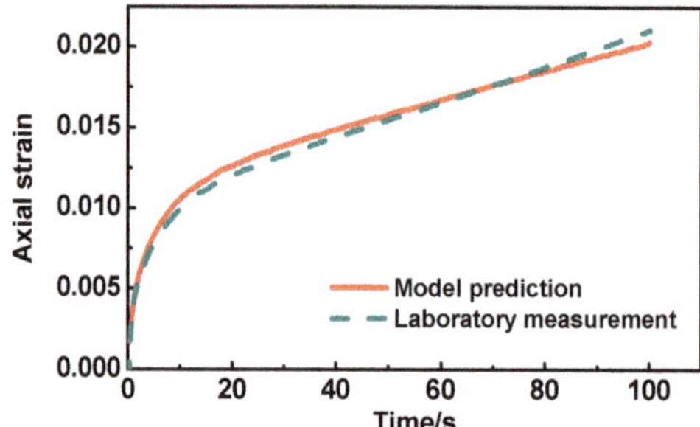

Figure 9. Results of discrete element model (DEM) prediction and laboratory measurement.

4. Results and Discussion

4.1. Effect of Aggregate Angularity

The aggregate angularity could represent the morphological characteristics of aggregates, and significantly affect the skeleton stability, interlocking force, and shear resistance of aggregates. It has been proven that the aggregate angularity is closely related to the strength of asphalt mixtures, especially the permanent deformation resistance and shear resistance [28,29].

Some previous studies about aggregate angularity have been carried out, and some indexes to describe the aggregate angularity have been proposed. These angularity indexes are mainly divided into two categories, one denoting the roundness of aggregate corners and the other one denoting the roundness of the overall outline of the aggregate [34]. However, most of these angularity indexes are in two dimensions. Moreover, the measurement and calculation of the first type of angularity index is very complex, while the second type could be affected by the overall outline of aggregates. Therefore, the angularity indexes mentioned above cannot fully represent the angularity characteristics of aggregates. In this study, the ratio of surface area of irregular aggregate to equivalent ellipsoid was proposed to characterize the aggregate angularity. The equivalent ellipsoid of an aggregate has the same volume as the aggregate. It was considered that the equivalent ellipsoid of aggregates could accurately reflect the overall outline, and this angularity index could significantly reduce the influence of aggregate outline on angularity index, as expressed in Equation (7). The larger *AI* is, the better the aggregate angularity is.

$$AI = \left(\frac{S}{Sellipse} \right)^3 \tag{7}$$

where *AI* is the angularity index of aggregates, *S* is surface area of aggregates, and $S_{ellipse}$ is surface area of equivalent ellipsoid.

As mentioned above, irregular polyhedral aggregates were generated by cutting a cube or a sphere with random planes. Thus, the equivalent ellipsoid of an irregular polyhedral aggregate can be approximated to an equivalent sphere, and then Equation (8) can be further expressed as follows:

$$AI = \left(\frac{S}{Ssphere} \right)^3 = \left(\frac{S}{4\pi \left(\sqrt[3]{3V/4\pi} \right)^2} \right)^3 = \frac{S^3}{36\pi V^2} \tag{8}$$

where S_{sphere} is surface area of equivalent sphere and *V* is the volume of aggregates.

Coarse aggregates with different angularities were generated by cutting a cube or a sphere with random planes. Some coarse aggregates with different angularities are shown in Figure 10, in which their angularity become better with the increasing number. DE models of asphalt mixtures with different aggregate angularities were then obtained.

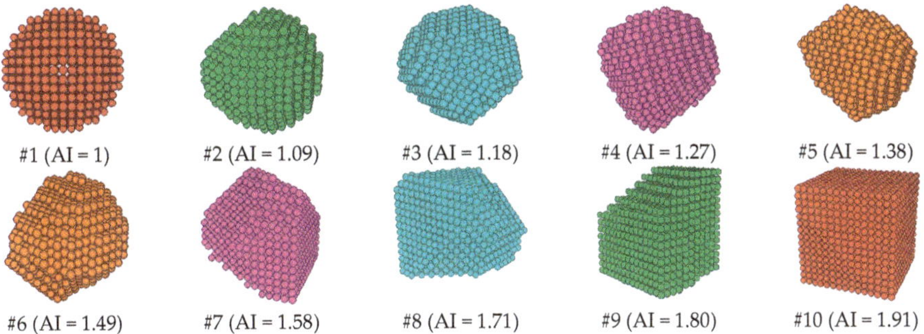

Figure 10. Irregular polyhedral aggregates with different angularities. AI: angularity index

In order to calculate the surface area of aggregates, the number of discrete elements on aggregate surface and the number of discrete elements that made up the aggregates were counted. The angularity index of the aggregates was then calculated by Equation (9). The method used to count the number of discrete elements on the aggregate surface and the number of discrete elements that make up aggregates is shown in Figure 11.

$$AI = \frac{\left(4R^2 n_S \right)^3}{36\pi \left(8R^3 n_V \right)^2} = \frac{n_S{}^3}{36\pi n_V{}^2} \tag{9}$$

where n_s is the number of discrete elements on the aggregate surface, n_v is the number of discrete elements that make up the aggregate, and *R* is radius of the discrete element.

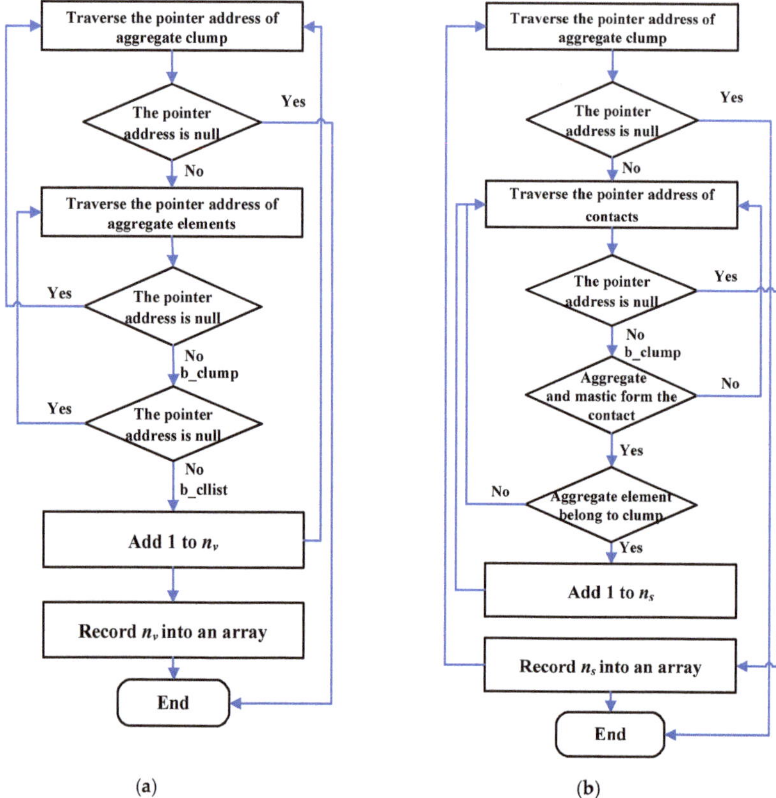

Figure 11. Procedure for quantifying the discrete element number of aggregates: (**a**) quantifying the number of aggregate surfaces; (**b**) quantifying the number that makes up aggregates.

The angularity of a single aggregate can be obtained by Equation (9). However, there are many aggregates within asphalt mixtures, and the angularities of each aggregate are different. It is necessary to evaluate the overall angularity of the aggregates within asphalt mixtures. In this study, the volumetrically weighted average of the angularities of each aggregate was used to describe the overall angularity of the aggregates within the asphalt mixtures, as expressed in Equation (10).

$$\overline{AI} = \frac{\sum_{i=1}^{n} Vi \times AIi}{\sum_{i=1}^{n} Vi} \tag{10}$$

where $\overline{AI}$ is the overall angularity of aggregates, AI_i is the angularity of aggregate i, V_i is the volume of aggregate i, and n is the number of aggregates.

Simulations of uniaxial static creep tests were then conducted on the DE models of asphalt mixtures with different overall aggregate angularities. The results are shown in Figure 12. The permanent deformation performance of asphalt mixtures improved with the increasing aggregate angularity, but the improvement weakened gradually. The axial deformation of the asphalt mixture was the smallest when the aggregates were closest to a cube. This is consistent with the requirement of selecting aggregates with shapes close to a cube. Therefore, aggregates with good angularities should be selected to improve the permanent deformation resistance of asphalt pavement.

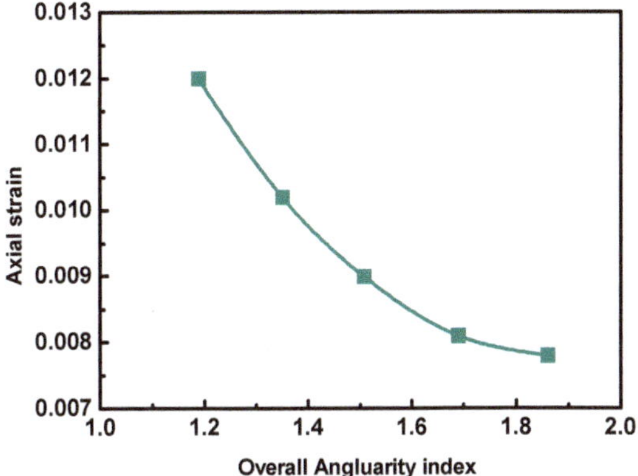

Figure 12. Axial strains with different overall aggregate angularities.

4.2. Effect of Aggregate Orientation

Previous studies about aggregate orientation have generally defined the long axis of the equivalent ellipse of aggregate in two dimensions as the long axis of aggregate, and used the angle between the long axis and the horizontal direction to represent the aggregate orientation [18–21]. In this study, the aggregate orientation was defined in three dimensions. The long axis of the equivalent ellipsoid of the aggregate was taken as the long axis, and the angles α, β, and γ between the long axis and X axis, Y axis, and Z axis were used to represent the aggregate orientation in three dimensions, as shown in Figure 13.

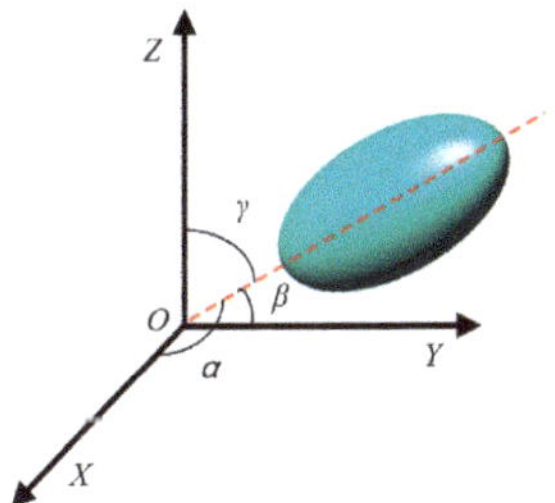

Figure 13. Orientation of ellipsoidal aggregate.

In order to investigate the effect of aggregate orientation on the permanent deformation performance of asphalt mixtures, a subroutine was written in PFC3D with the built-in "Fish" language. The long axis of the aggregate was set to be parallel to the YOZ surface, and the aggregate orientation was described by the angle between the long axis of the aggregate and the XOY surface. Some coarse aggregates with different orientations are shown in Figure 14.

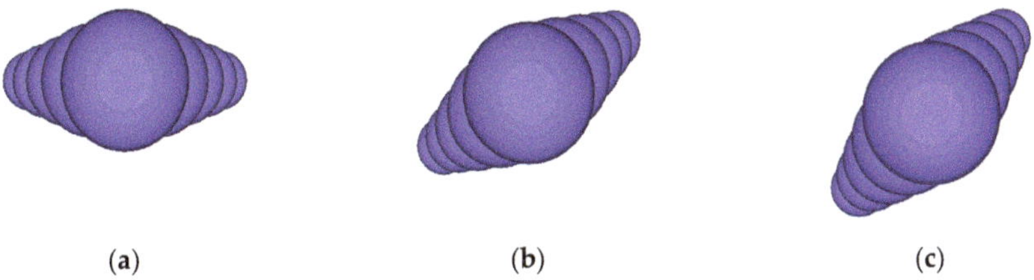

(a) (b) (c)

Figure 14. Ellipsoidal aggregates with different orientations: (**a**) 0°; (**b**) 30°; (**c**) 45°.

Graded coarse aggregates with different orientations were generated and randomly put into the cylinder space of the asphalt mixture model. The unbalanced forces between the aggregates were eliminated by the "cycle" command, so that the coarse aggregates within the asphalt mixtures could reach equilibrium. The angular velocities of the coarse aggregates were fixed to keep the aggregate orientations constant during modeling.

The effects of the orientations of aggregates of different sizes on the mechanical properties of the asphalt mixtures are different. To address the size effect, 30 cylinder DE models of asphalt mixtures with different orientations of aggregates with different sizes were developed. The aggregate orientations were varied from 0°, 30°, 45°, 60°, to 90°. The rest of the aggregates remained spherical. Figure 15 shows the coarse aggregates with different orientations.

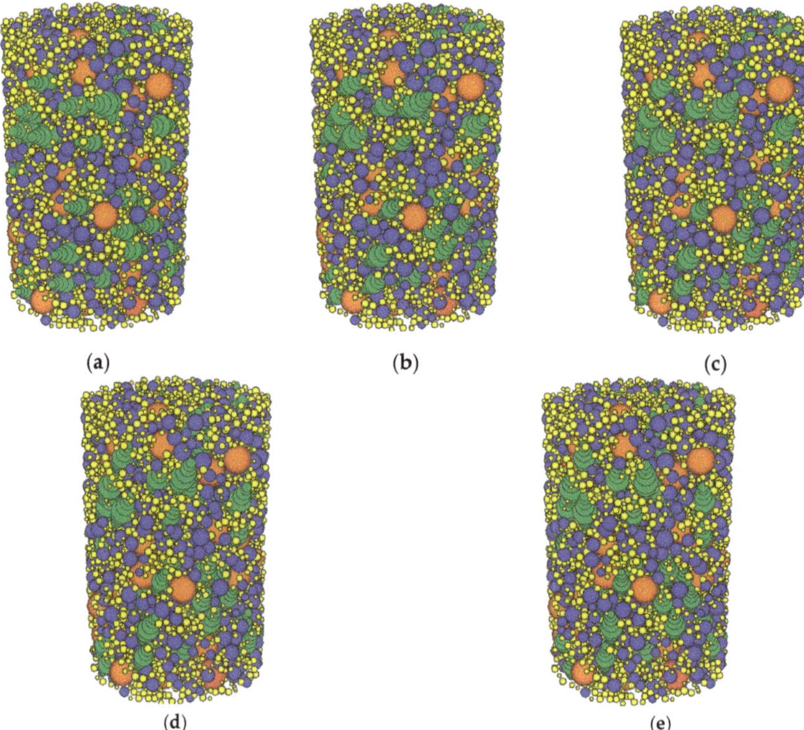

Figure 15. Coarse aggregates with different aggregate orientations within asphalt mixtures: (**a**) 0°; (**b**) 30°; (**c**) 45°; (**d**) 60°; (**e**) 90°.

Simulations of uniaxial static creep tests were then conducted on the prepared DE models. The results are shown in Figure 16.

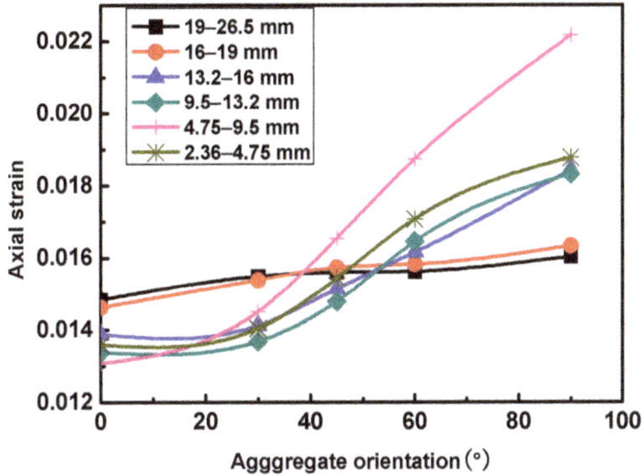

Figure 16. Axial strains with different aggregate orientations.

It can be seen from Figure 16 that the axial strains of the mixtures increased gradually with the increasing aggregate orientations. The orientations of 19–26.5mm and 16–19 mm aggregates had insignificant effects on the permanent deformation behavior of asphalt mixtures. The orientations of 13.2–16 mm, 9.5–13.2 mm, and 2.36–4.75 mm aggregates had more significant effects than those of 19–26.5 mm and 16–19 mm. The orientation of 4.75–9.5 mm aggregates had the greatest effect on the permanent deformation behavior of the asphalt mixtures. The reason is that the aggregates tended to rotate horizontally under vertical loads when the aggregate orientation was large. Thus, the aggregate skeleton was unstable and led to large axial deformation of the asphalt mixtures. When the aggregate orientation was closer to horizontal, the aggregate skeleton was more stable, and the axial deformation of the mixture specimens was smaller. It has been shown that the percentage pass of 4.75 mm plays an important role in controlling the aggregate skeleton, and the influence is next to that of air void content on the permanent deformation behavior of asphalt mixtures. This may be the reason why the orientation of 4.75–9.5 mm aggregates had such a great influence. Therefore, different kinds of rollers should be used during rolling compaction of asphalt pavement, which could force the aggregate orientation to be more horizontal. Thus, the asphalt pavement would reach a more stable state and its permanent deformation performance could be significantly improved.

4.3. Effect of Aggregate Surface Texture

The surface texture of aggregates reflects the roughness of aggregate surface. The rougher the aggregate surface, the better the surface texture. Internal friction between aggregates and adhesion between aggregate and asphalt binder are closely related to the surface texture of aggregates. It has been proven that the surface texture of aggregates has a significant impact on the performance of asphalt mixtures, especially the permanent deformation performance. The randomly created irregular polyhedral aggregates in the DE model could only reflect geometry and angularity, without characterizing the surface texture of aggregates. In this study, the friction coefficient between aggregates was used to indirectly represent the surface texture of aggregates. The higher the friction coefficient, the rougher the surface texture. A series of DE models of asphalt mixture specimens with different aggregate friction coefficients assigned were developed. The internal structures of the developed DE models were the same, and only the aggregate friction coefficients of polyhedron aggregates in the DE models were different. Simulations of uniaxial static creep tests were conducted on the prepared DE models. The results are shown in Figure 17.

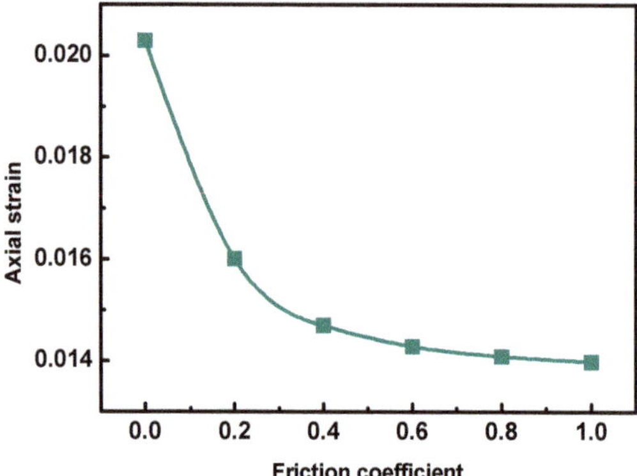

Figure 17. Axial strains with different aggregate surface textures.

It can be seen from Figure 17 that the permanent deformation performance of the asphalt mixtures was improved with the increasing friction coefficient, but the increment was gradually lowered. The permanent deformation resistance of the asphalt mixtures increased with rougher aggregate surface texture. Therefore, aggregates with good surface texture should be used to improve the permanent deformation performance of asphalt pavement.

4.4. Effect of Aggregate Distribution

Aggregates are ideally uniformly distributed within asphalt mixture for optimum performance. However, insufficient mixing force or mixing time may cause non-uniform distribution of aggregates. Aggregate segregation may also occur during paving. Early distresses of asphalt pavement such as cracking, pothole, blooding, and rutting are closely related to the material inhomogeneity caused by variable construction quality. The non-uniform distribution of aggregates greatly influences the internal structure of asphalt mixtures and consequently lowers the permanent deformation performance of asphalt mixtures.

To evaluate the uniformity of aggregate distribution, a cubic space containing asphalt mixtures was constructed by "wall" and the specimen was divided into three parts evenly in both the horizontal and vertical directions, as shown in Figure 18. Coarse aggregates with different sizes were added to each of the three parts, as shown in Figure 19 to create non-uniform distribution. The total volume of aggregates within the three parts remained constant. The distribution percentages of the aggregate volume within each part are shown in Table 4.

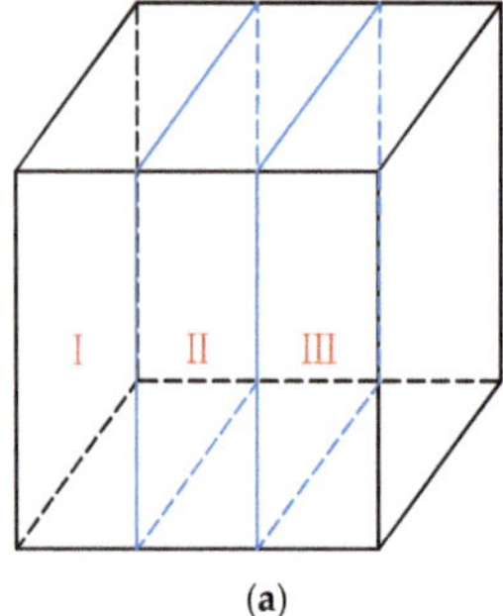
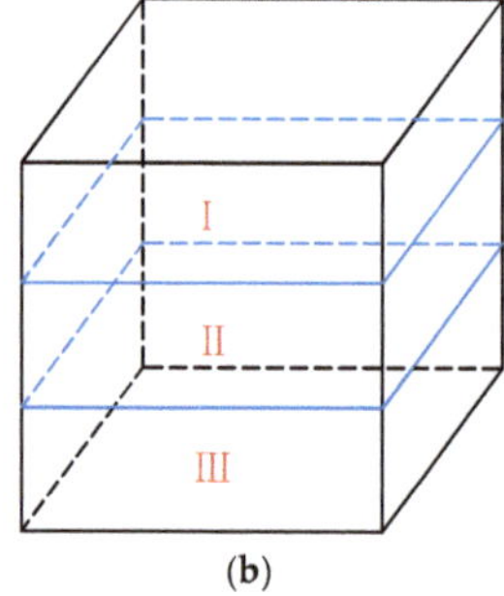

(a)   **(b)**

Figure 18. Partitioning within asphalt mixture specimen: (**a**) partitioning in the horizontal direction; (**b**) partitioning in the vertical direction.

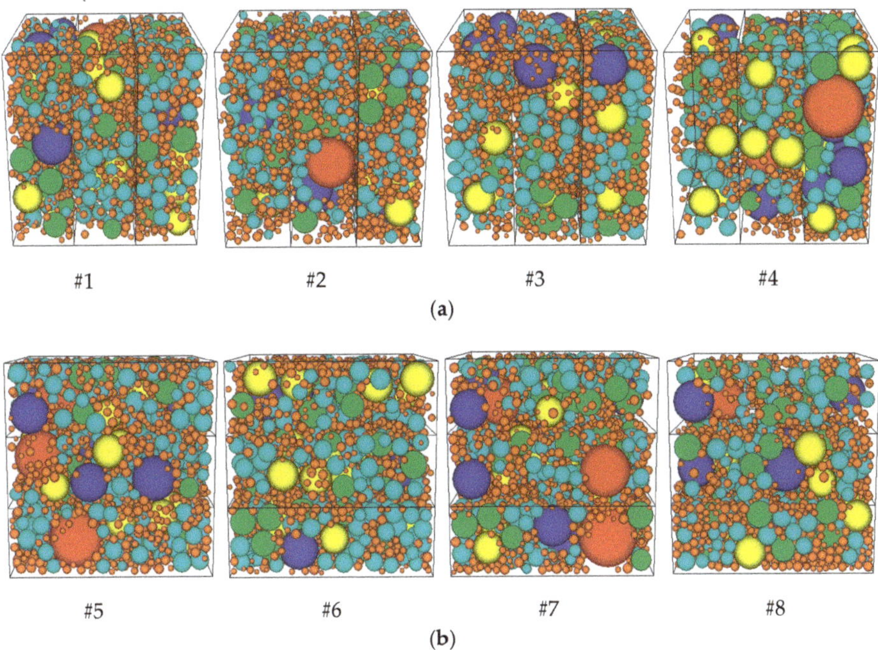

#1   #2   #3   #4

(**a**)

#5   #6   #7   #8

(**b**)

Figure 19. Aggregates distributed non-uniformly within DE models of asphalt mixtures: (**a**) aggregates distributed non-uniformly in the horizontal direction; (**b**) aggregates distributed non-uniformly in the vertical direction.

Table 4. Volume distribution percentages within different parts.

Model No.	Volume Distribution Percentage within Each Part (%)		
	Part I	Part II	Part III
Partitioned in the horizontal direction			
#1	30	30	40
#2	25	30	45
#3	20	30	50
#4	15	30	55
Partitioned in the vertical direction			
#5	30	30	40
#6	25	30	45
#7	20	30	50
#8	15	30	55

A quantitative indicator of aggregate distribution was proposed to measure the uniformity of distribution of aggregate. The volume of aggregates within each part was recorded and the coefficient of variation (*var*) of aggregate distribution was calculated as follows:

$$Vj = \sum_{i=1}^{m} Vi \tag{11}$$

$$\overline{V} = \sum_{j=1}^{n} Vj/n \tag{12}$$

$$var_j = \sqrt{\left(Vj - \overline{V}\right)^2 / (n-1)} / \overline{V} \tag{13}$$

$$var = \sum_{j=1}^{n} varj \tag{14}$$

where V_i is the volume of aggregate i within part j, V_j is the total volume of aggregates within part j, $\overline{V}$ is the average aggregate volume of all divided parts, var_j is the coefficient of variation of the total volume of aggregates within part j, and var is the coefficient of variation of the total volume of aggregates within specimen.

Figure 20 and Table 5 below present the results, in which the axial strain was calculated by measuring the axial deformation and dividing it by the initial axial length.

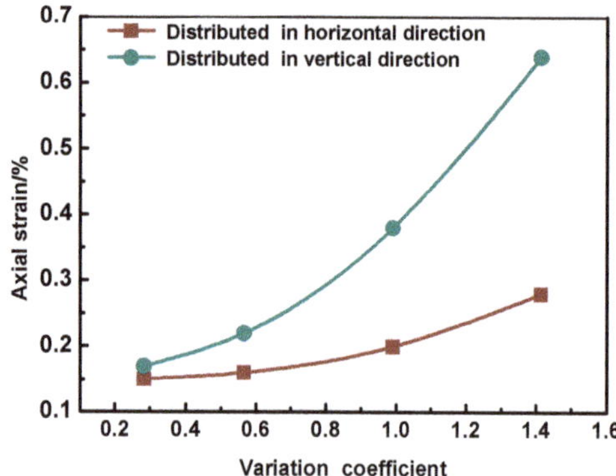

Figure 20. Axial strains with different coefficient of variation of aggregate distribution.

Table 5. Axial strains when aggregates were distributed non-uniformly.

Model No.	Coefficient of Variation	Axial Strain ε (%)	$\varepsilon/\varepsilon_0$
	Partitioned in the horizontal direction		
#1	0.283	0.15	1.12
#2	0.495	0.16	1.24
#3	0.707	0.20	1.56
#4	0.919	0.28	2.15
	Partitioned in the vertical direction		
#5	0.283	0.17	1.31
#6	0.495	0.22	1.70
#7	0.707	0.38	2.90
#8	0.919	0.64	4.90

As shown in Figure 20 and Table 5, axial deformation of the specimens increased as the coefficient of variation of aggregate distribution increased using both partition methods. Aggregate distribution had a significant impact on the deformation performance of the asphalt mixtures. Non-uniformity in the vertical direction had much more impact on the deformation of the specimen than that in the horizontal direction, as shown in Figure 20.

5. Conclusions

In this study, three-dimensional discrete element (DE) models of asphalt mixtures composed of coarse aggregates, asphalt mastic, and air voids was developed using PFC3D, and uniaxial static load creep tests were conducted on the prepared models to investigate the effects of aggregate mesostructure on the permanent deformation behavior of the asphalt mixtures. The effects of aggregate angularity, orientation, surface texture, and distribution on the permanent deformation behavior of asphalt mixtures were carefully analyzed. The following conclusions were drawn:

(1) The constructed DE models of asphalt mixtures could accurately capture the mesostructures of aggregates, such as angularity, orientation, surface texture, and distribution. The simulation of the uniaxial creep test conducted using the DEM could precisely estimate the permanent deformation behavior of asphalt mixtures.

(2) The permanent deformation performance of the asphalt mixtures was improved with increasing aggregate angularity. The axial deformation of the asphalt mixtures was smallest when the aggregates were close to cubical. Good aggregate angularity had a positive effect on the permanent deformation performances of asphalt mixtures, especially when the aggregates were nearly cubical.

(3) Aggregate packing was more stable when the orientations tended to be horizontal, which improved the permanent deformation performances of asphalt mixtures. The percentage pass of 4.75 mm played an important role in controlling the aggregate skeleton and had a great impact on the permanent deformation behaviors of the asphalt mixtures.

(4) The permanent deformation performance of the asphalt mixtures was improved with increasing friction coefficient, indicating that aggregates with good surface texture benefitted the permanent deformation performance of the asphalt mixtures.

(5) Axial deformation of the asphalt mixture specimen increased with increasing coefficient of variation of aggregate distribution. Non-uniform distribution of the aggregates had a negative impact on the permanent deformation performance of the asphalt mixtures, especially when the aggregates were distributed non-uniformly in the vertical direction.

In the future research steps, a viscoplastic mesomechanical model will be developed to further describe the viscoplastic behavior of the permanent deformation behavior of asphalt mixtures.

Author Contributions: Conceptualization, D.Z. and J.Z.; Data curation, D.Z., L.G.; Formal analysis, D.Z., L.G. and J.Z.; Funding acquisition, D.Z.; Methodology, D.Z. and J.Z.; Writing—original draft, D.Z.; Writing—review & editing, J.Z.

Funding: The study is financially supported by Postdoctoral Research funding Project of Jiangsu (2019K138), the Natural Science Foundation of Jiangsu (BK20140109) and the Research Foundation of Nanjing Institute of Technology (YKJ201512).

Conflicts of Interest: The authors declare no conflict of interest.

References

1. Masad, E.; Muhunthan, B.; Shashidhar, N.; Harman, T. Quantifying laboratory compaction effects on the internal structure of asphalt concrete. *Transp. Res. Rec.* **1999**, *1681*, 179–185. [CrossRef]

2. Masad, E.; Muhunthan, B.; Shashidhar, N.; Harman, T. Internal structure characterization of asphalt concrete using image analysis. *J. Comput. Civ. Eng.* **1999**, *13*, 88–95. [CrossRef]

3. Masad, E.; Somadevan, N. Microstructural finite-element analysis of influence of localized strain distribution on asphalt mix properties. *J. Eng. Mech.* **2002**, *128*, 1105–1114. [CrossRef]

4. Kose, S.; Guler, M.; Bahia, H.; Masad, E. Distribution of strains within hot-mix asphalt binders: Applying imaging and finite-element techniques. *Transp. Res. Rec.* **2000**, *1728*, 21–27. [CrossRef]

5. Wang, L.B.; Frost, J.D.; Shashidhar, N. Microstructure study of westrack mixes from x-ray tomography images. *Transp. Res. Rec. J. Transp. Res. Board* **2001**, *1767*, 85–94. [CrossRef]

6. Wang, L.; Wang, Y.; Mohammad, L.; Harman, T. Voids distribution and performance of asphalt concrete. *Int. J. Pavements* **2002**, *1*, 22–33.

7. You, Z.; Adhikari, S.; Kutay, M.E. Dynamic modulus simulation of the asphalt concrete using the X-ray computed tomography images. *Mater. Struct.* **2009**, *42*, 617–630. [CrossRef]

8. Ding, X.; Ma, T.; Gu, L.; Zhang, Y. Investigation of surface micro-crack growth behavior of asphalt mortar based on the designed innovative mesoscopic test. *Mater. Des.* **2020**, *185*, 108238. [CrossRef]

9. Chang, K.-N.G.; Meegoda, J.N. Micromechanical Ssimulation of hot mix asphalt. *J. Eng. Mech.* **1997**, *123*, 495–503. [CrossRef]

10. Rougier, E.; Knight, E.E.; Broome, S.T.; Sussman, A.J.; Munjiza, A. Validation of a three-dimensional finite-discrete element method using experimentalresults of the Split-Hopkinson pressure bar test. *Int. J. Rock Mech. Min. Sci.* **2014**, *70*, 101–108. [CrossRef]

11. Buttlar, W.G.; You, Z. Discrete element modeling of asphalt concrete: Microfabric approach. *Transp. Res. Rec. J. Transp. Res. Board* **2001**, *1757*, 111–118. [CrossRef]

12. You, Z.; Buttlar, W.G. Discrete element modeling to predict the modulus of asphalt concrete mixtures. *J. Mater. Civ. Eng.* **2004**, *16*, 140–146. [CrossRef]

13. You, Z.; Liu, Y.; Dai, Q. Three-dimensional microstructural-based discrete element viscoelastic modeling of creep compliance tests for asphalt mixtures. *J. Mater. Civ. Eng.* **2011**, *23*, 79–87. [CrossRef]

14. Abbas, A.; Masad, E.; Papagiannakis, T.; Harman, T. Micromechanical modeling of the viscoelastic behavior of asphalt mixtures using the discrete-element method. *Int. J. Géoméch.* **2007**, *7*, 131–139. [CrossRef]

15. Liu, Y.; Dai, Q.; You, Z. Viscoelastic model for discrete element simulation of asphalt mixtures. *J. Eng. Mech.* **2009**, *135*, 324–333. [CrossRef]

16. Chen, J.; Zhang, Q.; Wang, H.; Wang, L.; Huang, X. Numerical investigation into the effect of air voids on the anisotropy of asphalt mixtures. *J. Wuhan Univ. Technol.-Mater. Sci. Ed.* **2017**, *32*, 473–481. [CrossRef]

17. Chen, J.; Pan, T.; Huang, X. Numerical investigation into the stiffness anisotropy of asphalt concrete from a microstructural perspective. *Constr. Build. Mater.* **2011**, *25*, 3059–3065. [CrossRef]

18. Ma, T.; Zhang, D.; Zhang, Y.; Hong, J. Micromechanical response of aggregate skeleton within asphalt mixture based on virtual simulation of wheel tracking test. *Constr. Build. Mater.* **2016**, *111*, 153–163. [CrossRef]

19. Ma, T.; Zhang, D.; Zhang, Y.; Wang, S.; Huang, X. Simulation of wheel tracking test for asphalt mixture using discrete element modelling. *Road Mater. Pavement Des.* **2016**, *19*, 367–384. [CrossRef]

20. Ma, T.; Zhang, D.; Zhang, Y.; Zhao, Y.; Huang, X. Effect of air voids on the high-temperature creep behavior of asphalt mixture based on three-dimensional discrete element modeling. *Mater. Des.* **2016**, *89*, 304–313. [CrossRef]

21. Ding, X.; Ma, T.; Huang, X. Discrete-element contour-filling modeling method for micromechanical and macromechanical analysis of aggregate skeleton of asphalt mixture. *J. Transp. Eng. Part B: Pavements* **2019**, *145*, 04018056. [CrossRef]

22. Zhang, Y.; Ma, T.; Ling, M.; Huang, X. Mechanistic sieve-size classification of aggregate gradation by characterizing load-carrying capacity of inner structures. *J. Eng. Mech.* **2019**, *145*, 04019069. [CrossRef]

23. Zhang, Y.; Ma, T.; Ling, M.; Zhang, D.; Huang, X. Predicting dynamic shear modulus of asphalt mastics using discretized-element simulation and reinforcement mechanisms. *J. Mater. Civ. Eng.* **2019**, *31*, 04019163. [CrossRef]

24. Zhang, Y.; Ma, T.; Huang, X.; Zhao, Y.; Hu, P. Algorithms for generating air-void structures of idealized asphalt mixture based on three-dimensional discrete-element method. *J. Transp. Eng. Part B: Pavements* **2018**, *144*, 04018023. [CrossRef]

25. Zhang, Y.; Ma, T.; Ding, X.; Chen, T.; Huang, X.; Xu, G. Impacts of air-void structures on the rutting tests of asphalt concrete based on discretized emulation. *Constr. Build. Mater.* **2018**, *166*, 334–344. [CrossRef]

26. *JTG F40-Technical Specification for Construction of Highway Asphalt Pavement*; Ministry of Transport of the People's Republic of China: Beijing, China, 2011.

27. *JTG E20-Standard Test Methods of Bitumen and Bituminous Mixtures for Highway Engineering*; Ministry of Transport of the People's Republic of China: Beijing, China, 2011.

28. Pan, T.; Tutumluer, E.; Carpenter, S.H. Effect of coarse aggregate morphology on permanent deformation behavior of hot mix asphalt. *J. Transp. Eng.* **2015**, *132*, 580–589. [CrossRef]

29. Stakston, A.D.; Bahia, H.U.; Bushek, J.J. Effect of fine aggregate angularity on compaction and shearing resistance of asphalt mixtures. *Transp. Res. Rec. J. Transp. Res. Board* **2002**, *1789*, 14–24. [CrossRef]

30. Masad, E.; Jandhyala, V.K.; Dasgupta, N.; Somadevan, N.; Shashidhar, N. Characterization of air void distribution in asphalt mixes using x-ray computed tomography. *J. Mater. Civ. Eng.* **2002**, *14*, 122–129. [CrossRef]

31. Itasca Consulting Group. *PFC3D Version 4.0*; Itasca Consulting Group: Minneapolis, MN, USA, 2008.

32. Thornton, C. The conditions of failure of a face-centered cubic array of uniform rigid spheres. *Géotechnique* **2015**, *29*, 441–459. [CrossRef]

33. Dai, Q.; You, Z. Prediction of creep stiffness of asphalt mixture with micromechanical finite-element and discrete-element models. *J. Eng. Mech.* **2007**, *133*, 163–173. [CrossRef]
34. Erdoğan; Turhan, S.; Fowler, D.; Garboczi, E.J. *Determination of Aggregate Shape Properties Using X-ray Tomographic Methods and the Effect of Shape on Concrete Rheology*; Research Report ICAR 106-1; International Center for Aggregate Research: Austin, TX, USA, 2005.

 *materials* 

Article

Rheological and Interaction Analysis of Asphalt Binder, Mastic and Mortar

Meng Chen [1], Barugahare Javilla [1], Wei Hong [1], Changluan Pan [1], Martin Riara [1,2], Liantong Mo [1,*] and Meng Guo [3]

[1] State Key Laboratory of Silicate Materials for Architectures, Wuhan University of Technology, Wuhan 430070, China; 265797@whut.edu.cn (M.C.); makorogo@whut.edu.cn (B.J.); hw0238@whut.edu.cn (W.H.); panchangluan@whut.edu.cn (C.P.); m.m.riara@whut.edu.cn (M.R.)
[2] Department of Physical sciences, South Eastern Kenya University, Kitui 170-90200, Kenya
[3] College of Architecture and Civil Engineering, Beijing University of Technology, Beijing 100124, China; gm@bjut.edu.cn
* Correspondence: molt@whut.edu.cn; Tel.: +86-027-87162595; Fax: +86-027-87873892

Received: 6 December 2018; Accepted: 27 December 2018; Published: 2 January 2019

Abstract: This paper investigated the rheological properties of asphalt binder, asphalt mastic and asphalt mortar and the interaction between asphalt binder, mineral filler and fine aggregates. Asphalt binder, mastic and mortar can be regarded as the binding phase at different scales in asphalt concrete. Asphalt mastic is a blend of asphalt binder and mineral filler smaller than 0.075 mm while asphalt mortar consists of asphalt binder, mineral filler and fine aggregate smaller than 2.36 mm. The material compositions of mastic and mortar were determined from the commonly used asphalt mixtures. Dynamic shear rheometer was used to conduct rheological analysis on asphalt binder, mastic and mortar. The obtained test data on complex modulus and phase angle were used for the construction of rheological master curves and the investigation of asphalt-filler/aggregate interaction. Test results indicated a modulus increase of three- to five-fold with the addition of filler and a further increase of one to two orders of magnitude with cumulative addition of fine aggregates into asphalt binder. Fine aggregates resulted in a phase change for mortar at high temperatures and low frequencies. The filler had stronger physical interaction than fine aggregate with an interaction parameter of 1.8–2.8 and 1.15–1.35 respectively. Specific area could enhance asphalt-filler interaction. The mastic and mortar modulus can be well predicted based on asphalt binder modulus by using particle filling effect. Asphalt mortar had a significant modulus reinforcement and phase change and thus could be the closest subscale in terms of performance to that of asphalt mixtures. It could be a vital scale that bridges the gap between asphalt binder and asphalt mixtures in multiscale performance analysis.

Keywords: bitumen; mastic; mortar; rheological properties; physical interaction

1. Introduction

By the end of 2017, the total length of highways in China open to traffic reached 477 million km with a highway density of 19.5 km per 100 km^2. Among these highways, 136.5 thousand kilometers of expressways have been built according to advanced modern transportation standard. Up to 90% of high-grade highways in China are asphalt pavements. These asphalt pavements usually consist of a three-layer structure comprising the wearing course layer, intermediate binder course layer and base course layer from top to down respectively. Most of the asphalt pavements are designed to have a service life of about 15 years for high-grade highways. However, large amount of maintenance and repair works are needed before reaching the design life. Permanent deformation or rutting, moisture damage, reflective cracking, low temperature cracking and fatigue cracking are the main distresses addressed during maintenance [1–4]. Currently, it has become a big challenge to design more durable asphalt pavements to deal with the increasing traffic volume of modern transportation and extreme climate change in the future.

Asphalt pavements are usually paved with hot mix asphalt which consists of coarse aggregates, fine aggregates, filler and bitumen as the binder. Each of these components has a great influence on the performance of asphalt mixtures after they have been paved and compacted [5–7]. In general, coarse aggregates form the stone skeleton while fine aggregate, mineral filler and asphalt binder fill in the voids within the coarse aggregates and also hold the stone skeleton as a whole. Depending on the difference of scale, the binding phase is usually divided in to three levels: pure asphalt binder, asphaltic mastic (a blend of asphalt binder and mineral filler smaller than 0.075 mm) and asphaltic mortar (a matrix that consists of fine aggregate smaller than 2.36 mm, mineral filler and asphalt binder). The stone skeleton of coarse aggregates provides the main contribution of loading bearing and rutting resistance, while the binding phases strongly relate to the problems of aging, fatigue, cracking, and moisture damage [8,9].

In order to design better and more durable asphalt mixtures, the fundamental properties of asphalt mixtures should be predictable by an upscaling procedure based on the properties of asphalt binder, mastic and mortar [7,10]. This requires that a correlation exists between the test results of asphalt binder, mastic and mortar scales and those of asphalt mixtures. Because of the exclusion of mineral fractions, a gap exists between the properties of pure asphalt binder and those of asphalt mixtures. Asphalt mastic is usually considered as the actual binding component in the mix which coats the coarse aggregates [11]. Apart from coarse aggregates, asphalt mastic is further blended with fine fractions of aggregate to form the mortar that fills in the voids within the coarse aggregates. Therefore, asphalt mortar is the sub scale close to asphalt mixture [12]. In general, asphaltic mortar can bridge the gap of scale between bitumen and asphalt mixture.

Asphalt binder can be reinforced by the addition of mineral filler. Further reinforcement is obtained by the addition of fine aggregates [13–15]. In general, the mastic and mortar system can be regarded as two-phase composite materials, in which, asphalt binder is the matrix phase and mineral particles are the reinforcement phase. The interaction between asphalt binder, filler and fine fractions was found to be a physical process and could be explained by the mechanisms of particle reinforcement [14–16]. Fillers and fine aggregate had significant influences on rheological properties of the corresponding mastic and mortar. Finally, they affect the performances of asphalt mixture at both high and low temperatures.

In recent years, many researches have been done on asphalt mastic with various types of filler including Portland cement [17–19], hydrated lime [17,19–21], fly ash [20,22], natural and synthetic zeolites derived from fly ash [23–26], volcanic ash [27], oil shale ash [28], rice straw ash [29], red mud [30,31], limestone dust [19], glass powder [31], brick dust [31], carbide lime [31], copper tailings [21,31], natural bentonite clay [32], ladle furnace slag [33], silica fume [21], magnetite [34], waste stone sawdust [35] and steel slag [36]. Asphalt mixtures prepared using the same content of bitumen but different waste materials as fillers (glass powder, limestone dust, red mud, rice straw ash, brick dust, carbide lime and copper tailings) showed satisfactory mechanical and volumetric properties. In particular, fine fillers such as limestone dust and red mud had a significant positive effect on the stiffness and cracking resistance of the asphalt mixtures. Free energy of adhesion between bitumen and aggregates was improved by stone powder fillers but use of hydrated lime, calcium carbonate and Portland cement had a negative effect on the adhesion energy. Recycled ladle furnace slag is beneficial to the volumetric properties, stiffness, indirect tensile strength and resistance to dynamic loading of asphalt concrete. Active filler including hydrated lime and cement had the potential to improve the resistance to moisture damage [17–19]. The usage of micro filler and nano-clay had shown high reinforcing potential in bitumen mastics [27,32]. Many waste materials were used as substitutes to natural limestone filler due to the consideration of recycling and environmental protection. Magnetite filler could be exploited for induction or microwave healing of asphalt pavement cracks [34]. Regardless of the type of filler to be used, high interaction or compatibility between asphalt binder and filler is always desired [37–43]. Diab reported that there was no proof of chemical mechanisms between bitumen and various fillers including blast-furnace slag, silica fume, fly ash, and hydrated lime. Nonlinear rheological properties could differentiate the performance of different types of mastic [22]. Guo's research indicated that the interfacial interaction between asphalt binder and filler strongly depended on the diffusion behavior of asphalt binder components as well as the chemical composition of mineral fillers [15,43,44]. The reinforcement effect of filler was widely investigated by means of rheological properties including complex modulus, phase angle, creep recovery and stiffness. Various indices were proposed to evaluate the asphalt-filler interaction ability. Liu reported that the evaluation index based on phase angle was more sensitive than that based on complex modulus. Temperature and specific surface area were the two main factors effecting the interfacial interaction between asphalt binders and mineral fillers [14,45–47].

With respect to asphaltic mortar, recycled waste materials such as gneiss, ceramsite, ceramic, marble, redbrick ash, steel slag, fine dune sand, and river sand were reported as a possible substitute for fine aggregates [48,49]. It was found that the morphological characteristics of fine aggregates (i.e., shape factors, angularity, and surface texture) significantly affect the mechanical performance of asphalt mixtures [50]. The increase of fine aggregate content was found to jeopardize the resistance of asphalt mixture on low-temperature cracking [51]. Fine aggregates had a more significant impact on skid resistance on the macro-texture level than micro-texture level [52,53]. Li reported that the complex modulus values between asphaltic mastic and mortar were highly correlated and the dissipated energy method could well explain the fatigue performance of asphalt materials at different scales [6].

In the recent years, various types of filler and fine aggregates have been used as a substitute for traditional mineral components in asphalt mixtures. The associated asphalt mixtures were reported to have a comparable performance with control ones based on laboratory testing. However, the long-term pavement performance of these non-traditional asphalt mixtures is needed for validation. Rutting and moisture damage was frequently found among the premature distresses of asphalt pavements in China. Improper use of filler and fine aggregate worsened the resistance of rutting and moisture damage. This paper aimed to get insight into the correlation of the rheological property and the degree of physical interaction between asphalt binder, mastic and mortar over a wide range of temperatures (30 °C to 70 °C) and frequencies (0.1 rad/s to 400 rad/s). The sensitivity of common parameters used to analyze the interaction among asphalt binder, filler and fine aggregate was also assessed. In addition, the asphalt materials subscale whose interaction was more representative for a multiscale performance research was established. The material components of the studied asphaltic mastic and mortar were determined from the commonly used asphalt mixtures in asphalt pavements. This ensured that the used asphaltic mastic and mortar were consistent with those applied in actual asphalt pavements. The rheological properties of asphalt binder, mastic and mortar were investigated by means of dynamic shear rheometer. The interaction between asphalt binder, filler and fine aggregate was evaluated based on the change of complex modulus and phase angle. Several interaction parameters proposed by particle reinforcement theory were applied for interaction evaluation and validation.

2. Materials and Methods

2.1. Materials

Neat bitumen (90#) and SBS (Styrene-Butadiene-Styrene) polymer modified bitumen were used. Both asphalt binders were supplied by Panjin Northern Asphalt Co. Ltd, Panjin, China and the content of SBS polymer in the modified asphalt binder was 4.5%. SBS modified bitumen is commonly used in the construction of the surface wearing course and intermediate layer, while neat bitumen 90# is used in the bottom layer of most three-layer asphalt pavements in northern China. Table 1 shows the basic properties of the asphalt binders.

Limestone and basalt aggregates were selected for this study. AC-13 asphalt mixtures designed for the surface wearing course layer contained basalt aggregates while AC-20 asphalt mixture designed for the intermediate binder course layer and AC-25 asphalt mixture for the base course layer contained limestone aggregates. 13, 20 and 25 are the nominal maximum aggregate sizes. More detailed information on the related properties of the used aggregate and filler can be found in our previous paper [54]. The gradation limits of these three mixtures were designed according to JTG F40-2004 [55] and the combined aggregates grading can be found in Table 2. The optimum bitumen contents of AC-13, AC-20 and AC-25 asphalt mixtures were 4.7%, 4.3% and 3.9% respectively as determined by Marshall method with 75 blows per face. These bitumen contents were in agreement with the practical application values in the field.

Based on the mixture compositions listed in Table 2, the corresponding compositions for mastic and mortar fractions were determined as shown in Tables 3 and 4. Asphalt mastic was determined by the combination of bitumen content and filler content. Asphalt mortar was a mixture of fine aggregates, filler and asphalt binder. Depending on the maximum size of the fine aggregates, three different levels of the mortar scale were investigated. The highest, middle and lowest mortar scales contained fine aggregates with a maximum size of 2.36 mm, 1.18 mm and 0.6 mm respectively. Asphalt mastic and mortar specimens were made following an optimized protocol to obtain homogeneous mixtures. The compositions of these asphaltic materials were heated in the oven at 170 °C for 2 h, after which they were blended until they became homogeneous.

Table 1. Basic properties of the asphalt binders.

Index	Units	Asphalt Binder	
		Neat Bitumen (90#)	SBS Modified Bitumen
Penetration (25 °C, 100 g, 5 s)	0.1 mm	84	73
Ductility (15 °C, 5 cm/mm)	cm	98.1	>100
Viscosity (135 °C)	Pa·s	0.324	0.645
Softening point	°C	49.5	52

2.2. Test Methods

DSR (MCR101, Anton Paar, Graz, Germany) was used to test the rheological properties of asphalt binder, mastic and mortar specimens. DSR test samples were sandwiched between two circular plates of diameter 25 mm. The lower plate was fixed while the upper plate oscillated back and forth, at frequencies ranging from 0.01 rad/s to 400 rad/s and at temperatures between 30 °C and 70 °C. Asphalt mastic was tested based on the same protocols as bitumen. For mortar samples, due to the addition of fine aggregates, the gap height between the parallel plates was designed to be a minimum of three times the size of the largest fine aggregates in the sample. For that case, asphalt mortar samples containing 0.6, 1.18 and 2.36 mm fine aggregates needed a gap height of 2 mm, 4 mm and 7 mm respectively as illustrated in Table 5.

Circular mortar test specimens were made by pouring hot and well blending mortar into silicon rubber mounds with a diameter 25 mm and depth of 2, 4 and 7 mm depending on the maximum size of fine aggregates. In order to improve the interface adhesion between the mortar specimen and DSR plates, super glue was used for bonding specimens to the DSR plates. To do so, the zero gap calibration of the DSR plates was first carried out. Super glue was spread on the surfaces of the bottom and upper plates. The mortar specimens were removed from the silicon rubber mold and carefully placed on the bottom plate. The upper plate was lowered and the mortar specimens were squeezed to the desired thickness. DSR frequency sweep testing on asphalt binder, mastic and mortar samples was carried out on five test temperatures including 30 °C, 40 °C, 50 °C, 60 °C, and 70 °C and frequencies from 0.1 rad/s to 400 rad/s. The obtained complex modulus and phase angle was used to investigate the rheological properties of various asphaltic materials and the interaction between asphalt binder, filler and fine aggregates.

Table 2. Aggregate grading and compositions of various asphalt mixtures.

Sieve Size (mm)	Passing Percent (%)		
	AC-13	AC-20	AC-25
31.5	-	-	100
26.5	-	100	98
19	-	95	86
16	100	83	79
13.2	95	72	69
9.5	77	61	47
4.75	53	41	34
2.36	37	30	24
1.18	27	23	18
0.6	19	16	13
0.3	14	11	10
0.15	10	9	8
0.075	6	5	5
Aggregate	Basalt	Limestone	Limestone
Filler	Limestone	Limestone	Limestone
Asphalt binder	SBS	SBS	90#
Asphalt binder content	4.7	4.4	3.9

Table 3. Compositions of asphalt mastic contained filler and asphalt binder.

Specimen	Asphalt Binder (A)	Filler Type (F)	F/A Ratio
AC-13's mastic	SBS	Limestone powder	1.280
AC-20's mastic	SBS	Limestone powder	1.136
AC-25's mastic	90#	Limestone powder	0.850

Table 4. Compositions of asphalt mortar containing find aggregates, filler and asphalt binder.

Sieve Size (mm)	Percent Composition by Weight (%)								
	Basalt Aggregates			Limestone Aggregates			Limestone Aggregates		
	I	II	III	I	II	III	I	II	III
2.36	27.7	-	-	24.2	-	-	22.4	-	-
1.18	18.2	25.2	-	16.5	21.8	-	16.2	20.8	-
0.6	13.0	18.0	24.0	14.3	18.9	24.2	16.2	20.8	26.3
0.3	9.5	13.2	17.6	11.0	14.5	18.6	12.4	16.0	20.2
0.15	6.1	8.4	11.2	5.5	7.3	9.3	6.2	8.0	10.1
0.075	6.9	9.6	12.8	7.7	10.2	13.0	8.7	11.2	14.2
Filler	10.4	14.4	19.2	11.0	14.5	18.6	8.3	10.6	13.4
Asphalt binder content	8.1	11.3	15.1	9.7	12.8	16.4	9.7	12.5	15.8
Asphalt binder	-	SBS	-		SBS			90#	
Asphalt mixture		AC-13			AC-20			AC-25	

Note: I, II and III are mortar scales containing fine aggregates with a maximum size of 2.36 mm, 1.18 mm and 0.6 mm respectively.

Table 5. Designed height of DSR test specimens.

Type of Binder	Height (mm)
Asphalt	1.0
Mastic	1.0
Mortar with fine aggregates smaller than 0.6 mm (0.6 mortar)	2.0
Mortar with fine aggregates smaller than 1.18 mm (1.18 mortar)	4.0
Mortar with fine aggregates smaller than 2.36 mm (2.36 mortar)	7.0

3. Results and Discussion

3.1. Master Curves of Various Asphalt Materials

Figures 1–6 present the master curves of various asphalt materials at a reference temperature of 60 °C. The master curves were constructed by means of the time-temperature superposition principle (TTSP). The complex modulus and phase angle data obtained at different temperatures including 30 °C, 40 °C, 50 °C, 60 °C and 70 °C, were horizontally shifted to the reference temperature of 60 °C to form a smooth master curve. By doing this, the rheological properties indicated by complex modulus and phase angle were explained over a wide range of reduced frequencies. Data at low frequencies represent the properties at high temperature (e.g., 70 °C). On the contrary, those at high frequencies indicate the properties at low temperature (e.g., 30 °C).

Figure 1 shows various master curves of complex modulus obtained from AC-25 based asphaltic materials including neat asphalt binder, mastic and mortar. The differences are distinct between neat asphalt binder, mastic and mortars containing different sizes of fine aggregates. Compared with the neat asphalt binder, the addition of filler made an obvious modulus increase of the resultant mastic. Further addition of fine aggregate led to a higher modulus for mortar. Increasing the maximum size of fine aggregates also resulted in an modulus increase of mortar. In general, mastic had a threefold increase in complex modulus compared with neat asphalt binder. However, depending on the size of fine aggregate, mortar could have an increase of one to two orders of magnitude on complex modulus. The huge increase in complex modulus of mortar helps to improve the rutting resistance of asphalt mixtures at high temperature.

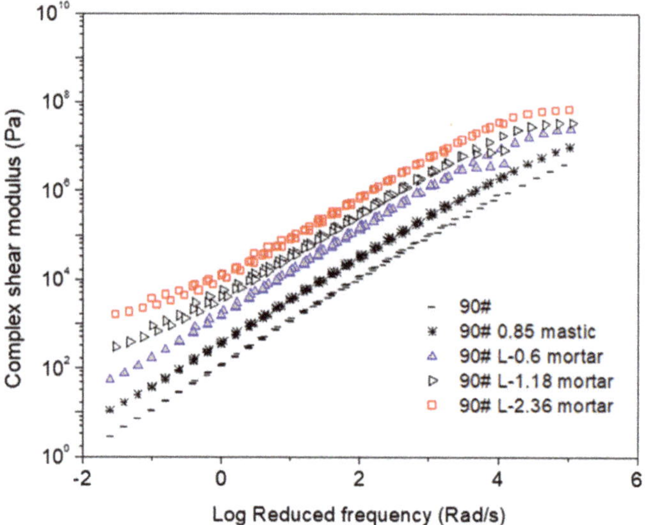

Figure 1. Complex modulus master curves of AC-25 based asphaltic materials including neat asphalt binder, mastic and mortar at a reference temperature of 60 °C.

Figure 2 presents the master curves of phase angle obtained from AC-25 based asphaltic materials including neat asphalt binder, mastic and mortar. Among these asphalt materials, a significant change can be seen at low reduced frequencies, which is equal to high temperature of 70 °C. The phase angle tended to drop down with the addition of filler and fine aggregates. Larger size of fine aggregates led to a more obvious reduction on phase angle at the range of high frequency. It should be noted that phase angle is an index for the ratio between viscosity and elasticity. A purely viscous liquid and an ideal elastic solid have a phase angle of 90° and 0°, respectively. Neat asphalt binder behaved like a purely viscous liquid with a phase angle close to 90° at very high temperatures. Asphalt mastic only deviated a little from neat asphalt binder. However, the addition of fine aggregate could result in a significant change on phase angle, which was strongly indicated by a phase change from a viscous to viscoelastic range. In general, low phase angle and viscoelastic behavior is desired for improving rutting resistance of asphalt mixtures. Combining the data obtained on complex modulus and phase angle, it can be seen that the corresponding material components in asphalt mixture including asphalt binder, mastic and mortar behaved differently at high temperatures. The addition of filler and fine aggregates could result in higher complex moduli and lower phase angles as well as phase change from a viscous to viscoelastic range.

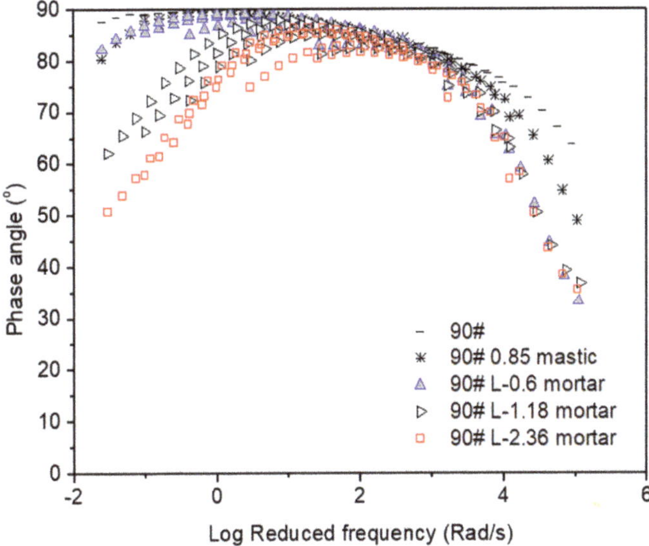

Figure 2. Phase angle master curves of AC-25 based asphaltic materials including neat asphalt binder, mastic and mortar at a reference temperature of 60 °C.

Figure 3 presents the master curves of complex modulus obtained from AC-20 based asphalt materials. AC-20 asphalt mixture is commonly used as the intermediate layer in typical three-layer asphalt pavements in China. The intermediate layer is considered to be subjected to disadvantageous shear stress under wheal loadings and thus susceptible to rutting at high temperatures. Field inspection demonstrated that the intermediate layer contributed most of permanent deformation of asphalt pavements with a semi-rigid base. In order to increase the rutting resistance, SBS modified asphalt binder is commonly used instead of neat asphalt binder together with a stronger coarse stone skeleton containing fewer fine fractions for the intermediate layer. In Figure 3, the material composition of the studied mastic and mortar was determined from the aggregate gradation and asphalt binder content of a typical AC-20 mixture. Similarly, SBS modified asphalt mastic and mortars all have higher complex modulus compared with SBS modified asphalt binder. In general, the addition of filler into SBS modified asphalt binder quadrupled the complex modulus of mastic. Further addition of fine aggregates resulted in a more significant improvement on complex modulus of mortar. Increasing the size of fine aggregate from 0.6 mm to 2.36 mm led to three-fold increase in complex modulus.

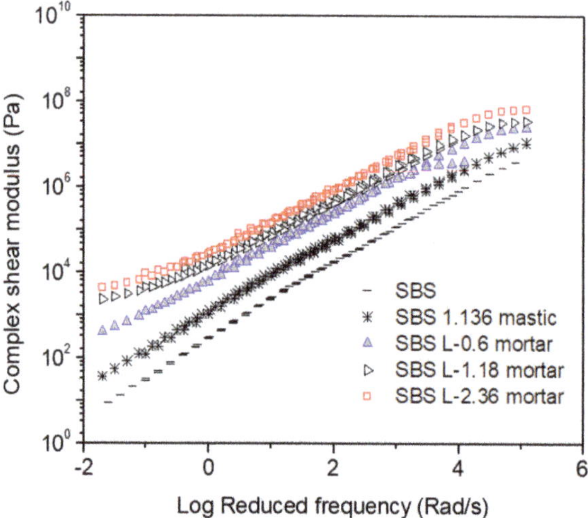

Figure 3. Complex modulus master curves of AC-20 based asphaltic materials including SBS modified asphalt binder, mastic and mortar at a reference temperature of 60 °C.

Figure 4 gives a summary of the master curves of phase angle obtained from AC-20 based asphalt materials. SBS modified asphalt binder, mastic and mortars containing different fine aggregate maximum sizes were included. The change tendency of asphalt mastic is close to that of SBS modified asphalt binder with a slight reduction over the wide range of reduced frequencies. The addition of fine aggregates resulted in an obvious shift of phase angle, thus confer more elastic property on asphalt mortar. This helped to transfer the visco characteristic of bitumen to a viscoelastic characteristic of mortar at high temperatures. Therefore, the actual fine fractions improve the rutting resistance of the AC-20 mixture.

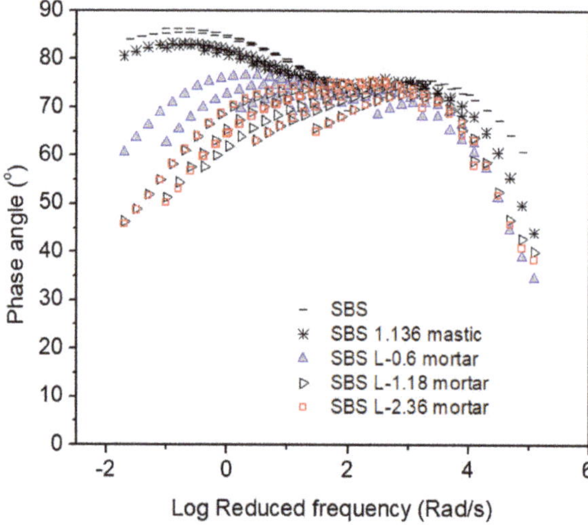

Figure 4. Phase angle master curves of AC-20 based asphaltic materials including neat asphalt binder, mastic and mortar at a reference temperature of 60 °C.

Figures 5 and 6 give the master curves of complex modulus and phase angle of AC-13 based asphalt materials including neat asphalt binder, mastic and mortar. AC-13 asphalt mixture is widely used in the construction of surface wearing course in China. Basalt aggregates are commonly used instead of limestone aggregates in order to improve surface abrasion resistance. It is observed that the change tendencies of AC-13 based asphalt materials were similar to those of AC-20 and AC-25 based asphalt mixtures, which were presented in Figures 1 and 3 respectively. The hardening effect on complex modulus can be well distinguished by the addition of filler and fine basalt aggregates. With respect to phase angle, the addition of basalt fine aggregates can result in a similar phase change at low reduced frequencies, which is an equivalent to high temperature.

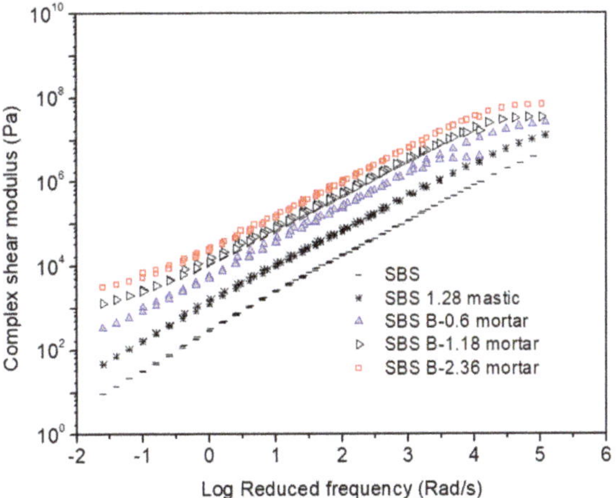

Figure 5. Complex modulus master curves of AC-13 based asphaltic materials including SBS modified asphalt binder, mastic and mortar at a reference temperature of 60 °C.

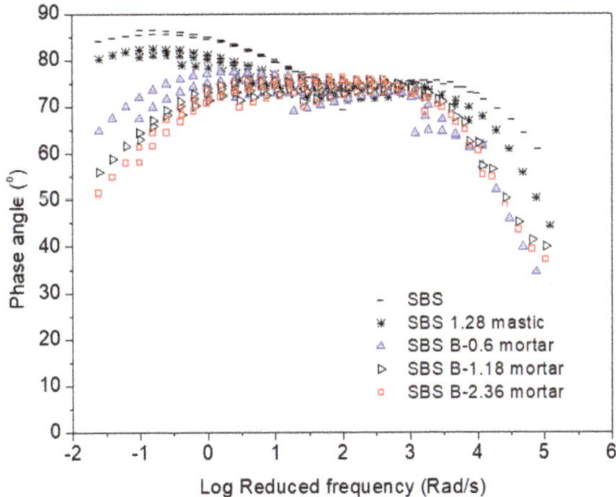

Figure 6. Phase angle master curves of AC-13 based asphalt materials including SBS modified asphalt binder, mastic and mortar at a reference temperature of 60 °C.

Figures 7 and 8 show the influence of type of asphalt binder on complex modulus and phase angle, respectively. Compared with neat asphalt binder, SBS modified asphalt materials had an advantage on complex modulus at low reduced frequencies regardless of the scale of material. It also indicated that mortar made by neat asphalt binder combined with larger fine aggregates (for example, 2.36 mm) could be comparable to mortar that consists of SBS modified asphalt binder and smaller fine aggregate (for example, 0.6 mm). This indicated that well-designed material composition of asphalt mortar could be important for performance improvement. With respect to phase angle, a plateau was observed for SBS modified asphalt binder, but not for neat asphalt binder materials. This strongly indicated the existence of the polymer network. In general, SBS modification could result in a small reduction of phase angle. This effect was relatively limited when compared with neat asphalt binder at low reduced frequencies. The effect of filler is not obvious for either neat asphalt binder or SBS modified asphalt binder. However, fine aggregates could be decisive for asphalt mortar. The interaction between fine aggregates tended to result in a phase change, which is considered to benefit the rutting resistance of asphalt mixtures at high temperatures. Based on the analysis above, it could be concluded that the use of polymer modified asphalt binder and coarser fine aggregate should be of interest to improve the rheological properties of mortar at high-temperatures.

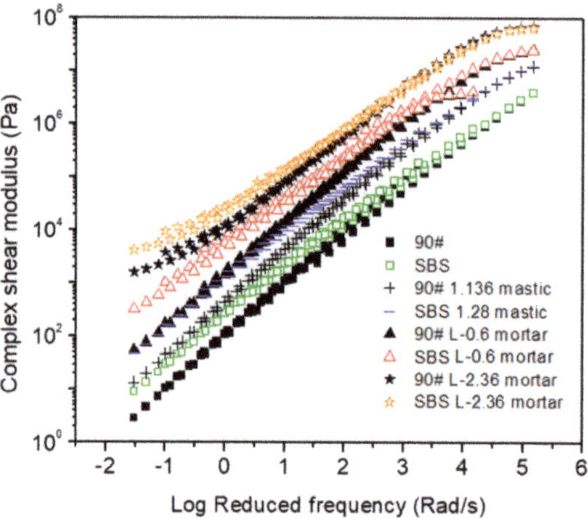

Figure 7. Complex modulus comparison between 90# neat asphalt and SBS modified asphalt based materials at a reference temperature of 60 °C.

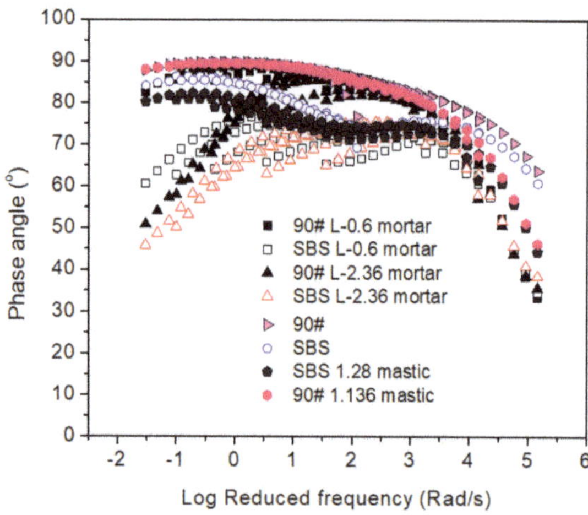

Figure 8. Phase angle comparison between 90# neat asphalt and SBS modified asphalt based materials at a reference temperature of 60 °C.

3.2. Interaction Analysis between Asphalt Binder, Filler and Fine Aggregate

Asphalt mastic and mortar can be regarded as two-phase particle reinforced composite materials. In this system, asphalt binder is the matrix phase, while filler and fine aggregates acts as the reinforcement phase. Both the filler and fine aggregates can be defined as rigid particles when considering the huge modulus difference between asphalt binder and mineral stone. The addition of filler and fine aggregates into asphalt binder results in modulus reinforcement. The interaction between asphalt binder and mineral particles could be expressed in terms of phase angle using the following equation:

$$\tan \delta_c = (1 - \varnothing_f)(1 + A)\tan \delta_m, \tag{1}$$

where: δ_c and δ_m are the phase angles of asphalt based composite material and the asphalt matrix, respectively; $\varnothing_f$ is the volume fraction of the reinforcement phase. A is the interaction parameter between the asphalt matrix phase and reinforcement phase.

Since the phase angle of mastic/mortar and asphalt binder can be tested, the interaction parameter between asphalt binder and mineral particles can be calculated by using:

$$A = \frac{\tan \delta_c}{\tan \delta_m (1 - \varnothing_f)} - 1, \tag{2}$$

It should be noted that a lower value of the A interaction parameter indicates a stronger interaction between asphalt binder and mineral particles.

K.D. Ziegel et al. proposed Equation (3) to estimate the loss tangent of composite materials as the function of filler volume fraction [16]:

$$\tan \delta_c = \frac{\tan \delta_m}{1 + 1.5 \varnothing_f B'} \tag{3}$$

where: B is the interaction parameter that represents the matrix phase and reinforcement phase interaction ability. A higher value of B indicates a stronger interaction between the matrix phase and reinforcement phase.

Similarly, with the known phase angle of mastic/mortar and asphalt binder, the interaction parameter of B can be determined by using the equation below:

$$B = (\frac{\tan \delta_m}{\tan \delta_c} - 1)/(1.5 \varnothing_f), \tag{4}$$

According to Palierne model [15], the modulus of two-phase composite materials can be estimated by using Equation (5):

$$G_c^* = G_m^* \times (\frac{1 + 1.5 \varnothing_f C}{1 - \varnothing_f C}), \tag{5}$$

where: G_C^* and G_m^* are the modulus of the composite material and the asphalt matrix, respectively; C is the interaction parameter that represents the matrix-particle reinforcement interaction.

With respect to the system of asphalt mastic and mortar, the interaction parameter of C value can be determined by using the modulus of asphalt binder, mastic and mortar:

$$C = \frac{\frac{G_c^*}{G_m^*} - 1}{(1.5 + \frac{G_c^*}{G_m^*}) \times \varnothing_f}, \tag{6}$$

A greater C value indicates a stronger asphalt and mineral interaction. In Ziegel's study [16], it was found the polymer-filler interaction parameter of C values determined from storage modulus ranged from 1.1 to 6.5 depending on the combination of polymer and filler, which indicated that C value is of fundamental importance in selecting fillers for various commercial end uses.

An extensive research on asphalt-filler interaction ability done by Liu and Zhao, showed that the sensitivity of interaction parameters A, B, C from high to low is in the order of B, A and C. The change of B and A was complex while C tended to be relatively constant [14]. The C value determined by using complex modulus varied from 0.8 to 3.0 for asphalt mastics containing limestone, Portland cement and hydrated lime. Hydrated lime showed the strongest interaction between asphalt binder and filler, followed by Portland cement, and limestone had the least interaction among these three types of filler. The interaction ability of asphalt binder and filler increases with an increase in temperature and decreases with an increase in loading frequency. Guo's study indicated that the interaction parameter C was the least sensitive compared with A and B yet effective to evaluate the effect of temperature, components of filler and specific surface area of filler on interfacial interaction. The C value ranged from 1.45 to 1.85 for mastic containing andesite, granite and limestone. Increasing temperature from 15 °C to 35 °C led to a higher C value. Guo concluded that temperature and the specific surface area of fillers were the two main factors effecting the interfacial interaction between asphalt binder and mineral fillers [15].

Figures 9–11 gives the values of interaction parameter A for various types of mastic and mortar determined from various asphalt mixtures. According to these figures, A value changed significantly over a wide range of reduced frequency or temperature (30 °C to 70 °C). A peak value tended to occur at a reduced frequency of around 300 rad/min at a reference temperature of 60 °C. Asphalt mastic tended to have lower A value compared with the corresponding mortar, indicating that asphalt-filler interaction was stronger than those between asphalt binder and fine aggregate. Increasing the size of fine aggregates could result in higher A values, thus weak asphalt-aggregate interaction. It was found that neat asphalt-based materials in Figure 9 tended to have a lower A value than SBS modified asphalt-based materials in Figure 10. The difference between basalt and limestone fine aggregate is marginal as indicated in Figures 10 and 11, respectively.

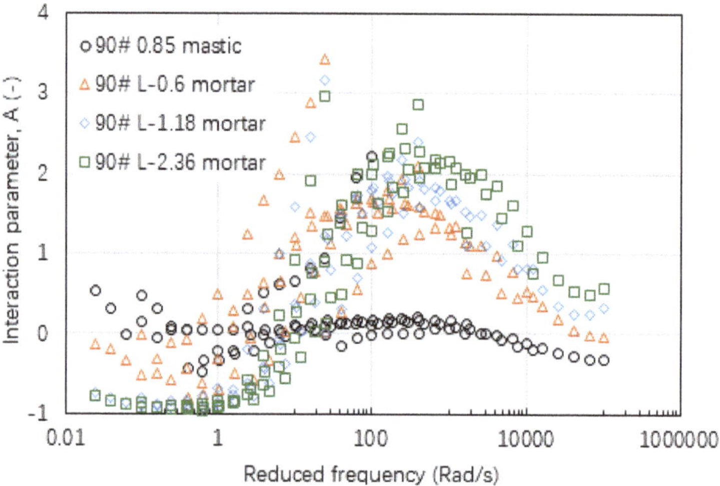

Figure 9. Interaction parameter of A values of mastic and mortar based on AC-25 mixture.

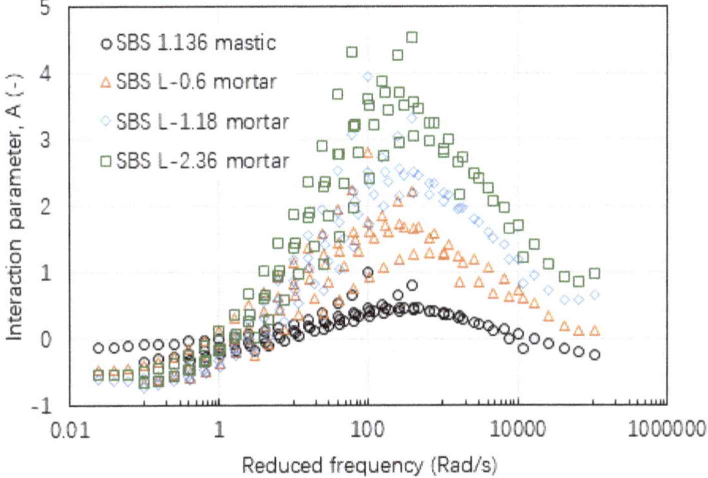

Figure 10. Interaction parameter of A values of mastic and mortars based on AC-20 mixture.

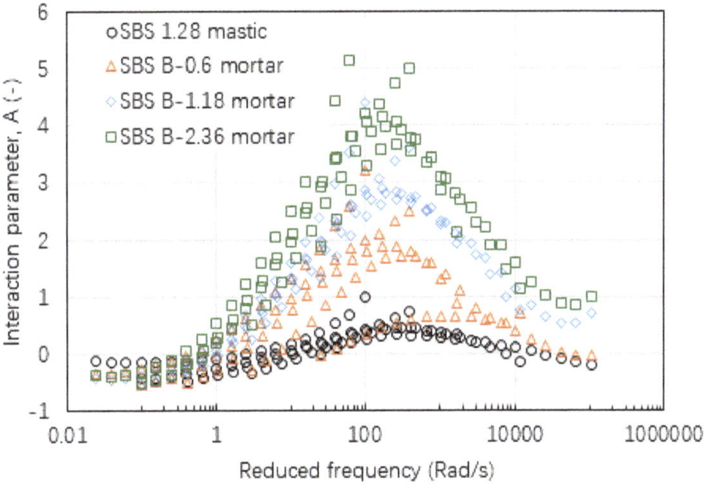

Figure 11. Interaction parameter of A values of mastic and mortar based on AC-13 mixture.

Figures 12–14 show the values of interaction parameter B for various types of mastic and mortar. The change tendency of interaction parameter B was contrary to interaction parameter A. Increasing frequency or reducing temperature decreased the B value to a minimum at a frequency around 200rad/s with a reference temperature of 60 °C. After reaching this critical frequency, further increase in frequency resulted in an increase in B value. It was observed that B value could not distinguish well the difference between mastic and mortar as well the effect of aggregate size of mortar. Regardless of the type of asphalt material, larger scatter of B values was found at the low range of reduced frequency, or high temperatures. At high frequencies or low temperatures, asphalt mastic and mortar had a close value of interaction parameter B.

The analysis based on interaction parameters, A and B pointed to a phenomenon that asphalt-filler/aggregate interaction varies over a wide range of frequency or temperature (30 °C to 70 °C in this study). Furthermore, there was a critical frequency or temperature that may show poor interaction ability between asphalt binder and mineral particle. Since temperature has a great influence on asphalt binder's viscoelastic properties, it indirectly hinted that asphalt-filler/aggregate interaction may be sensitive to the viscoelastic properties of asphalt binder.

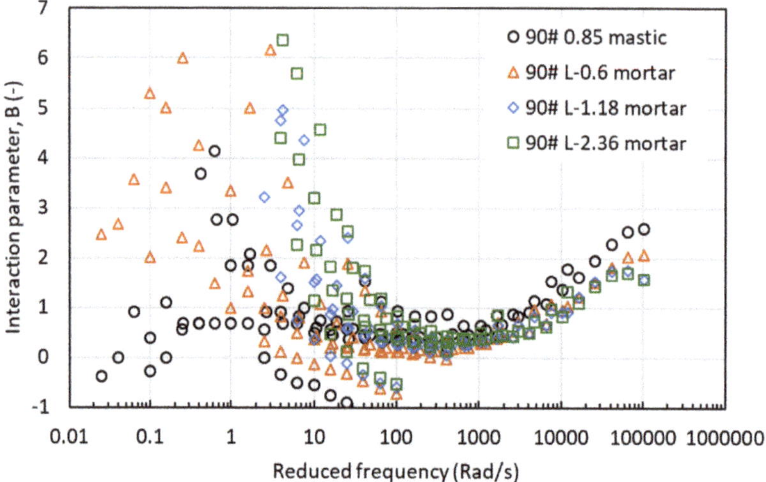

Figure 12. Interaction parameter of B values for mastic and mortar based on AC-25 mixture.

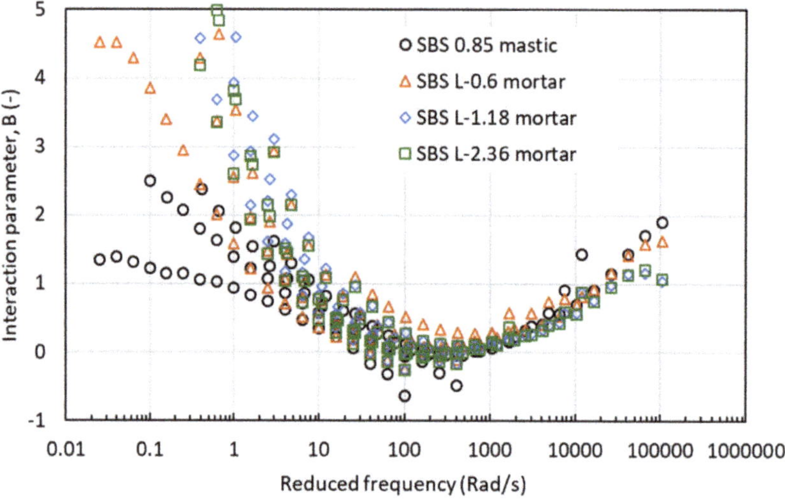

Figure 13. Interaction parameter of B values for asphalt mastic and mortar based on AC-20 mixture.

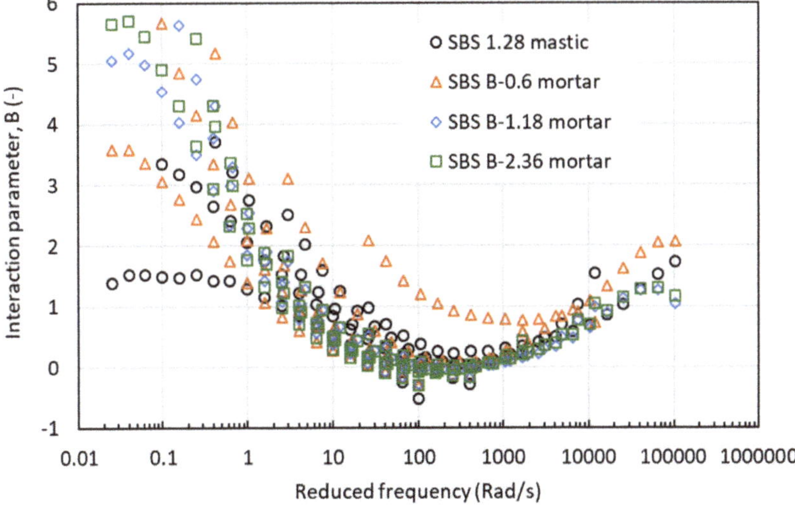

Figure 14. Interaction parameter of B values for asphalt mastic and mortar based on AC-13 mixture.

Figures 15–17 show the values of interaction parameter C for various types of mastic and mortar. Compared with interaction parameters A and B, C tended to be relatively constant with a small change over a wide range of frequency or temperature. The higher value of C indicated that mastic had a stronger interaction ability compared with mortar. Bigger sizes of fine aggregate resulted in a very slight reduction on C value. Table 6 lists a summary of C values of various asphalt materials. Neat asphalt binder had a stronger interaction with filler compared with SBS modified asphalt binder. Increasing filler-asphalt binder ratio could slightly increase C value. All of the mortar tended to have a relatively constant C value ranging from 1.20 to 1.31 independent of the type of asphalt binder, aggregate and grading.

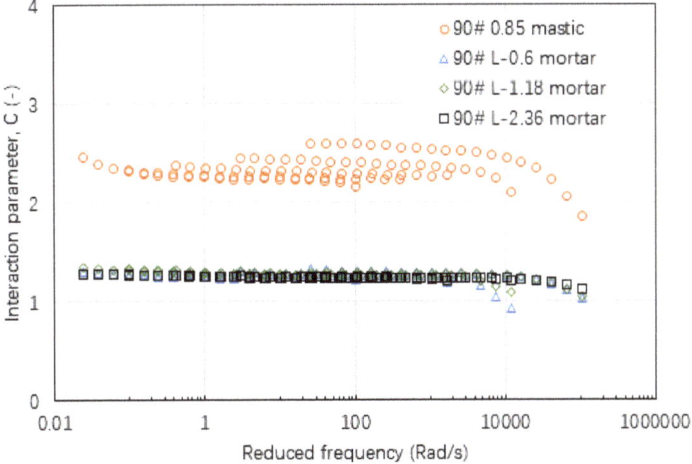

Figure 15. Interaction parameter of C values for asphalt mastic and mortar based on AC-25 mixture.

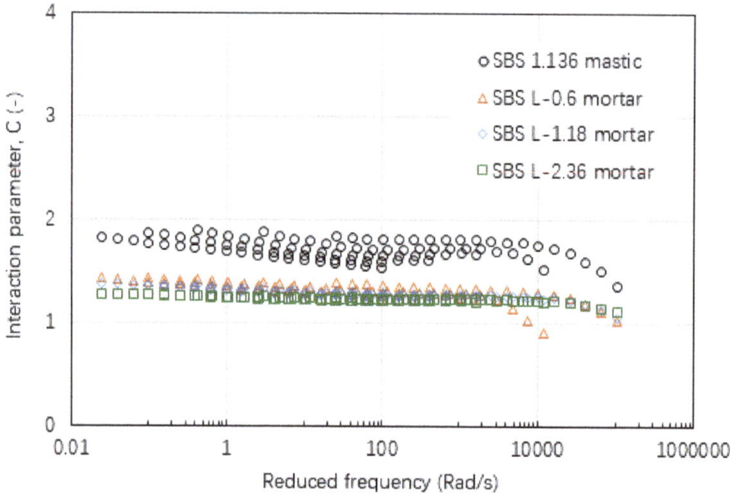

Figure 16. Interaction parameter of C values for asphalt mastic and mortar based on AC-20 mixture.

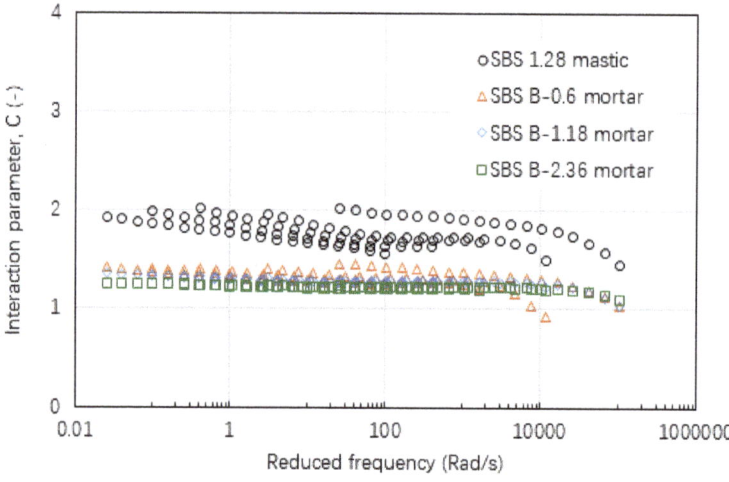

Figure 17. Interaction parameter of C values for asphalt mastic and mortar based on AC-13 mixture.

Table 6. A summary of interaction parameter, C values of various asphalt materials.

Material Scale	Asphalt Binder	Filler	Fine Aggregate	Interaction Parameter, C Value
Mastic	90#	Limestone (0.85:1)	-	2.523
	SBS	Limestone (1.136:1)	-	1.806
	SBS	Limestone (1.28:1)	-	1.877
Mortar	90#	Limestone	0.6 mm limestone	1.281
	90#	Limestone	1.18 mm limestone	1.272
	90#	Limestone	2.36 mm limestone	1.243
	SBS	Limestone	0.6 mm limestone	1.310
	SBS	Limestone	1.18 mm limestone	1.288
	SBS	Limestone	2.36 mm limestone	1.233
	SBS	Basalt	0.6 mm limestone	1.308
	SBS	Basalt	1.18 mm limestone	1.277
	SBS	Basalt	2.36 mm limestone	1.217

The above analysis indicated that the interaction parameters A and B were determined based on phase angle and they showed a large scatter due to frequency and temperature dependency. This may make the property prediction difficult. A relatively constant value of interaction parameter C was useful for modulus prediction. The validation of C value was carried out by using different materials and combinations. For this purpose, an AH-70 neat asphalt binder with a penetration of 67, Portland cement as filler as well as river sand as fine aggregate were applied. Two types of mastic were prepared by using AH-70 neat asphalt binder and SBS modified asphalt binder plus cement with a filler-asphalt binder ratio of 1:1. Two types of mortar were checked by AH-70 neat asphalt binder and SBS modified asphalt binder together with limestone filler and river sand as a substitution for fine aggregates. The compositions of these two types of asphalt mortar were determined according to the mortar with a maximum size of 1.18 mm and 0.6 mm based on AC-20 asphalt mixture as listed in Table 4. Figure 18 shows the tendency of C values of these four asphalt materials that were used for the purpose of validation. The C values of mastic containing cement tended to decline with increasing frequency while mortar containing river sand remained constant. Table 7 shows the validation results based on C values. It could be seen that AH-70 neat asphalt binder had a slightly higher C value compared with SBS modified asphalt binder. Similarly, cement filler had stronger interaction ability than limestone filler as listed in Table 6. Mortar containing river sand showed a comparable result with different types of mortar as listed in Table 6. It indicated that the interaction ability between asphalt binder and filler was stronger than those between asphalt binder and fine aggregate. The specific area of filler was found to significantly influence the interaction ability. In general, the C values remained relatively constant, independent of the type of materials and composition, thus could be better for the prediction of mortar modulus based on asphalt binder modulus. The above analysis indicated that the interaction between asphalt binder, filler and fine aggregate was a physical effect, which could be explained by particle filling effect. On the contrary with physical effect, the chemical interaction effect was reported by Singh in the crumb rubber-asphalt binder system [56]. Singh reported that the chemical interaction effect between crumb rubber and asphalt binder and the filler nature of rubber particles can be well identified. The effect of particle filler nature was found to be significantly greater than the effect of chemical interaction on viscosity, $G^*/Sin\delta$, recovery, and fatigue life, while the chemical interaction effect was obvious on nonrecoverable creep compliance response. Increased crumb rubber particle size may cause changes in the filler nature of crumb rubber modified asphalt composite. Compared with crumb rubber, it is clear that filler and fine aggregate lack chemical interaction with asphalt binder. Due to the simple particle filling effect, specific area could be significant for asphalt binder-filler mastic system, while the size of fine aggregate could be important for mortar system. Nevertheless, further testing could be carried out to reveal the microscopic bitumen-filler/aggregate interaction in the asphalt mixture composites.

Table 7. Interaction validation by using cement as filler and river sand as a substitution for fine aggregate.

Material Scale	Bitumen	Filler	Fine Aggregate	Interaction Parameter, C Value
Mastic	AH-70	Cement (1:1)	-	2.864
	SBS	Cement (1:1)	-	2.678
Mortar	AH-70	Limestone	1.18 mm sand	1.312
	SBS	Limestone	0.6 mm sand	1.340

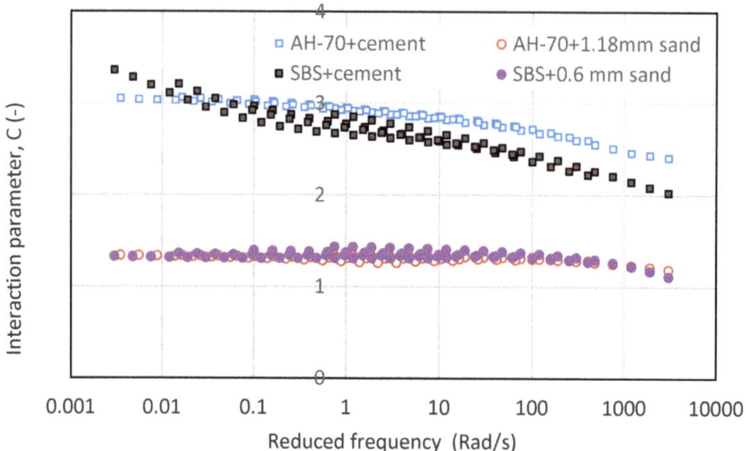

Figure 18. Interaction parameters, C values of mastic made by using cement as filler and mortar using river sand as a substitution for fine aggregate.

4. Conclusions

The rheological properties of asphalt binder, asphalt mastic and asphalt mortar were investigated by means of dynamic shear rheometer. The interaction between asphalt binder, mineral filler and fine aggregates was evaluated by using the data obtained from complex modulus and phase angle. Based on the experimental results and the analysis done, the following conclusions were drawn:

i. The modulus reinforcement with the addition of filler and fine aggregates into asphalt binder can be well explained by particle reinforcement mechanics. Compared with asphalt binder, asphalt mastic could result in a three- to five-fold increase while asphalt mortar could have a modulus increase of one to two orders of magnitude. The phase change was identified by the significant reduction on phase angle at low frequencies or high temperatures. The combined effect of increasing modulus and reducing phase angle can contribute to high rutting resistance of asphalt mixtures.

ii. Three interaction parameters were used to analyze the interaction among asphalt binder, filler and fine aggregate. The sensitivity of these parameters was different and did not give a consistent result. The parameters based on phase angle showed large scatter; however, the parameter based on complex modulus was relatively constant.

iii. The interaction between the asphalt binder and the filler was stronger than its interaction with fine aggregates. Therefore, the specific area could be important for enhancing asphalt binder-filler interaction.

iv. Asphalt mortar tended to have a constant C value irrespective of type of asphalt binder, fine aggregate and material composition. This allowed for a better prediction of mortar modulus based on asphalt binder modulus. In general, the interaction between asphalt binder, filler and fine aggregate was physical and could be explained by particle filling effect. The temperature sensitivity of mastic and mortar was thus controlled by the type of asphalt binder. The effect of fine aggregate, for example, aggregate contact and friction on rheological properties was prominent at high temperatures. Considering the significant effect of modulus reinforcement and phase change, asphalt mortar could be a vital scale to bridge the gap between asphalt binder and asphalt mixture in a multiscale performance research.

Author Contributions: Conceptualization, M.C., C.P. and L.M.; Methodology, M.C., W.H. and M.R.; Software, B.J. and M.G.; Validation, M.C., C.P. and M.R.; Formal Analysis, W.H. and M.G.; Investigation, L.M.; Resources,

B.J. and M.G.; Data Curation, M.C. and M.R.; Writing—Original Draft Preparation, M.C. and B.J.; Writing—Review & Editing, L.M. and M.R.; Visualization, L.M. and M.R.; Supervision, L.M.

Funding: This research was funded by National Natural Science Foundation of China (No. U1733121) and the China Scholarship Council.

Acknowledgments: The authors are grateful for the cooperation between the People's Republic of China and the Governments of Uganda, Kenya and express their desire to see a prolonged and stronger cooperation between the two states.

Conflicts of Interest: The authors declare no conflict of interest.

References

1. Li, Q.; Yang, H.; Ni, F.; Ma, X.; Luo, L. Cause analysis on permanent deformation for asphalt pavements using field cores. *Constr. Build. Mater.* **2015**, *100*, 40–51. [CrossRef]
2. Wu, S.; Zhao, Z.; Xiao, Y.; Yi, M.; Li, M. Evaluation of mechanical properties and aging index of 10-year field aged asphalt materials. *Constr. Build. Mater.* **2017**, *155*, 1158–1167. [CrossRef]
3. Xu, T.; Huang, X. Investigation into causes of in-place rutting in asphalt pavement. *Constr. Build. Mater.* **2012**, *28*, 525–530. [CrossRef]
4. Liu, J.; Zhao, S.; Li, L.; Lic, P.; Saboundjian, S. Low temperature cracking analysis of asphalt binders and mixtures. *Cold Reg. Sci. Technol.* **2017**, *141*, 78–85. [CrossRef]
5. Underwood, B.S.; Kim, Y.R. Microstructural investigation of asphalt concrete for performing multiscale experimental studies. *Int. J. Pavement Eng.* **2013**, *14*, 498–516. [CrossRef]
6. Li, Q.; Chen, X.; Li, G.; Zhang, S. Fatigue resistance investigation of warm-mix recycled asphalt binder, mastic, and fine aggregate matrix. *Fatigue Fract. Eng. Mater.* **2018**, *41*, 400–411. [CrossRef]
7. Füssl, J.; Lackner, R. Multiscale fatigue model for bituminous mixtures. *Int. J. Fatigue* **2011**, *33*, 1435–1450. [CrossRef]
8. Reza Pouranian, M.; Haddock John, E. Determination of voids in the mineral aggregate and aggregate skeleton characteristics of asphalt mixtures using a linear-mixture packing model. *Constr. Build. Mater.* **2018**, *188*, 292–304. [CrossRef]
9. Sreedhar, E.; Coleri, E.; Sadat, S. Selection of a performance test to assess the cracking resistance of asphalt concrete materials. *Constr. Build. Mater.* **2018**, *179*, 285–293. [CrossRef]
10. Gudipudi, P.P.; Underwood, B.S. Use of Fine Aggregate Matrix Experimental Data in Improving Reliability of Fatigue Life Prediction of Asphalt Concrete Sensitivity of This Approach to Variation in Input Parameters. *Transp. Res. Rec.* **2017**, *2631*, 65–73. [CrossRef]
11. Hospodka, M.; Hofko, B.; Blab, R. Introducing a new specimen shape to assess the fatigue performance of asphalt mastic by dynamic shear rheometer testing. *Mater. Struct.* **2018**, *51*, 46. [CrossRef]
12. Mo, L.; Huurman, M.; Wu, S.; Molenaar, A. Ravelling investigation of porous asphalt concrete based on fatigue characteristics of bitumen–stone adhesion and mortar. *Mater. Des.* **2009**, *30*, 170–179. [CrossRef]
13. Lesueur, D.; Blazquez, M.L.; Garcia, D.A.; Rubio, A.R. On the impact of the filler on the complex modulus of asphalt mixtures. *Road Mater. Pavement* **2018**, *19*, 1057–1071. [CrossRef]
14. Liu, G.; Zhao, Y.; Zhou, J.; Li, J.; Yang, T.; Zhang, J. Applicability of evaluation indices for asphalt and filler interaction ability. *Constr. Build. Mater.* **2017**, *148*, 599–609. [CrossRef]
15. Guo, M.; Tan, Y.; Hou, Y.; Wang, L.; Wang, Y. Improvement of evaluation indicator of interfacial interaction between asphalt binder and mineral fillers. *Constr. Build. Mater.* **2017**, *151*, 236–245. [CrossRef]
16. Ziegel, K.D.; Romanov, A. Modulus reinforcement in elastomer composites. I. Inorganic fillers. *J. Appl. Polym.* **1973**, *17*, 1119–1131.
17. Kakade, V.B.; Reddy, M.A.; Reddy, K.S. Rutting performance of hydrated lime modified bituminous mixes. *Constr. Build. Mater.* **2018**, *186*, 1–10. [CrossRef]
18. Huang, B.; Shu, X.; Dong, Q.; Shen, J. Laboratory evaluation of moisture susceptibility of hot-mix asphalt containing cementitious fillers. *J. Mater. Civ. Eng.* **2010**, *22*, 667–673. [CrossRef]
19. Sakanlou, F.; Shirmoha mmadi, H.; Hamedi, G. Investigating the effect of filler types on thermodynamic parameters and their relationship with moisture sensitivity of asphalt mixes. *Mater. Struct.* **2018**, *51*, 39. [CrossRef]

20. Jamshidi, A.; Hasan, M.; Lee, M. Comparative study on engineering properties and energy efficiency of asphalt mixes incorporating fly ash and cement. *Constr. Build. Mater.* **2018**, *168*, 295–304. [CrossRef]

21. Das, A.; Singh, D. Effects of Basalt and Hydrated Lime Fillers on Rheological and Fracture Cracking Behavior of Polymer Modified Asphalt Mastic. *J. Mater. Civ. Eng.* **2018**, *30*, 04018011. [CrossRef]

22. Diab, A.; You, Z. Linear and Nonlinear Rheological Properties of Bituminous Mastics under Large Amplitude Oscillatory Shear Testing. *J. Mater. Civ. Eng.* **2018**, *30*, 04017303. [CrossRef]

23. Zhang, Y.; Leng, Z.; Zou, F.; Wang, L.; Chen, S.; Tsang, D. Synthesis of zeolite a using sewage sludge ash for application in warm mix asphalt. *J. Clean. Prod.* **2018**, *172*, 686–695. [CrossRef]

24. Woszuk, A. Application of fly ash derived zeolites in warm-mix asphalt technology. *Materials* **2018**, *11*, 1542. [CrossRef] [PubMed]

25. Woszuk, A.; Panek, R.; Madej, J.; Zofka, A.; Franus, W. Mesoporous silica material MCM-41: Novel additive for warm mix. *Constr. Build. Mater.* **2018**, *183*, 270–274. [CrossRef]

26. Topal, A.; Sengoz, B.; Kok, B.V.; Yilmaz, M.; Dokandari, P.A.; Oner, J.; Kaya, D. Evaluation of mixture characteristics of warm mix asphalt involving natural and synthetic zeolite additives. *Constr. Build. Mater.* **2014**, *57*, 38–44. [CrossRef]

27. Liu, X.; Liu, W.; Wang, S.; Wang, Z.; Shao, L. Performance Evaluation of Asphalt Mixture with Nanosized Volcanic Ash Filler. *J. Transp. Eng. Part B-Pavements* **2018**, *144*, 04018028. [CrossRef]

28. Wang, W.; Cheng, Y.; Tan, G.; Liu, Z.; Shi, C. Laboratory investigation on high- and low-temperature performances of asphalt mastics modified by waste oil shale ash. *J. Mater. Cycles Waste* **2018**, *20*, 1710–1723. [CrossRef]

29. Sargin, S.; Saltan, M.; Morova, N.; Serin, S.; Terzi, S. Evaluation of rice husk ash as filler in hot mix asphalt concrete. *Constr. Build. Mater.* **2013**, *48*, 390–397. [CrossRef]

30. Zhang, J.; Liu, S.; Yao, Z.; Wu, S.; Jiang, H.; Liang, M.; Qiao, Y. Environmental aspects and pavement properties of red mud waste as the replacement of mineral filler in asphalt mixture. *Constr. Build. Mater.* **2018**, *180*, 605–613. [CrossRef]

31. Choudhary, J.; Kumar, B.; Gupta, A. Application of waste materials as fillers in bituminous mixes. *Waste Manag.* **2018**, *78*, 417–425. [CrossRef]

32. Roman, C.; Garcia-Morales, M. Comparative assessment of the effect of micro- and nano-fillers on the microstructure and linear viscoelasticity of polyethylene-bitumen mastics. *Constr. Build. Mater.* **2018**, *169*, 83–92. [CrossRef]

33. Bocci, E. Use of ladle furnace slag as filler in hot asphalt mixtures. *Constr. Build. Mater.* **2018**, *161*, 156–164. [CrossRef]

34. Giustozzi, F.; Mansour, K.; Patti, F.; Pannirselvam, M.; Fiori, F. Shear rheology and microstructure of mining material-bitumen composites as filler replacement in asphalt mastics. *Constr. Build. Mater.* **2018**, *171*, 726–735. [CrossRef]

35. Al-Khateeb, G.G.; Khedaywi, T.S.; Irfaeya, M.F. Mechanical Behavior of Asphalt Mastics Produced Using Waste Stone Sawdust. *Adv. Mater. Sci. Eng.* **2018**, *5362397*. [CrossRef]

36. Topini, D.; Toraldo, E.; Andena, L.; Mariani, E. Use of recycled fillers in bituminous mixtures for road pavements. *Constr. Build. Mater.* **2018**, *159*, 189–197. [CrossRef]

37. Craus, J.; Ishai, I.; Sides, A. Some physico-chemical aspects of the effect and the role of the filler in bituminous paving mixtures. *J. Assoc. Asph. Paving Technol.* **1978**, *47*, 558–588.

38. Zhou, S.; Liu, S.; Xiang, Y. Effects of Filler Characteristics on the Performance of Asphalt Mastic: A Statistical Analysis of the Laboratory Testing Results. *Int. J. Civ. Eng.* **2018**, *16*, 1175–1183. [CrossRef]

39. Clopotel, C.; Velasquez, R.; Bahia, H. Measuring physico-chemical interaction in mastics using glass transition. *Road Mater. Pavement* **2012**, *13*, 304–320. [CrossRef]

40. Davis, C.; Castorena, C. Implications of physico-chemical interactions in asphalt mastics on asphalt microstructure. *Constr. Build. Mater.* **2015**, *94*, 83–89. [CrossRef]

41. Hesami, E.; Birgisson, B.; Kringos, N. Numerical and experimental evaluation of the influence of the filler-bitumen interface in mastics. *Mater. Struct.* **2014**, *47*, 1325–1337. [CrossRef]

42. Tan, Y.; Li, X.; Wu, J. Internal influence factors of asphalt-aggregate filler interactions based on rheological characteristics. *J. Mater. Civ. Eng.* **2012**, *24*, 1520–1528.

43. Tan, Y.; Guo, M. Interfacial thickness and interaction between asphalt and mineral fillers. *Mater. Struct.* **2014**, *47*, 605–614. [CrossRef]

44. Luo, D.; Guo, M.; Tan, Y. Molecular Simulation of Minerals-Asphalt Interfacial Interaction. *Minerals* **2018**, *8*, 176. [CrossRef]

45. Cheng, Y.; Tao, J.; Jiao, Y.; Tan, G.; Guo, Q.; Wang, S.; Ni, P. Influence of the properties of filler on high and medium temperature performances of asphalt mastic. *Constr. Build. Mater.* **2016**, *118*, 268–275. [CrossRef]

46. Zhang, J.; Fan, Z.; Hu, D.; Hu, Z.; Pei, J.; Kong, W. Evaluation of asphalt–aggregate interaction based on the rheological properties. *Int. J. Pavement Eng.* **2016**, 1–7. [CrossRef]

47. Liu, G.; Yang, T.; Li, J.; Jia, Y.; Zhao, Y.; Zhang, J. Effects of aging on rheological properties of asphalt materials and asphalt-filler interaction ability. *Constr. Build. Mater.* **2018**, *168*, 501–511. [CrossRef]

48. Kofteci, S.; Nazary, M. Experimental study on usability of various construction wastes as fine aggregate in asphalt mixture. *Constr. Build. Mater.* **2018**, *185*, 369–379. [CrossRef]

49. Rondon-Quintana, H.A.; Ruge-Cardenas, J.C.; Patino-Sanchez, D.F.; Vacca-Gamez, H.A.; Reyes-Lizcano, F.A.; de Farias, M.M. Blast Furnace Slag as a Substitute for the Fine Fraction of Aggregates in an Asphalt Mixture. *J. Mater. Civ. Eng.* **2018**, *30*, 04018244. [CrossRef]

50. Xie, X.; Lu, G.; Liu, P.; Wang, D.; Fan, Q.; Oeser, M. Evaluation of morphological characteristics of fine aggregate in asphalt pavement. *Constr. Build. Mater.* **2017**, *139*, 1–8. [CrossRef]

51. Ma, H.; Zhou, C.; Feng, D.; Sun, L. Influence of Fine Aggregate Content on Low-Temperature Cracking of Asphalt Pavements. *J. Mater. Civ. Eng.* **2017**, *45*, 835–842. [CrossRef]

52. Cong, L.; Wang, T. Effect of fine aggregate angularity on skid-resistance of asphalt pavement using accelerated pavement testing. *Constr. Build. Mater.* **2018**, *168*, 41–46.

53. Xiao, Y.; Wang, F.; Cui, P.; Lei, L.; Lin, J.; Yi, M. Evaluation of Fine Aggregate Morphology by Image Method and Its Effect on Skid Resistance of Micro-Surfacing. *Materials* **2018**, *11*, 920. [CrossRef] [PubMed]

54. Javilla, B.; Mo, L.; Fang, H.; Shu, B.; Wu, S. Multi-stress loading effect on rutting performance of asphalt mixtures based on wheel tracking testing. *Constr. Build. Mater.* **2017**, *148*, 1–9. [CrossRef]

55. *JTG F40-2004 Technical Specification for Construction of Highway Asphalt Pavements*; People's Republic of China, Ministry of Transport: Beijing, China, 2004.

56. Singh, D.; Ashish, P.K.; Jagadeesh, A. Influence of Particle and Interaction Effects of Different Sizes of Crumb Rubber on Rheological Performance Parameters of Binders. *J. Mater. Civ. Eng.* **2018**, *30*, 04018066. [CrossRef]

Article

Unified Strength Model of Asphalt Mixture under Various Loading Modes

Chengdong Xia [1], Songtao Lv [1,*], Lingyun You [2], Dong Chen [1,3], Yipeng Li [1] and Jianlong Zheng [1]

[1] National Engineering Laboratory of Highway Maintenance Technology, Changsha University of Science & Technology, Changsha 410004, China; xiachengdong@stu.csust.edu.cn (C.X.); 15878778765@163.com (D.C.); lypcsust@163.com (Y.L.); zjl@csust.edu.cn (J.Z.)

[2] Department of Civil and Environmental Engineering, Michigan Technological University, 1400 Townsend Drive, Houghton, MI 49931-1295, USA; liyou@mtu.edu

[3] Guangxi Communications Design Group Co., Ltd., Nanning 530000, China

*   Correspondence: lst@csust.edu.cn

Received: 21 February 2019; Accepted: 14 March 2019; Published: 17 March 2019

Abstract: Although the rutting resistance, fatigue cracking, and the resistance to water and frost are important for the asphalt pavement, the strength of asphalt mixture is also an important factor for the asphalt mixture design. The strength of asphalt mixture is directly associated with the overall performance of asphalt mixture. As a top layer material of asphalt pavement, the strength of asphalt mixture plays an indispensable role in the top structural bearing layer. In the present design system, the strength of asphalt pavement is usually achieved via the laboratory tests. The stress states are usually different for the different laboratory approaches. Even at the same stress level, the laboratory strengths of asphalt mixture obtained are significantly different, which leads to misunderstanding of the asphalt mixtures used in asphalt pavement structure design. The arbitrariness of strength determinations affects the effectiveness of the asphalt pavement structure design in civil engineering. Therefore, in order to overcome the design deviation caused by the randomness of the laboratory strength of asphalt mixtures, in this study, the direct tension, indirect tension, and unconfined compression tests were implemented on the specimens under different loading rates. The strength model of asphalt mixture under different loading modes was established. The relationship between the strength ratio and loading rate of direct tension, indirect tension, and unconfined compression tests was adopted separately. Then, one unified strength model of asphalt mixture with different loading modes was established. The preliminary results show that the proposed unified strength model could be applied to improve the accurate degree of laboratory strength. The effectiveness of laboratory-based asphalt pavement structure design can therefore be promoted.

Keywords: structure design; asphalt mixture; laboratory strength; unified strength model; loading modes

---

1. Introduction

The flexible and rigid pavements are the two most important roads or highways. Over 95% of the roads in the world are flexible asphalt pavements [1,2] because of its good driving comfort [1,3,4], durability [5–8], and resistance to water damage [9–11]. The main material component of asphalt pavement structure is asphalt mixture [12,13]. However, under the dual influence of vehicle load and environmental factors [14–16], asphalt pavement will produce different types of diseases. There are three main types of diseases: rutting, low temperature cracking, and fatigue cracking. Rutting is the result of excessive shear deformation due to insufficient shear strength of asphalt mixture, which is related to the high temperature performance of the asphalt mixture [17–20]. Low temperature cracking is related to low temperature fracture strength of the asphalt mixture [21,22]. Fatigue cracking is mainly related to fatigue strength of the asphalt mixture [8,23]. Especially in recent years, the dramatic increase of heavy-duty vehicles has put forward higher requirements for the structural strength design of asphalt pavement. How to reduce rutting, low temperature cracking, and fatigue cracking of asphalt pavement is an urgent problem to be solved. The structural design of the asphalt pavement belongs to the mechanical–empirical method in China, where the elastic layered half-space is employed to calculate the mechanical response of pavement. The theories of maximum tensile stress and strain are implemented as the failure criterion of asphalt pavement [24,25]. The laboratory strength test methods mainly include direct tension [26–29], unconfined compression [30–32], bending [33–35], indirect tension [36–39], shear [40,41], and triaxial tests [42,43], which are performed to evaluate the tensile, compressive, bending, and shear properties of materials. For the unconfined compression test, the applied strain rate in height direction of specimens is 1.3 mm/(min · 25 mm) stipulated by AASHTO (American Association of State Highway and Transportation Officials) T 167 and ASTM (American Society for Testing Materials) D 1074. The standard size of Marshall Specimens for indirect tension test is Φ101.6 mm $\times$ 63.5 mm, which is adopted in the specifications or standards of United States, Japan, and Australia, while the field-drilling specimen with Φ150 mm is employed in the British standards. Moreover, in these standards, AASHTO T 283, BS EN (British European) 12697-23: 2012, and the Specifications and Test Methods of Asphalt and Asphalts Mixtures for Highway Engineering (JTG E20-2011) used the loading rate of 50 mm/min for the indirect tension test. The unconfined compression test and direct tension test are one-dimensional stress states. The bending test is divided into upper compression and lower tension with the neutral surface as the boundary. The stress at a certain point is the one-dimensional stress state, but the overall stress state is more in line with the stress characteristics of the pavement structure. The center point of the indirect tension test is under vertical compression and horizontal tension, which is in a two-dimensional stress state and conforms to the stress state of the pavement structure. The triaxial test is mainly aimed at the stress characteristics of asphalt pavement under a complex stress state. However, the strengths obtained by different test methods are usually quite discrepant, and it is difficult to compare between them. Therefore, the arbitrariness of the strengths of asphalt mixture under different loading modes should be considered during the asphalt pavement structure design and the related risk management.

The strength of asphalt mixture and other related mechanical parameters have always been a common topic for the civil engineers and researchers. Su et al. [44] used the superpave indirect tensile (IDT) strength test to evaluate the concrete strength of reclaimed asphalt pavement (RAP). It was found that the IDT strength of concrete decreased with the increase of percentage of RAP and temperature. Saride et al. [45] studied the RAP/VA (Reclaimed Asphalt Pavements/Virgin Aggregates) mixture stabilized by alkali activated fly ash. It was found that the strength of the mixture meets the strength requirements of the specification. Ji, X et al. [46] stated that UPT-NSM (Uniaxial Penetration Test—Numerical Simulation Method) can be utilized to optimize the gradation better than the step-filling test to improve the shear strength and rutting resistance of an asphalt mixture. It was noted that the anti-shear strength and dynamic stability of graded asphalt mixture optimized by UPT-NSM are 25.5% and 27.0% higher than the specified gradation, respectively. Li et al. [47] studied the influence of production conditions on the indirect tensile strength characteristics of foamed asphalt mixture. The strength characteristics of the same graded foamed asphalt mixture were mainly affected by curing time, cement dosage, and asphalt content, and it almost had nothing to do with the foam characteristics. Gaus et al. [48] explored the use of button granular asphalt (BGA) instead of petroleum asphalt to produce asphalt concrete bearing course (AC-BC) mixes. Compared with an AC-BC mixture without BGA, the application of BGA partly replacing petroleum asphalt in AC-BC mixture improves its compressive strength and elastic modulus. There was no significant difference in the Poisson ratio of all mixtures.

At present, numerous factors affecting the strength of asphalt mixtures have been reported by laboratory and field researchers and there were many useful conclusions that were established. However, the strengths that were obtained using different test methods are still tough to compare, which leads to the randomness of the strength indexes in asphalt mixture design. The negative impacts on the design and analysis of asphalt mixtures form these randomness indexes are self-evident. Fortunately, there are many studies have been reported by the researchers for unified parameter model of materials. For example, Yu, M.H et al. [49,50] proposed a unified strength criterion for rock considering the effect of intermediate principal stress. Its strength parameters can be determined using conventional triaxial compression tests. It is found that the unified strength theory can be used to describe various types of rock. You, M et al. [51] put forward that the unified strength theory of linearity and nonlinearity is constructed directly in the form of principal stress. Based on the test results of true triaxial compression, conventional triaxial compression and elongation of rocks, the material parameters and fitting deviations in several strength criteria and their applicability are determined. Danni et al. [52] found that the strength properties of high-strength concrete (HSC) under multi-axial stress may be conducted via shear-type four-parameter unified strength theory (STFP-UST) and nonlinear unified strength theory (N-UST) through analysis of the failure surfaces of several twin shear strength criteria. Wu et al. [53] introduced a shape factor that is expressed as a function of the corner radius ratio, $\rho = 2r/b\rho = 2r/b$. Doing so, a unified model for the concrete strength of FRP-confined columns that have an arbitrary corner radius is described. This model can be degenerated into two special cases for circular columns and sharp cornered square columns when $\rho = 1\rho = 1$ and 0, respectively. Through collecting all of the available experimental results on both circular and square columns from the open literature for model evaluation, a comprehensive and updated database has been established. A better correlation of the proposed model has been demonstrated by comparing between the test results and the model predictions. Wu, Y.F et al. [54] also proposed a new model based on the Hoek–Brown failure criterion. The existing strength models for FRP (Fiber-Reinforced Polymer)-confined circular and square concrete columns are reviewed, evaluated, and compared with the proposed model. Then, using an updated database, a large number of test data is to evaluate the models. A comparison between the models and the test results is used to demonstrate the accuracy of the proposed model. In addition, the model has a unified form for both circular and square columns. It can be used to predict the strength of columns that have existing damage or cracks. Wei et al. [55,56] presented a new stress–strain model for FRP-confined concrete columns. One of advantages of the model is its unified form (mathematical expression). Compared with the test results, the model can be predicted the ultimate stress and strain more accurately, particularly the strain. You et al. [57,58] established a three-dimensional (3D) microstructure-based computational model through applying a coupled thermo-viscoelastic, thermo-viscoplastic, and thermo-viscodamage constitutive model. The result reflected that the generated 3D microstructure model and the presented constitutive model could be implemented effectively to predict the overall thermo-mechanical response of asphalt concrete. Hajj et al. [59] proposed a unified permanent deformation model, which uses response measurements of two tests. The new model quantifies the accumulated permanent shear strain as a function of the number of load cycles and factor of safety (FOS). The safety factor is defined in q-p space and evaluated according to the applied stress and triaxial compressive strength characteristics (cohesive force and internal friction angle). For specific mixtures used in this study, there is a good correlation between cumulative permanent shear strain and FOS level regardless of stress conditions and test types.

The above researches laid a foundation for establishing the unified strength model of asphalt mixture. However, the mentioned-above research on the unified strength model of materials mainly applied on cement concrete and rock materials, while the related unified parameter model employed on asphalt mixture is usually based on the computation model revealed the thermos-mechanical response and permanent deformation of asphalt mixture [60,61]. Owing to the complex composition and structure of asphalt mixtures and the various destruction forms, there is little research on the unified strength model of asphalt mixtures.

Therefore, direct tension, unconfined compression, and indirect tension strength tests under different loading rates were carried out in this paper. A unified strength model of asphalt mixture under different loading modes was established by using the relationship between the strength ratio of direct tension, unconfined compression, and indirect tension tests and the loading rate ratio.

The main objectives of this study are to reveal the strength rate characteristics of asphalt mixtures under various loading modes and establish a unified strength model to solve the uncertainty of the strength parameters of asphalt mixtures under various loading modes. The direct tension, indirect tension, and unconfined compression tests were applied in the study.

2. Materials and Sample Preparations

2.1. Materials

In this paper, strength tests of direct tension, indirect tensile, and unconfined compression were implemented separately to establish a unified strength model of asphalt mixture under different loading modes. The dense gradation asphalt mixture AC-13C that was composed of SBS (styrene-butadiene-styrene) modified asphalt made of Xiamen Huate group Co., Ltd, Xiamen, China and limestone aggregates produced in Shizichang, Niujiaowu, Foshan City, China was chosen. The performance indexes of SBS modified asphalt are shown in Table 1, the densities of limestone aggregate are shown in Table 2, and the properties of the aggregate are shown in Table 3.

Table 1. Test results of SBS (I-D) modified asphalt.

Test Projects	Test Standard: JTG F40-2004 (China) [62]		
	Technical Requirements	Test Results	Test Methods
Penetration (25 °C,100 g, 5 s) (0.1 mm)	40~60	55.9	T 0604-2000
Penetration index PI	≥ 0	0.533 ($R^2 = 0.997$)	T 0604-2000
Ductility (5 cm/min, 5 °C) (cm)	≥ 20	35.1	T 0605-1993
Softening point (Ring ball) (°C)	≥ 60	70.5	T 0606-2000
135°C dynamic viscosity (Pa s)	≤ 3	2.36	T 0615-2000
Flash point (°C)	≥ 230	264	T 0611-1993
Solubility (%)	99	99.8	T 0607-1993
Density (15 °C)	—	1.03	T 0603-1993
Rolling Thin Film Oven Test (RTFOT) (163 °C, 85 min) — Mass loss (%)	$\leq \pm 1.0$	0.22	T 0609-1993
Residual penetration ratio (25 °C) (%)	≥ 65	75.1	T 0604-2000
Residual ductility(5 °C) (cm)	≥ 15	23.2	T 0605-1993

Table 2. Densities of limestone aggregate.

Sizes of Sieve (mm)	Apparent Density (g/cm^3)	Bulk Density (g/cm^3)	Skin Drying Density (g/cm^3)	Water Absorption (%)
16–13.2	2.671	2.577	2.611	1.32
13.2–9.5	2.673	2.569	2.608	1.53
9.5–4.75	2.661	2.572	2.607	1.35
4.75–2.36	2.649			
2.36–1.18	2.642			
1.18–0.6	2.606			
0.6–0.3	2.592	-	-	-
0.3–0.15	2.586			
0.15–0.075	2.615			

Table 3. Properties of aggregate.

Test Item	Technical Requirements	Test Results	Test Methods
Crushed stone value (%)	$\leq$26	17.9	T 0316-2005
Apparent relative density (g/cm^3)	$\geq$2.6	2.71	T 0321-2005
Content of flat and elongated particles in coarse aggregate (%)	$\leq$15	9	T 0312-2005
Content of SiO$_2$ (%)	/	1.81	/

Through the above test results, it was shown that SBS modified asphalt and the aggregate satisfied the requirements of JTG F40-2004 [62], which were the technical specifications for asphalt pavement construction in China. The aggregate gradation curve of dense graded asphalt mixture (AC-13C) and the target gradation of the asphalt mixture is shown Figure 1. The optimum asphalt content was determined using the Marshall tests, and the test results are displayed in Table 4.

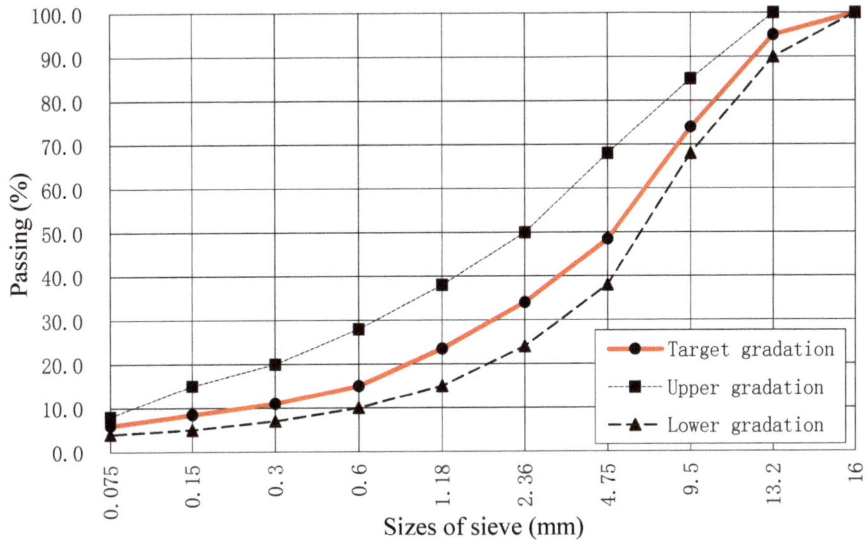

Figure 1. Aggregate gradation curve of dense graded asphalt mixture (AC-13C).

Table 4. Results of Marshall test at the optimal asphalt-aggregate ratio.

Asphalt Aggregate Ratio (%)	Bulk Specific Gravity (g cm^{-3})	Volume of Air Voids VV (%)	Voids Filled with Asphalt VFA (%)	Marshall Stability (kN)	Flow Value (0.1 mm)
5.2	2.44	4.51	67.20	12.71	27.89
/	/	3–5	65–75	>8	20–40

2.2. Sample Preparations

According to the Specifications and Test Methods of Asphalt and Asphalts Mixtures for Highway Engineering (JTG E20-2011) [63], the block samples of asphalt mixture plates were made through the method of vibration wheel grinding. Along the rolling direction, each beam was cut to a length, width, and height of 250 mm, 50 mm, and 50 mm, respectively, for direct tension specimens. The cylindrical specimens for the unconfined compressive fatigue test were made using an SGC (Superpave Gyratory Compactor) with a size of Φ100 mm $\times$ 100 mm, and the indirect tensile specimens were prepared by slicing the top and bottom surface of the specimens of unconfined compressive moduli test to the size. The cylindrical specimens with height of 100 $\pm$ 2mm and diameter of 100 $\pm$ 2 mm of asphalt mixture made using a SGC gyratory compactor were prepared for unconfined compression. In addition, the indirect tensile specimens were prepared by slicing the top and bottom surface of the specimens of unconfined compressive moduli test to the height of 60 ± 2 mm and diameter of 100 ± 2 mm. Then, the asphalt mixture specimens were put in the environment chamber at 15 °C for 4–5 h. Subsequently, it was placed on the strength test support of MTS (Material Testing System)-Landmark. The preliminary contact between the indenter of the strength test and the specimen was adjusted to start the test, and the test process was completed in environment chamber. The set-up details of the strength tests are shown in Figure 2.

Figure 2. Testing process of strength under different loading modes: (**a**) direct tension test, (**b**) indirect tension test, (**c**) unconfined compression test, and (**d**) the environmental chamber.

3. Test Results and Analysis

The displacement control mode for China, the United States, and Europe was adopted in the loading rate control mode of strength test. Among them, the loading rate of the unconfined compressive strength test with the size of Φ 100 mm × 100 mm was 2 mm/min and 5.08 mm/min for AASHTO T167 and JTG E20-2011, respectively. The loading rate for indirect tensile strength was 50 mm/min for AASHTO T 283, BS EN 12697-23: 2012, and JTG E20-2011. The loading rate of flexural strength was also 50mm/min for JTG E20-2011. However, the test methods of direct tensile strength were not clearly given in this regulation. In order to explore the loading rate characteristics of the asphalt mixture strength under different loading modes, the displacement loading rate of test regulations could not be unified, so the stress loading rate control mode was adopted, and the test temperature was unified at 15 °C.

3.1. Direct Tensile Strength Test at Different Loading Rates

Under the stress control mode, the direct tensile strength tests of asphalt mixtures at different loading rates were carried out, and the results are shown in Table 5.

Table 5. Test results of direct tensile strength of asphalt mixture.

Number	Loading Rate v (MPa/s)	Section Area of Specimen A (mm^2)	Failure Loading F (kN)	Strength R_D (MPa)	Average Value of Strength R_D (MPa)	Coefficient of Variation
1		2631.1	7.317	2.781		
2	5	2596.8	8.159	3.142	2.95	0.050
3		2581.9	7.557	2.927		

Table 5. *Cont.*

Number	Loading Rate v (MPa/s)	Section Area of Specimen A (mm²)	Failure Loading F (kN)	Strength R_D (MPa)	Average Value of Strength R_D (MPa)	Coefficient of Variation
4	10	2560.2	9.398	3.671	3.487	0.055
5		2588.9	9.235	3.567		
6		2611.3	8.416	3.223		
7	20	2621.5	10.772	4.109	4.158	0.054
8		2594.5	11.561	4.456		
9		2617.7	10.233	3.909		
10	30	2600.1	11.586	4.456	4.552	0.025
11		2597.3	12.236	4.711		
12		2559.6	11.490	4.489		
13	40	2579	12.887	4.997	4.821	0.039
14		2599.8	11.858	4.561		
15		2630.1	12.901	4.905		
16	50	2671	13.040	4.882	5.012	0.019
17		2599.1	13.276	5.108		
18		2611.8	13.179	5.046		
19	60	2567.7	13.080	5.094	5.13	0.013
20		2598.1	13.567	5.222		
21		2666.3	13.529	5.074		
22	70	2621.5	13.608	5.191	5.197	0.007
23		2617.3	13.723	5.243		
24		2613.5	13.478	5.157		

The strength values in Table 5 were fitted with the loading rate. The fitted curve is shown in Figure 3.

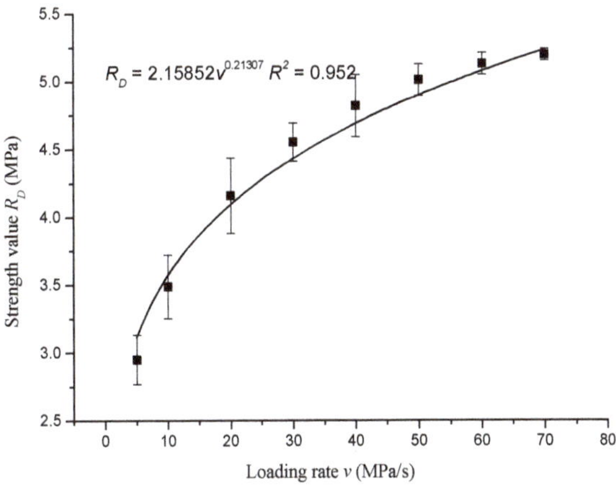

Figure 3. The curve of direct tensile strength of asphalt mixture with loading rate.

The fitting equation was as follows:

$$R_D = 2.15852v^{0.21307}, R^2 = 0.952 \tag{1}$$

According to the fitting results, the direct tensile strength R_D of the asphalt mixture varied with the loading rate v as a power function. The strength increased with the increase of loading rate, and the rate of strength increase slowed down with the increase of loading rate.

3.2. Indirect Tensile Strength Test at Different Loading Rates

According to Chinese Standard Test Methods of Bituminous and Bituminous Mixtures for Highway Engineering (JTG E20-2011) [63], the indirect tensile strength tests of asphalt mixtures under different loading rates were performed. The results of the tests are shown in Table 6.

Table 6. Indirect tensile strength test of asphalt mixture.

Number	Loading Rate v (MPa/s)	Section Area of Specimen A (mm²)	Failure Loading F (kN)	Strength R_D (MPa)	Average Value of Strength R_D (MPa)	Coefficient of Variation
1		58.5	28.948	3.111		
2	5	60.4	33.164	3.452	3.258	0.044
3		60.5	30.900	3.211		
4		60.1	34.997	3.661		
5	10	58.9	35.778	3.819	3.704	0.022
6		59.1	34.142	3.632		
7		59.3	40.709	4.316		
8	20	59.6	40.299	4.251	4.41	0.041
9		58.8	43.611	4.663		
10		59.2	46.808	4.971		
11	30	59.7	44.307	4.666	4.837	0.026
12		59.6	46.205	4.874		
13		59	47.542	5.066		
14	40	59.8	50.621	5.322	5.185	0.020
15		60.1	49.393	5.167		
16		59.7	53.129	5.595		
17	50	61.2	55.106	5.661	5.487	0.037
18		59.6	49.343	5.205		
19		60.5	55.592	5.777		
20	60	61.4	56.810	5.817	5.658	0.035
21		60.1	51.430	5.38		
22		61.7	57.814	5.891		
23	70	60.3	53.241	5.551	5.784	0.029
24		60.8	57.154	5.91		

The strength values in Table 6 were fitted with the loading rate. The fitted curve is shown in Figure 4.

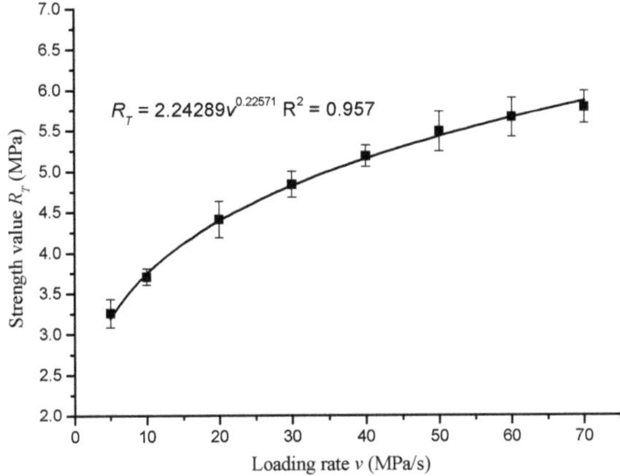

Figure 4. Indirect tensile strength curve of asphalt mixture with loading rate.

The fitting equation was as follows:

$$R_T = 2.24289v^{0.22571}, R^2 = 0.957 \tag{2}$$

According to the fitting results, the indirect tensile strength R_T of asphalt mixture varied with the loading rate v as a power function. The strength increased with the increase of the loading rate, and the rate of strength increase slowed down with the increase of the loading rate.

3.3. Unconfined Compressive Strength Test at Different Loading Rates

Considering the test threshold of employed material testing system (MTS) was 100 kN, through the tentative experiments it was found that the unconfined compression failure load exceeded 100 kN when the loading rate was greater than 3 MPa/s. For the sake of safety and operability of the test, the strength values at the loading rates that exceeded the threshold of MTS were obtained using the delay prediction method in this study. The proposed delay prediction method was performed based on enough laboratory test data within the threshold of MTS, such that the strength values at the loading rate that exceeded the threshold of MTS could be output from the fitting curve. The test results are shown in Table 7.

Table 7. Test results of unconfined compressive strength of asphalt mixture.

Number	Loading Rate v (MPa/s)	Section Area of Specimen A (mm²)	Failure Loading F (kN)	Strength R_D (MPa)	Average Value of Strength R_D (MPa)	Coefficient of Variation
1		34.862	34.862	4.441		
2	0.02	32.169	32.169	4.098	4.134	0.057
3		30.325	30.325	3.863		
4		38.473	38.473	4.901		
5	0.05	38.795	38.795	4.942	5.062	0.039
6		41.943	41.943	5.343		
7		47.249	47.249	6.019		
8	0.1	46.998	46.998	5.987	5.901	0.025
9		44.721	44.721	5.697		

Table 7. *Cont.*

Number	Loading Rate v (MPa/s)	Section Area of Specimen A (mm^2)	Failure Loading F (kN)	Strength R_D (MPa)	Average Value of Strength R_D (MPa)	Coefficient of Variation
10		63.773	63.773	8.124		
11	0.5	66.851	66.851	8.516	8.421	0.025
12		67.691	67.691	8.623		
13		78.429	78.429	9.991		
14	1	76.255	76.255	9.714	9.816	0.013
15		76.483	76.483	9.743		
16		87.064	87.064	11.091		
17	2	90.636	90.636	11.546	11.441	0.022
18		91.735	91.735	11.686		

The strength values in Table 7 are fitted with the loading rate. The fitted curve is shown in Figure 5.

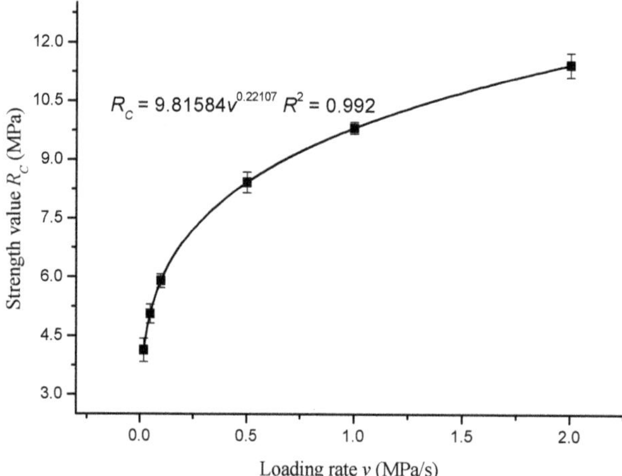

Figure 5. Unconfined compressive strength versus loading rate of asphalt mixture.

The fitting equation was as follows:

$$R_C = 9.81584v^{0.22107}, \quad R_2 = 0.992 \tag{3}$$

According to the fitting results ($R_C = 9.81584v^{0.22107}$ and $R_2 = 0.992$), the unconfined compressive strength R_C of the asphalt mixture varied with the loading rate v yielding to a power function. The strength values of asphalt mixture at the loading rates of 5 MPa/s, 10 MPa/s, 20 MPa/s, 30 MPa/s, 40 MPa/s, 50 MPa/s, 60 MPa/s, and 70 MPa/s were predicted as 14.01 MPa, 16.33 MPa, 19.035 MPa, 20.82 MPa, 22.187 MPa, 23.309 MPa, 24.267 MPa, and 25.109 MPa, respectively.

3.4. Research on Strength Parameters based on Mohr–Coulomb Theory

Asphalt mixture is mainly composed of asphalt and aggregate. The cohesive force is mainly provided by asphalt. The internal friction angle can occur when aggregates are embedded. At present,

Mohr–Coulomb theory can be widely used to analyze the strength parameters of asphalt mixtures when the strength characteristics of asphalt mixtures are researched. Based on Mohr–Coulomb theory, cohesive force C and internal friction angle φ can be obtained using a triaxial test, and a tension and compression test. Triaxial test equipment is complex, expensive, and difficult to operate. Although the real stress state of pavement can be well simulated by it, it has certain limitations to get actual application and project popularization.

It is convenient to determine the cohesive force C and internal friction angle φ of asphalt mixture through a direct tensile strength test and unconfined compressive strength test. The material and mechanical assumptions are that the material composition variables, the mechanical excitation variables, and the intrinsic parameters of the two tests are the same. After the R_C and R_D obtained from unconfined compressive and direct tensile strength tests, the two parameters can be calculated according to the conversion Equations (6) and (8) given below. The conversion relations can be derived from a Mohr circle.

When direct tensile test was carried out, $\sigma_1 = R_t$ and $\sigma_3 = 0$; when unconfined compression test was implemented, $\sigma_1 = 0$ and $\sigma_3 = -R_C$. According to the geometric relationship in Figure 6:

$$\frac{l + \sigma_1/2}{l + \sigma_1 + |\sigma_3|/2} = \frac{\sigma_1}{|\sigma_3|} \tag{4}$$

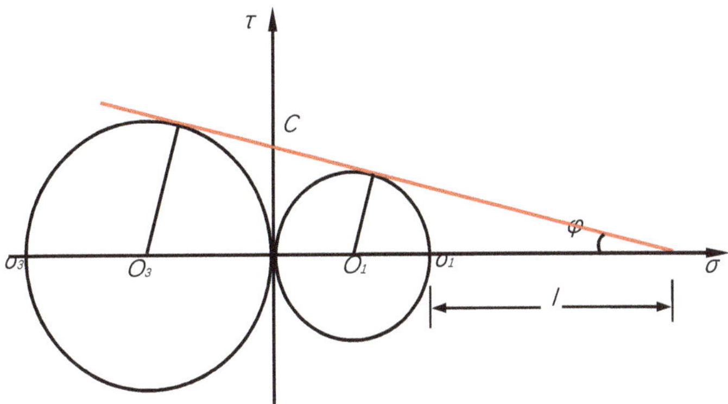

Figure 6. Mohr circle diagram for determining the values of C and φ through unconfined compressive and direct tensile strength.

Substituting the above conditions into Equation (4) to obtain:

$$l = \frac{R_D^2}{R_C - R_D} \tag{5}$$

In the right triangle:

$$\sin\varphi = \frac{\sigma_1/2}{l + \sigma_1/2} = \frac{R_D}{2l + R_D} = \frac{R_C - R_D}{R_C + R_D} \tag{6}$$

C is the intercept between straight line and ordinate. From Equation (6), $\tan\varphi$ is calculated as follows:

$$\tan\varphi = \frac{R_C - R_D}{2\sqrt{R_C R_D}} = \frac{C}{l + R_D} \tag{7}$$

The solution C is obtained below:

$$C = \frac{\sqrt{R_C R_D}}{2} \tag{8}$$

The results of unconfined compressive and direct tensile strength tests in Tables 5 and 7 are substituted for Equations (7) and (8) to calculate the cohesive force C and internal friction angle φ at different loading rates, as shown in Table 8.

Table 8. Cohesive force and internal friction angles at different loading velocities.

Loading Rate (MPa/s)	Unconfined Compressive Strength (MPa)	Direct Tensile Strength (MPa)	Cohesive Force (MPa)	Internal Friction Angle (°)
5	14.01	2.95	3.214	40.702
10	16.33	3.487	3.773	40.397
20	19.035	4.158	4.448	39.900
30	20.82	4.552	4.868	39.880
40	22.187	4.821	5.171	40.015
50	23.309	5.012	5.404	40.245
60	24.267	5.13	5.579	40.616
70	25.109	5.197	5.712	41.074

The variation of cohesive force and internal friction angle with loading rate is shown in Figure 7.

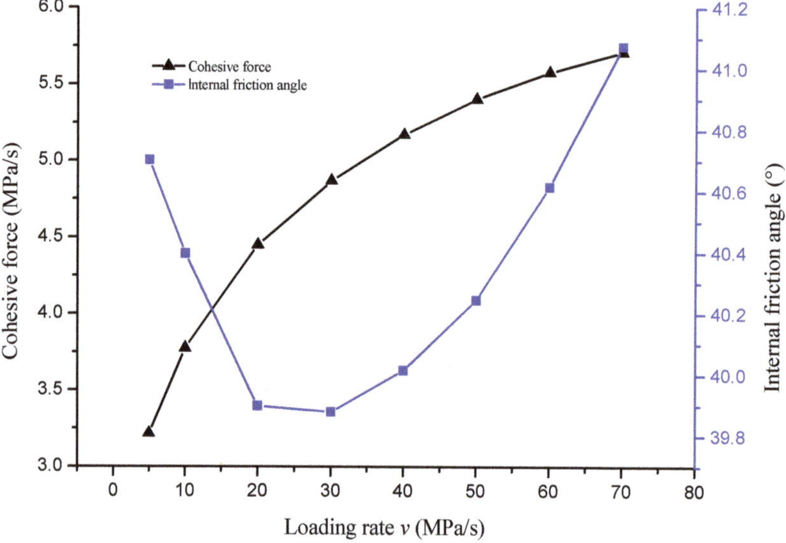

Figure 7. The variation of cohesive force and internal friction angle with loading rate.

Figure 7 shows that the cohesion increased sharply with the increase of loading rate, and then the growth rate tended to be gentle. Equation (8) also shows that cohesive force C is half of the geometric average value of unconfined compressive strength R_C and direct tensile strength R_D, so the loading pattern of cohesive force was consistent with that of unconfined compressive strength and direct tensile strength. The cohesive force of mixtures was mainly provided by the cementation between asphalt and aggregate. When the loading rate was high, the material exhibited more low-temperature morphology and the cohesive force was greater. The pattern of variation of internal friction angle with loading rate was not as clear as for the cohesive force. The pattern of variation decreased first and then increased in the experiment. The pattern of variation was that the internal friction angle decreased with the loading rate and tended to be flat in theory. The hypothesis of the theoretical pattern of variation was that the material and mechanical parameters of unconfined compression and direct tension were the same. In fact, it was difficult to satisfy the hypothesis under the experimental conditions.

3.5. Preliminary Explanation of Strength Discrepancy of Various Loading Modes

Asphalt mixture is usually used as the surface material of asphalt pavement, which directly bears various vehicle loads and environmental factors [64]. It is a kind of composite material, which mainly consists of asphalt, coarse aggregate, fine aggregate, and filler. These materials of different quality and quantity are mixed to form different structures, which have different mechanical properties.

The research on the composition and structure of asphalt mixture mainly includes surface theory and mortar theory. Surface theory holds coarse aggregate, fine aggregate, and filler to form a mineral skeleton. Asphalts binders with bonding ability are distributed on the surface of mineral skeleton and cemented into a whole structure. According to the theory of mortar, the mixture is a kind of dispersed system with a multi-level spatial network cementitious structure. The dispersed system includes a coarse dispersed system (asphalt mixture), subdivided dispersed system (asphalt mortar), and differential dispersed system (asphalt mastic).

The size and distribution of mineral aggregates in asphalt mixture, the position of aggregates, and the ratio of closed voids to connected voids of asphalt mixture are all important parts of its structure. Their differences will have a great impact on the mechanical properties of asphalt mixture. The properties of asphalt mixture improved its structure, especially the interaction between aggregate and cementing material, which made the chemical bond between the two materials and the mixture become a cohesive structure with high strength. Usually, the spatial structure of asphalt mixture is a cementitious structure. In this structure, the main factors determining the anti-destruction performance of asphalt mixture are the cohesive force between aggregates under the action of asphalt mortar, the embedding effect between aggregates, the internal friction resistance between coarse aggregates and fine aggregates, etc.

Under the condition of direct tensile test, asphalt mixture specimens are subjected to tensile stress. The deformation of asphalt mixture makes the aggregate pull apart, and the binder filled between aggregates plays a good bonding role, which is mainly borne by the cohesive force between asphalt and aggregate and the cohesive force of asphalt. The direct tensile strength is the smallest compared with the strength values under the other two test conditions. Under the condition of indirect tensile test, the asphalt mixture is in a bidirectional stress state. The compressive properties depend on the embedding effect of the aggregate, and the transverse tensile properties depend on the cohesive force and internal friction between asphalt mortar and aggregate or asphalt. Under the condition of unconfined compressive strength test, asphalt mixture specimens are subjected to compressive stress. Under the action of compressive stress, the aggregate particles are close to each other, and the skeleton formed by the aggregate particles begins to play a role. The loading is mainly borne by the internal friction resistance and the embedding force formed by the aggregates, so the compressive strength is greater than that under the other two tests. In summary, when the material is in different loading modes, the external factors that determine the anti-destruction performance of materials is inconsistent, which is the main reason for the differences of the tensile, compressive, and indirect tensile strength parameters of the material, and the fundamental reason why the compressive strength is greater than the tensile strength in general.

At the same time, it can be seen from Figure 6 that the growth rates of the strength of direct tension, indirect tension and unconfined compression with the loading rate were varied. Among the three loading modes, the growth rate of the unconfined compressive strength was the largest, followed by the indirect tensile strength, which was related to the stress state and failure mode of the three.

3.6. Unification of the Relation Between Strength and Loading Rate under Different Stress Conditions

From Tables 6 and 7, the relationship between strength and loading rate under various loading modes can be determined. The average strength values of various loading modes at eight different loading rates from 5 MPa/s to 70 MPa/s were compared, as shown in Table 9. It should be noted that the relationship between the loading rate and vehicle speed can be expressed using the equation $v = p/(l/t)$, where v is the loading rate, t is the vehicle speed, p is the tire–pavement contact pressure, and l is the length of the tire ground. In general, the length (l) of the tire ground is 0.1 m, and the tire-pavement contact pressure (p) is 0.4 MPa. In this study, the loading rates of 5 MPa/s, 10 MPa/s, 20 MPa/s, 30 MPa/s, 40 MPa/s, 50 MPa/s, 60 MPa/s, and 70 MPa/s were adopted in the strength test of asphalt mixtures. Their corresponding vehicle speeds are 4.5 km/h, 9 km/h, 18 km/h, 27 km/h, 36 km/h, 45 km/h, 54 km/h, and 63 km/h, respectively.

Table 9. Strength values of different loading rates under various loading modes.

Loading Rates v (MPa/s)	Direct Tensile Strength R_D (MPa)	Indirect Tensile Strength R_T (MPa)	Unconfined Compressive Strength R_C (MPa)
5	2.95	3.258	14.01
10	3.487	3.704	16.33
20	4.158	4.41	19.035
30	4.552	4.837	20.82
40	4.821	5.185	22.187
50	5.012	5.487	23.309
60	5.13	5.658	24.267
70	5.197	5.784	25.109

The relationship between strength and loading rate under various loading modes was compared, as shown in Figure 8.

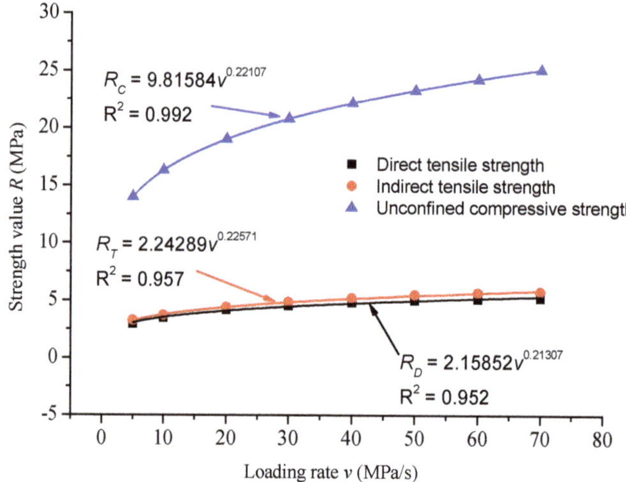

Figure 8. Comparison of the relationship between strength and loading rate under various loading modes.

The fitting curve parameters of the relationship between strength and loading rates are summarized in Table 10.

Table 10. Fitting curve equations of the relation between strength and loading rate under various loading modes.

Fitting Equation	$R = \alpha \times v^\beta$		
	α	β	R^2
Direct Tensile Test	2.15852	0.21307	0.952
Indirect Tensile Test	2.24289	0.22571	0.957
Unconfined Compression Test	9.81584	0.22107	0.992

It can be seen that the strength–loading rate curves of various loading modes increased with the increase of loading rate, and the strength decreased with the increase of loading rate. Under the same loading rate, the direct tensile strength was close to the indirect tensile strength, and the unconfined compressive strength was far greater than the direct tensile strength and the indirect tensile strength. The parameters of the strength–loading rate curves under various loading modes were quite different, which brings a lot of inconvenience to the experimental research. This section will use standardized methods to unify the relationship between strength and loading rate, as shown in Table 11.

Table 11. Relation between strength ratio and loading rate ratio.

Loading Rate Ratio v/v_s	Direct Tensile Strength Ratio	Indirect Tensile Strength Ratio	Unconfined Compressive Strength Ratio
0.071	0.568	0.563	0.558
0.143	0.671	0.640	0.650
0.286	0.800	0.762	0.758
0.429	0.876	0.836	0.829
0.571	0.928	0.896	0.884
0.714	0.964	0.949	0.928
0.857	0.987	0.978	0.966
1	1	1	1

The strength ratio and loading rate ratio of direct tension, unconfined compression, and indirect tension in Table 10 are fitted in Figure 9.

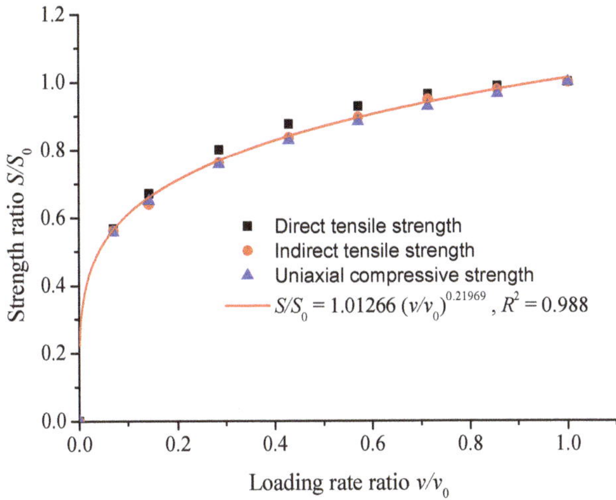

Figure 9. Strength ratio versus loading rate ratio under various loading modes.

The fitting equation was as follows:

$$S/S_0 = 1.01266 \, (v/v_0)^{0.21969}, R^2 = 0.988 \tag{9}$$

where, S is the strength values of various loading modes at different loading rates, MPa; S_0 is the strength values of various loading modes at 70 MPa/s loading rate, MPa; v_0 is the set loading rate, 70 MPa/s.

As shown in Figure 6, the relationship between strength and loading rate under various loading modes can be obtained by using the traditional power function equation, but the parameters of the equation could not be unified, and the difference was large. As shown in Figure 7, the relationship between strength and loading rate of direct tension, unconfined compression, and indirect tension could be unified by using the equation of strength ratio and loading rate ratio, and the correlation coefficient was better. By standardizing the curve function of strength ratio and loading rate ratio under various loading modes, the strength values of various loading modes under other loading rates could be predicted through the strength values of one stress state, and the strength values of different loading rates under other stress states could be predicted.

4. Summary and Conclusions

In this study, considering the main objective of this study was to solve the uncertainty of the strength performance of asphalt mixture under various loading models, only one of the dense gradation asphalt mixture AC-13C that were composed of SBS modified asphalt was applied in the laboratory tests. The strength of used asphalt mixture with various loading modes and various loading rates were investigated. The whole curves of strength with the variation of loading rates under three loading modes were obtained, and based on these, the strength mechanism with loading rates under three loading modes was explained preliminarily from the mechanical view. It is noteworthy that the achieved strength and loading rate of asphalt mixtures were treated using a standardized method. According to the mentioned research works, the main conclusions of this study are as follows:

1. Loading rate had a significant effect on the strength of the asphalt mixtures. The pattern of variation of the direct tensile strength, indirect tensile strength, and unconfined compressive strength vary with loading rate are shown in the equations $R_D = 2.15852v^{0.21307}$, $R^2 = 0.952$; $R_T = 2.24289v^{0.22571}$, $R^2 = 0.957$; and $R_C = 9.81584v^{0.22107}$, $R^2 = 0.992$, respectively.

2. Under the same laboratory conditions, the strength of the asphalt mixture was affected by the loading mode. Among the three loading modes, the value of the unconfined compressive strength was the largest, followed by the indirect tensile strength. The strength difference under different loading modes was explained by the structural composition of the asphalt mixture.

3. Unified strength models of asphalt mixtures under different loading modes could be depicted as $S/S_0 = 1.01266(v/v_0)^{0.21969}$, $R^2 = 0.988$. The proposed model could be applied to remove the uncertainties of strength parameters under different loading modes. Through the unified strength model, as long as the strength value under one loading mode was achieved, the strength values under the other two modes can be obtained, which greatly improved the efficiency of the laboratory test.

This study's main aim was to report the methodology to develop a unified strength model based on one special materials. However, in order to further confirm the reliability and effectiveness of the proposed model, more types of materials should be applied in later works.

Author Contributions: Conceptualization: C.X., S.L., D.C., and Y.L.; Data curation: C.X., S.L., L.Y., D.C., and Y.L.; Funding acquisition: S.L. and J.Z.; Methodology: C.X. and S.L.; Software: S.L.; Supervision: L.Y. and J.Z.; Writing—original draft: C.X., S.L., L.Y., and D.C.; Writing—review and editing: C.X. and S.L.

Funding: This work was supported by National Natural Science Foundation of China (51578081, 51608058); The Ministry of Transport Construction Projects of Science and Technology (2015318825120); Key Projects of Hunan Province-Technological Innovation Project in Industry (2016GK2096); The Inner Mongolia Autonomous Region Traffic and Transportation Department Transportation Projects of Science and Technology (NJ-2016-35, HMJSKJ-201801); and The Hunan Province Transport Construction Projects of Science and Technology (201701).

Conflicts of Interest: The authors declare no conflict of interest.

References

1. Aziz, M.M.A.; Rahman, M.T.; Hainin, M.R.; Bakar, W.A.W.A. An overview on alternative binders for flexible pavement. *Constr. Build. Mater.* **2015**, *84*, 315–319. [CrossRef]
2. Lv, S.; Liu, C.; Zheng, J.; You, Z.; You, L. Viscoelastic Fatigue Damage Properties of Asphalt Mixture with Different Aging Degrees. *Ksce J. Civ. Eng.* **2018**, *22*, 2073–2081. [CrossRef]

3. Zhang, J.; Wang, M.; Wang, D.; Li, X.; Song, B.; Liu, P. Feasibility study on measurement of a physiological index value with an electrocardiogram tester to evaluate the pavement evenness and driving comfort. *Measurement* **2018**, *117*, 1–7. [CrossRef]

4. Lv, S.; Liu, C.; Yao, H.; Zheng, J. Comparisons of synchronous measurement methods on various moduli of asphalt mixtures. *Constr. Build. Mater.* **2018**, *158*, 1035–1045. [CrossRef]

5. Mannan, U.A.; Islam, M.R.; Tarefder, R.A. Effects of recycled asphalt pavements on the fatigue life of asphalt under different strain levels and loading frequencies. *Int. J. Fatigue* **2015**, *78*, 72–80. [CrossRef]

6. Lv, S.; Fan, X.; Xia, C.; Zheng, J.; Chen, D.; You, L. Characteristics of Moduli Decay for the Asphalt Mixture under Different Loading Conditions. *Appl. Sci.* **2018**, *8*, 840. [CrossRef]

7. Wang, H.; Liu, X.; Zhang, H.; Apostolidis, P.; Scarpas, T.; Erkens, S. Asphalt-rubber interaction and performance evaluation of rubberised asphalt binders containing non-foaming warm-mix additives. *Road Mater. Pavement Des.* **2018**, 1–22. [CrossRef]

8. Yi, J.; Shen, S.; Muhunthan, B.; Feng, D. Viscoelastic–plastic damage model for porous asphalt mixtures: Application to uniaxial compression and freeze–thaw damage. *Mech. Mater.* **2014**, *70*, 67–75. [CrossRef]

9. Wang, W.; Wang, L.; Xiong, H.; Luo, R. A review and perspective for research on moisture damage in asphalt pavement induced by dynamic pore water pressure. *Constr. Buil. Mater.* **2019**, *204*, 631–642. [CrossRef]

10. Chen, S.; You, Z.; Sharifi, N.P.; Yao, H.; Gong, F. Material selections in asphalt pavement for wet-freeze climate zones: A review. *Constr. Buil. Mater.* **2019**, *201*, 510–525. [CrossRef]

11. You, L.; You, Z.; Yang, X.; Ge, D.; Lv, S. Laboratory Testing of Rheological Behavior of Water-Foamed Bitumen. *J. Mater. Civ. Eng.* **2018**, *30*, 04018153. [CrossRef]

12. Li, X.; Chen, S.; Xiong, K.; Liu, X. Gradation Segregation Analysis of Warm Mix Asphalt Mixture. *J. Mater. Civ. Eng.* **2018**, *30*, 04018027. [CrossRef]

13. Woszuk, A.; Franus, W. A Review of the Application of Zeolite Materials in Warm Mix Asphalt Technologies. *Appl. Sci.* **2017**, *7*, 293. [CrossRef]

14. Li, X.; Zhou, Z.; Lv, X.; Xiong, K.; Wang, X.; You, Z. Temperature segregation of warm mix asphalt pavement: Laboratory and field evaluations. *Constr. Buil. Mater.* **2017**, *136*, 436–445. [CrossRef]

15. Jin, J.; Xiao, T.; Zheng, J.; Liu, R.; Qian, G.; Xie, J.; Wei, H.; Zhang, J.; Liu, H. Preparation and thermal properties of encapsulated ceramsite-supported phase change materials used in asphalt pavements. *Constr. Buil. Mater.* **2018**, *190*, 235–245. [CrossRef]

16. Jin, J.; Xiao, T.; Tan, Y.; Zheng, J.; Liu, R.; Qian, G.; Wei, H.; Zhang, J. Effects of TiO$_2$ pillared montmorillonite nanocomposites on the properties of asphalt with exhaust catalytic capacity. *J. Clean. Prod.* **2018**, *205*, 339–349. [CrossRef]

17. Qin, X.; Zhu, S.; He, X.; Jiang, Y. High temperature properties of high viscosity asphalt based on rheological methods. *Constr. Buil. Mater.* **2018**, *186*, 476–483. [CrossRef]

18. Javilla, B.; Fang, H.; Mo, L.; Shu, B.; Wu, S. Test evaluation of rutting performance indicators of asphalt mixtures. *Constr. Buil. Mater.* **2017**, *155*, 1215–1223. [CrossRef]

19. Jin, J.; Tan, Y.; Liu, R.; Zheng, J.; Zhang, J. Synergy Effect of Attapulgite, Rubber, and Diatomite on Organic Montmorillonite-Modified Asphalt. *J. Mater. Civ. Eng.* **2019**, *31*, 04018388. [CrossRef]

20. Woszuk, A.; Franus, W. Properties of the Warm Mix Asphalt involving clinoptilolite and Na-P1 zeolite additives. *Constr. Buil. Mater.* **2016**, *114*, 556–563. [CrossRef]

21. Tan, Y.; Sun, Z.; Gong, X.; Xu, H.; Zhang, L.; Bi, Y. Design parameter of low-temperature performance for asphalt mixtures in cold regions. *Constr. Buil. Mater.* **2017**, *155*, 1179–1187. [CrossRef]

22. Sanchez-Alonso, E.; Vega-Zamanillo, A.; Castro-Fresno, D.; DelRio-Prat, M. Evaluation of compactability and mechanical properties of bituminous mixes with warm additives. *Constr. Buil. Mater.* **2011**, *25*, 2304–2311. [CrossRef]

23. Ameri, M.; Yeganeh, S.; Erfani Valipour, P. Experimental evaluation of fatigue resistance of asphalt mixtures containing waste elastomeric polymers. *Constr. Buil. Mater.* **2019**, *198*, 638–649. [CrossRef]

24. Huang, T.; Zheng, J.L.; Lv, S.T.; Zhang, J.H.; Wen, P.H.; Bailey, C.G. Failure criterion of an asphalt mixture under three-dimensional stress state. *Constr. Buil. Mater.* **2018**, *170*, 708–715. [CrossRef]

25. You, L.; Yan, K.; Hu, Y.; Ma, W. Impact of interlayer on the anisotropic multi-layered medium overlaying viscoelastic layer under axisymmetric loading. *Appl. Math. Model.* **2018**, *61*, 726–743. [CrossRef]

26. López, C.; González, A.; Thenoux, G.; Sandoval, G.; Marcobal, J. Stabilized emulsions to produce warm asphalt mixtures with reclaimed asphalt pavements. *J. Clean. Prod.* **2019**, *209*, 1461–1472. [CrossRef]

27. Goh, S.W.; Akin, M.; You, Z.; Shi, X. Effect of deicing solutions on the tensile strength of micro- or nano-modified asphalt mixture. *Constr. Buil. Mater.* **2011**, *25*, 195–200. [CrossRef]

28. You, L.; You, Z.; Dai, Q.; Xie, X.; Washko, S.; Gao, J. Investigation of adhesion and interface bond strength for pavements underlying chip-seal: Effect of asphalt-aggregate combinations and freeze-thaw cycles on chip-seal. *Constr. Buil. Mater.* **2019**, *203*, 322–330. [CrossRef]

29. Lv, S.; Liu, C.; Chen, D.; Zheng, J.; You, Z.; You, L. Normalization of fatigue characteristics for asphalt mixtures under different stress states. *Constr. Buil. Mater.* **2018**, *177*, 33–42. [CrossRef]

30. Hoy, M.; Horpibulsuk, S.; Arulrajah, A.; Mohajerani, A. Strength and Microstructural Study of Recycled Asphalt Pavement: Slag Geopolymer as a Pavement Base Material. *J. Mater. Civ. Eng.* **2018**, *30*, 04018177. [CrossRef]

31. Hoy, M.; Horpibulsuk, S.; Arulrajah, A. Strength development of Recycled Asphalt Pavement—Fly ash geopolymer as a road construction material. *Constr. Buil. Mater.* **2016**, *117*, 209–219. [CrossRef]

32. Chávez-Valencia, L.E.; Alonso, E.; Manzano, A.; Pérez, J.; Contreras, M.E.; Signoret, C. Improving the compressive strengths of cold-mix asphalt using asphalt emulsion modified by polyvinyl acetate. *Constr. Buil. Mater.* **2007**, *21*, 583–589. [CrossRef]

33. Falchetto, A.C.; Moon, K.H.; Wang, D.; Riccardi, C.; Wistuba, M.P. Comparison of low-temperature fracture and strength properties of asphalt mixture obtained from IDT and SCB under different testing configurations. *Road Mater. Pavement Des.* **2018**, *19*, 591–604. [CrossRef]

34. Kim, M.; Mohammad, L.; Elseifi, M. Characterization of fracture properties of asphalt mixtures as measured by semicircular bend test and indirect tension test. *Transp. Res. Rec. J. Transp. Res. Board* **2012**, *2296*, 115–124. [CrossRef]

35. Lv, S.; Wang, X.; Liu, C.; Wang, S. Fatigue Damage Characteristics Considering the Difference of Tensile-Compression Modulus for Asphalt Mixture. *J. Test. Eval.* **2018**, *46*, 2470–2482. [CrossRef]

36. Li, X.; Lv, X.; Liu, X.; Ye, J. Discrete Element Analysis of Indirect Tensile Fatigue Test of Asphalt Mixture. *Appl. Sci.* **2019**, *9*, 327. [CrossRef]

37. You, L.; You, Z.; Dai, Q.; Guo, S.; Wang, J.; Schultz, M. Characteristics of water-foamed asphalt mixture under multiple freeze-thaw cycles: Laboratory evaluation. *J. Mater. Civ. Eng.* **2018**, *30*, 04018270. [CrossRef]

38. Lv, S.; Wang, S.; Liu, C.; Zheng, J.; Li, Y.; Peng, X. Synchronous Testing Method for Tension and Compression Moduli of Asphalt Mixture under Dynamic and Static Loading States. *J. Mater. Civ. Eng.* **2018**, *30*, 04018268. [CrossRef]

39. Tao, M.; Mallick, R.B. Effects of Warm-Mix Asphalt Additives on Workability and Mechanical Properties of Reclaimed Asphalt Pavement Material. *Transp. Res. Rec. J. Transp. Res. Board* **2009**, *2126*, 151–160. [CrossRef]

40. Mohammad, L.N.; Raqib, M.; Huang, B. Influence of Asphalt Tack Coat Materials on Interface Shear Strength. *Transp. Res. Rec.* **2002**, *1789*, 56–65. [CrossRef]

41. Zhou, L.; Huang, W.; Xiao, F.; Lv, Q. Shear adhesion evaluation of various modified asphalt binders by an innovative testing method. *Constr. Buil. Mater.* **2018**, *183*, 253–263. [CrossRef]

42. Folino, P.; Xargay, H. Recycled aggregate concrete—Mechanical behavior under uniaxial and triaxial compression. *Constr. Buil. Mater.* **2014**, *56*, 21–31. [CrossRef]

43. Shang, H.-S.; Song, Y.-P.; Qin, L.-K. Experimental study on strength and deformation of plain concrete under triaxial compression after freeze-thaw cycles. *Build. Environ.* **2008**, *43*, 1197–1204. [CrossRef]

44. Su, Y.M.; Hossiney, N.; Tia, M. Indirect Tensile Strength of Concrete Containing Reclaimed Asphalt Pavement Using the Superpave Indirect Tensile Test. *Adv. Mater. Res.* **2013**, *723*, 368–375. [CrossRef]

45. Saride, S.; Avirneni, D. Strength Characteristics of Geopolymer Fly Ash-Stabilized Reclaimed Asphalt Pavement Base Courses. In *Geoenvironmental Practices and Sustainability*; Springer: Singapore, 2017; pp. 267–275.

46. Ji, X.; Jiang, Y.; Zou, H.; Cao, F.; Hou, Y. Application of numerical simulation method to improve shear strength and rutting resistance of asphalt mixture. *Int. J. Pavement Eng.* **2018**, 1–10. [CrossRef]

47. Li, L.H.; Wang, T.X. Study on Strength Characteristics of Foamed Asphalt Mixture. *J. Build. Mater.* **2009**, *12*, 549–553.

48. Gaus, A.; Tjaronge, M.W.; Ali, N.; Djamaluddin, R. Compressive Strength of Asphalt Concrete Binder Course (AC-BC) Mixture Using Buton Granular Asphalt (BGA). *Procedia Eng.* **2015**, *125*, 657–662. [CrossRef]

49. Yu, M.H.; Zan, Y.W.; Zhao, J.; Yoshimine, M. A Unified Strength criterion for rock material. *Int. J. Rock Mech. Min. Sci.* **2002**, *39*, 975–989. [CrossRef]

50. Michelis, P. Polyaxial yielding of granular rock. *J. Eng. Mech.* **1985**, *111*, 1049–1066. [CrossRef]
51. You, M. Discussion on unified strength theories for rocks. *Chin. J. Rock Mech. Eng.* **2013**, *32*, 258–265.
52. Danni, L.; Qinbin, L.; Yu, H.; Tao, C.; University, T. Strength criterion for high-strength concrete based on the unified strength theory. *J. Hydraul. Eng.* **2015**, *46*, 74–82. [CrossRef]
53. Wu, Y.F.; Wang, L.M. Unified Strength Model for Square and Circular Concrete Columns Confined by External Jacket. *J. Struct. Eng.* **2009**, *135*, 253–261. [CrossRef]
54. Wu, Y.F.; Zhou, Y.W. Unified Strength Model Based on Hoek-Brown Failure Criterion for Circular and Square Concrete Columns Confined by FRP. *J. Compos. Constr.* **2010**, *14*, 175–184. [CrossRef]
55. Wei, Y.Y.; Wu, Y.F. Unified stress–strain model of concrete for FRP-confined columns. *Constr. Build. Mater.* **2012**, *26*, 381–392. [CrossRef]
56. Youssef, M.N.; Feng, M.Q.; Mosallam, A.S. Stress–strain model for concrete confined by FRP composites. *Compos. Part B Eng.* **2007**, *38*, 614–628. [CrossRef]
57. You, T.; Abu Al-Rub, R.K.; Darabi, M.K.; Masad, E.A.; Little, D.N. Three-dimensional microstructural modeling of asphalt concrete using a unified viscoelastic–viscoplastic–viscodamage model. *Constr. Build. Mater.* **2012**, *28*, 531–548. [CrossRef]
58. You, T.; Al-Rub, R.K.A.; Masad, E.A.; Kassem, E.; Little, D.N. Three-dimensional microstructural modeling framework for dense-graded asphalt concrete using a coupled viscoelastic, viscoplastic, and viscodamage model. *J. Mat. Civ. Eng.* **2013**, *26*, 607–621. [CrossRef]
59. Hajj, E.; Siddharthan, R.; Sebaaly, P.; Weitzel, D. Laboratory-based Unified Permanent Deformation Model for Hot-Mix Asphalt Mixtures. *J. Test. Eval.* **2007**, *35*, 272–280. [CrossRef]
60. Pasetto, M.; Baldo, N. Numerical visco-elastoplastic constitutive modelization of creep recovery tests on hot mix asphalt. *J. Traffic Transp. Eng.* **2016**, *3*, 390–397. [CrossRef]
61. Costanzi, M.; Cebon, D. Generalized Phenomenological Model for the Viscoelasticity of Idealized Asphalts. *J. Mater. Civ. Eng.* **2014**, *26*, 399–410. [CrossRef]
62. *(JTG F40-2004) Technical Specifications for Construction of Highway Asphalt Pavement*; Renmin Communication Press: Beijing, China, 2004.
63. *(JTG E20-2011) Specifications and Test Methods of Bitumen and Biminous Mixtures for Highway Engineering*; Renmin Communication Press: Beijing, China, 2011.
64. Lv, S.; Wang, S.; Guo, T.; Xia, C.; Li, J.; Hou, G. Laboratory Evaluation on Performance of Compound-Modified Asphalt for Rock Asphalt/Styrene–Butadiene Rubber (SBR) and Rock Asphalt/Nano-CaCO3. *Appl. Sci.* **2018**, *8*, 1009. [CrossRef]

 *materials*

Article

Characterization of Asphalt Mixture Moduli under Different Stress States

Xiyan Fan, Songtao Lv *, Naitian Zhang *, Chengdong Xia and Yipeng Li

National Engineering Laboratory of Highway Maintenance Technology, Changsha University of Science and Technology, Changsha 410114, China; fxy@stu.csust.edu.cn (X.F.); xiachengdong@stu.csust.edu.cn (C.X.); lypcsust@163.com (Y.L.)
* Correspondence: lst@csust.edu.cn (S.L.); agnore@foxmail.com (N.Z.)

Received: 3 January 2019; Accepted: 22 January 2019; Published: 27 January 2019

Abstract: Modulus testing methods under various test conditions have a large influence on modulus test results, which hinders the accurate evaluation of the stiffness of asphalt mixtures. In order to decrease the uncertainty in the stiffness characteristics of asphalt mixtures under various stress states, the traditional unconfined compression test, direct tensile test, and the synchronous test method, based on the indirect tension and four-point bending tests, were carried out for different loading frequencies. Results showed that modulus test results were highly sensitive to the shape, size, and stress state of the specimen. Additionally, existing modulus characteristics did not reduce these differences. There is a certain correlation between the elastic modulus ratio and the frequency ratio for asphalt under multiple stress states. The modulus, under multiple stress states, was processed using min–max normalization. Then, the standardization model for tensile and compressive characteristics of asphalt under diverse stress states was established based on the sample preparation, modulus ratio variations, and loading frequency ratio. A method for deriving other moduli from one modulus was realized. It is difficult to evaluate the stiffness performance in diverse stress states for asphalt by only using conventional compressive and tensile tests. However, taking into account the effects of stress states and loading frequencies, standardized models can be used to reduce or even eliminate these effects. The model realizes the unification of different modulus test results, and provides a theoretical, methodological, and technical basis for objectively evaluating moduli.

Keywords: road engineering; asphalt mixture; modulus test; synchronous test method; tensile moduli; compressive moduli; indirect tensile moduli; four-point bending moduli; standardized

1. Introduction

The multiplayer elastic system theory is the basis of asphalt pavement design in various countries [1]. The modulus is the primary input of pavement design methodology, and is used to predict and understand the performance of asphalt mixes [2], as it can characterize the ability of asphalt mix to disperse loads and control traffic levels [3].

The modulus is a critical parameter in asphalt pavement structure design, and directly affects the accuracy of a structural design. Analysis results can be used to calculate reasonable pavement thicknesses and evaluate the long-term performance characteristics of the asphalt pavement, such as fatigue, permanent deformation, and cracking [4–6]. Compared with Marshall-compacted specimens, the dynamic stability, flexural strength, and water stability of rotary-compacted asphalt mixture specimens have been greatly improved. Specimens formed by vertical vibration are denser and stronger than those formed by Marshall compaction. The moduli test methods of pavement materials are generally divided into four-point bending, direct tensile, indirect tensile, and unconfined compression [7–9]. The stiffness modulus is measured by the material testing system (MTS). Under the stress-control mode, the specimens are loaded, and the real-time data of load and displacement are collected. The slope of the corresponding stress–strain curve is the stiffness modulus [10,11].

Different design methods have different definitions of asphalt mixture modulus: US AASHTO2002 uses dynamic compression modulus, whereas the Australian Pavement Structure Design Guide and the Japanese method use an indirect tensile test to determine the resilient modulus of the asphalt mixture. Due to differences in testing methods, the standard test method for asphalt mixture modulus in each country is different. For example, ASTM D4123-82 and AASHTO TP31-96 are used to determine the indirect tensile resilience modulus of asphalt mixtures; AASHTO TP8 and SHRP M-009 are used to determine the flexural tensile modulus of asphalt mixtures; and ASTM 3496-99, ASTM D3497-79, AASHTO TP62-03, and NCHRP 9-19/9-29/Report 513 are used to determine the compression modulus. European Standards EN 12697-26:2018 determine the modulus of bituminous mixes by the indirect tensile method. The principles of these standard testing methods are basically the same, but there are differences in the preparation process, test temperature, frequency, placement of LVDT (linear Variable Differential Transducers), loading time, failure judgement, and modulus calculation method. Different test methods give dissimilar stiffness moduli of a given asphalt mixture [12]. Varying moduli values greatly influence the stress, strain, and displacement results for the pavement layer, which will seriously affect asphalt pavement design. The parameter which accurately reflects the mechanical properties of an asphalt mixture is still unknown [13]. The current importance of reasonable modulus parameters in the structural design and behavior analysis of asphalt pavement is largely undetermined.

The traditional model for the indirect tensile modulus and flexural modulus is based on the assumption that the tension and compression moduli are similar. The measured modulus cannot accurately reflect the mechanical properties of a material while the difference between the tension and compression modulus is not accounted for [13]. Therefore, in several existing modulus testing methods, it is difficult to determine which parameters reflect the mechanical properties of the material itself.

On the basis of the traditional direct tensile modulus and unconfined compression modulus, Lv's derivation of indirect tensile resilience modulus and four-point bending modulus was used in this contribution [13]. A large number of tests under different loading frequencies were carried out using a material test system, including the direct tensile modulus, unconfined compression modulus, indirect tensile test, and four-point bending synchronous test. Based on the standardization idea, the modulus under different stress states was processed using min–max normalization measures. Experimental results were standardized and mapped to a 0–1 range. After standardization, the asphalt mixture modulus was converted to a dimensionless parameter under different stress states, and the comprehensive analysis was carried out. The standardized model for tension and compression moduli and the standardization model for the compressive, flexural, tensile, and indirect tensile moduli were obtained. The moduli under different loading states and different loading frequencies can be converted to simplify the modulus process. The modulus under different stress states was obtained by standardized treatment. Compared with the traditional method, the new method reduces the influence of loading mode, specimen shape, and specimen size on modulus test results, realizes the unification of different modulus test results, and provides a theoretical, methodological, and technical basis for objectively evaluating the modulus.

2. Sample Preparation

2.1. Materials

AC-13C (fine-grained asphalt mixtures) consist of limestone aggregates, SBS-modified asphalt binders, and limestone powders. A large number of direct tensile, indirect tensile, unconfined compression, and four-point bending tests were separately performed to ensure the significance of the model. The SBS-modified asphalt performance is illustrated in Table 1. The chemical and physical properties of aggregates are illustrated in Table 2. The dense skeleton-type gradation of aggregates was selected in accordance with the "Specifications for Design of Highway Asphalt Pavement" (Figure 1) [14]. Marshall tests were used to determine the optimum asphalt ratio (Table 3).

Table 1. The SBS-(I-D) modified asphalt performance.

Technical Indexes	Value	Specification
Penetration 25 °C, 100 g, 5 s (0.1 mm)	55.91	30–60
Ductility 5 cm/min, 5 °C (cm)	34.22	$\geq$20
Softening point TR&B (°C)	79.39	$\geq$60

Table 2. Chemical and physical properties of aggregates.

Items	Crushing Value	Content of Needle-Like Particles	Content of SiO$_2$	Apparent Density
Value	10.8	7.8	1.79	2.578
Specification	$\leq$26	$\leq$20	-	-

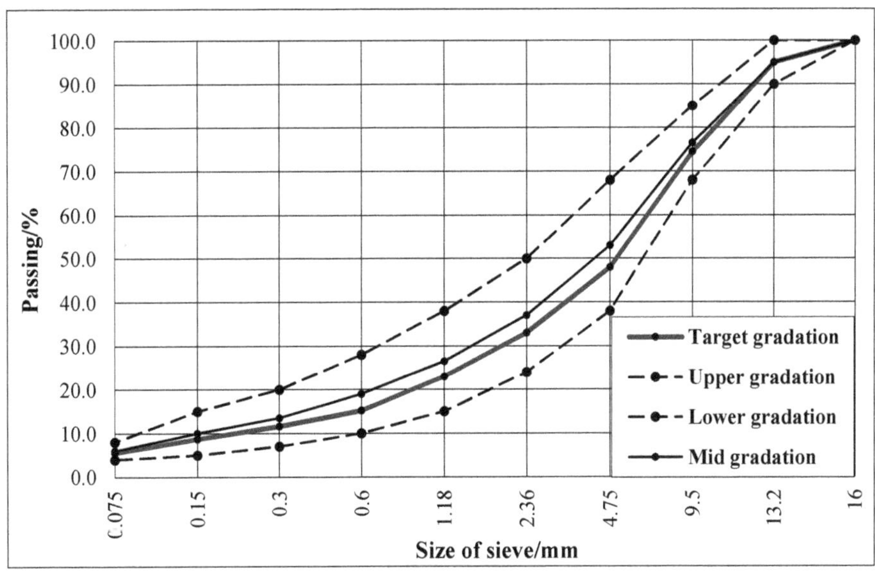

Figure 1. Aggregate gradation.

Table 3. Marshall test results at optimal asphalt content.

Asphalt Aggregate Ratio (%)	Bulk Specific Gravity (g·cm^{-3})	Volume of Air Voids VV (%)	Voids Filled with Asphalt VFA (%)	Voids in Mineral Aggregate VMA (%)	Marshall Stability (kN)	Flow Value (0.1 mm)
5.2	2.54	4.51	67.20	16.11	12.71	27.89

2.2. Specimen Manufacturing

When shaping the specimen, the weighed mineral was preheated in the oven for 4 h so that the mineral was fully dried, and then stirred in a stirring pot at room temperature for 90 s to ensure the mixture was stirred evenly. The uniaxial compression test adopts the method of rotating compaction. The specimens for direct tension and bending tension tests were obtained by cutting a rutting plate. Rutting plates were obtained by the wheel rolling method. The splitting test was done by a Marshall compactor.

Five specimens were constructed for each of the four tests, in accordance with the specifications and test methods (JTG E20-2011) [15]. Samples with dimensions of 400 mm × 300 mm × 80 mm were constructed for the vibrating compaction experiment. Beam specimens were divided from rutting plates to a size of 250 mm × 50 mm × 50 mm for direct tensile and the four-point-bending tests (Figure 2a,c). Cylinder specimens were drilled from rotating compaction samples to a dimension of Φ100 mm × 80 mm for the unconfined compressive test (Figure 2b). Indirect tensile specimens were prefabricated from a Marshall specimen to a dimension of Φ100 mm × 60 mm (Figure 2d).

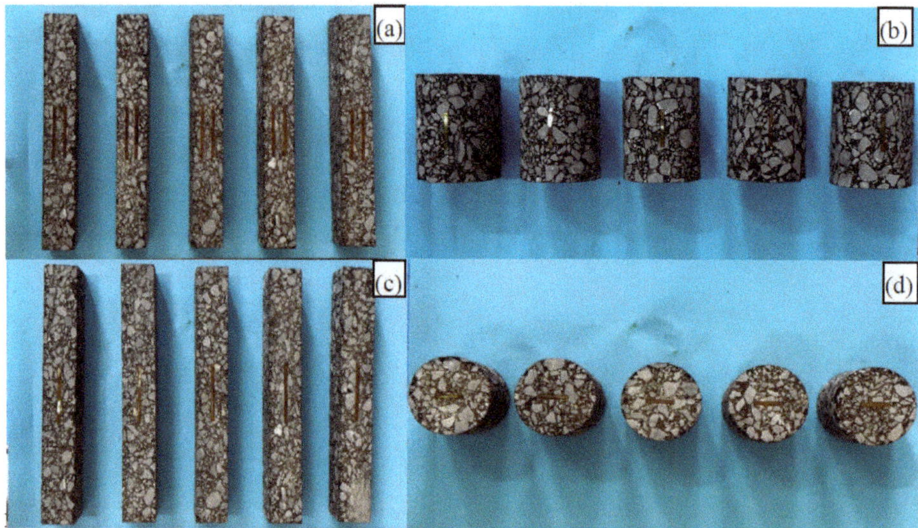

Figure 2. Samples for various tests for the: (**a**) four-point bending moduli test, (**b**) indirect tensile moduli test, (**c**) direct tensile moduli test, (**d**) unconfined compressive moduli test.

3. Moduli Tests

3.1. Experiment Conditions and Procedures

In order to ensure that the specimen is in the linear elastic stage, different loading frequencies corresponding to different load cycles should be used. If too few cycles are used, the rebound deformation will not be visible [16]. If too many cycles are used, the material will permanently deform and make the rebound deformation test results larger. Choosing the appropriate number of load cycles can meet the requirements of the rebound deformation test, and can also maximize the use of the specimen for the dynamic modulus test at different frequencies, which can effectively reduce the impact of test piece differences on the test results. The deformation ability of asphalt mixture is related to the load action time. Under one kind of test temperature condition, a higher loading frequency corresponds to a shorter load-action time. The deformation ability of an asphalt mixture cannot be brought into full play at high frequencies because the overall stiffness of the mixture is larger, and the deformation in each load cycle, as well as the cumulative deformation, are smaller. Low-frequency deformation has the opposite effect on asphalt specimens. Therefore, high-frequency tests are performed before low-frequency tests. For the purpose of mutual comparison, the moduli achieved by various methods, the same progression of experimental frequencies, and the number of loading cycles were used for all specimens (Table 4). The design of bitumen road surface structure in China sustains 100 kN as a standard axle load [14], and the corresponding tire compressive stress is 0.7 MPa. The dynamic modulus is designed for pavement structure design, so the test stress amplitude was set to 0.7 MPa. The use of a single loading frequency neglects the viscoelastic (i.e., time or frequency dependency) nature of the asphalt mix, inherited from the asphalt binder [17], so the loading frequencies were 0.1, 1, 10, 20, and 50 Hz. The test temperature for the dynamic modulus was 15 °C, and the stress ratio of the dynamic modulus test was 0.1. Each test was run five separate times to ensure accuracy.

The modulus tests were carried out by the material testing system (MTS), and strain measurement was gauged using strain gauges, LVDT, and a strain data collection system. The temperature of the test was controlled using an environmental chamber. Photographs of each modulus test are shown in Figure 3. The material test system (MTS) was used to carry out the unconfined compression test and the direct tensile dynamic modulus test. Based on the flexural modulus test [15], two strain gauges were attached to the upper and lower surface centers of the specimens to measure the tensile and compressive strains, respectively. The mid-span deflection of the test piece was gauged by LDVT. The indirect tensile test drew from the foundations of the classical Brazilian indirect tensile test. Strain gauges with horizontal radial and vertical diameters were separately bonded at the front and rear center of the specimen.

Figure 3. Photographs of the test equipment: (**a**) four-point bending test, (**b**) indirect tensile test, (**c**) unconfined compressive test, (**d**) direct tensile test.

Table 4. Loading cycles at different loading frequencies.

Loading Frequency (Hz)	0.1	1	10	20	50
Cycles	20	50	150	200	300

3.2. Calculation Formula

(1) Compressive and tensile resilient moduli from four-point bending test

According to the strength test for asphalt mixtures (JTG E20-2011) [15] under a four-point bending load, Pb is 1.523 kN. Two strains gauges were attached to the upper and lower surface centers of the specimens to measure the tensile and compressive strains, respectively (Figure 3). The mid-span deflection of the test piece was gauged by LDVT.

The formulae of tensile and compressive moduli using the four-point bending method are as follows [17]:

$$E_c = \frac{\sigma_{cmax}}{\varepsilon_{cmax}} = \frac{3M}{bh_2h\varepsilon_c} = \frac{PL(\varepsilon_t + \varepsilon_c)}{2b\varepsilon_c{}^2h^2},$$

(1)

$$E_t = \frac{\sigma_{tmax}}{\varepsilon_{tmax}} = \frac{3M}{bh_2h\varepsilon_t} = \frac{PL(\varepsilon_t + \varepsilon_c)}{2b\varepsilon_t{}^2h^2}.$$

(2)

In the formulae, σ_c, σ_t represent the values of compressive stress and tensile stress, respectively; $\varepsilon_c, \varepsilon_t$ represent the compressive strain at the upper surface center and the tensile strain at the lower surface center, respectively; and E_c, E_t represent compressive modulus and tensile modulus of specimens, respectively.

The formula for calculating the flexural modulus is provided in the specifications and test methods (JTG E20-2011) [15]:

$$E_f = \frac{23PL^3}{108bh^3w},$$

(3)

where E_f denotes the value of bending modulus, and w is the deflection.

(2) Compressive and tensile resilient modulus from the indirect tensile test

The formulae for the indirect tensile and compressive moduli test are as follows [10]:

$$\begin{cases} E_x = \frac{4P}{\pi L} \times \frac{a \times b + c \times d \times \mu^2}{b \times \Delta u - \mu \times d \times \Delta v} \\ E_y = \frac{4P}{\pi L} \times \frac{a \times b + c \times d \times \mu^2}{\mu \times c \times \Delta u + a \times \Delta v} \end{cases}$$

(4)

where $a, b, c,$ and d are calculated using

$$\begin{cases} a = \frac{Dl}{D^2 + l^2} - \arctan\frac{l}{D} + \frac{l}{2D} \\ b = \frac{l}{2D} - \ln\frac{D-l}{D+l} \\ c = \frac{l}{2D} \\ d = \frac{Dl}{D^2 + l^2} + \arctan\frac{l}{D} - \frac{l}{2D} \\ \Delta u = \varepsilon_h l \\ \Delta v = \varepsilon_v l \end{cases}$$

(5)

E_x is the horizontal tensile resilient modulus; E_y is the vertical compressive resilient modulus; P is the indirect tensile load; D indicates the diameter of the sample; L indicates the thickness of the sample; l indicates the strain gauge effective length; ε_h is the average tensile resilient strain measured by horizontal radial strain gauge; ε_v is the average compressive resilient strain measured by the vertical radial strain gauge; and $a, b, c,$ and d are all intermediate calculation variables.

4. Test Results and Analysis

Asphalt is a representative viscoelastic material that has different mechanical performances under diverse temperatures, loading frequencies, and stress states [18]. In order to accurately analyze the difference in asphalt tensile and compressive resilient modulus under different loading frequencies, modulus tests were carried out at different loading frequencies.

4.1. Contrastive Analysis of Modulus Test Results

The average value of the moduli from the last five cycles was defined as the dynamic resilient modulus of the asphalt. Dynamic moduli obtained by specified test methods are plotted in Figure 4. The comparison of six kinds of dynamic loading state under different frequencies is shown in Figure 5. The fitting curve of different dynamic resilient moduli under different loading frequencies is shown in

Figure 6. Fitting parameters of different dynamic resilient moduli under different loading frequencies is shown in Table 5.

The dynamic resilient modulus from the four test methods increased with loading frequency. As the loading frequency increased, the increase rate of each modulus slowed down. The modulus value rose with the increase of frequency under different stress states. The following phenomena were found under the same frequency: unconfined compressive modulus > indirect tensile modulus > direct tensile modulus > four-point bending flexural modulus (Figure 4).

The dynamic tension and compression-resilient modulus from the indirect tensile test and four-point bending test increased with increasing loading frequency, and the fastest increase was within 0.1–1 Hz.

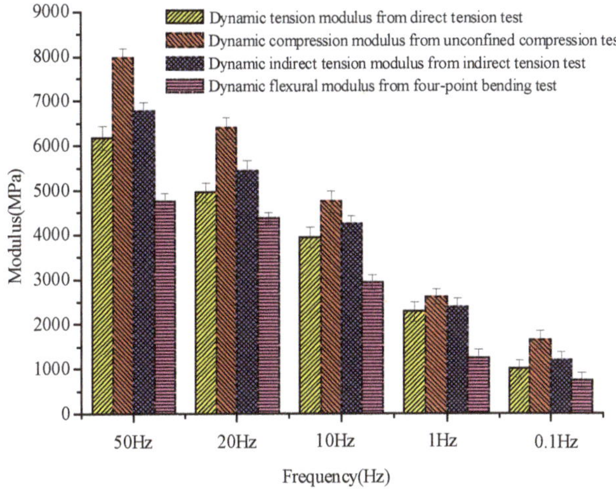

Figure 4. Dynamic moduli obtained by specified test methods.

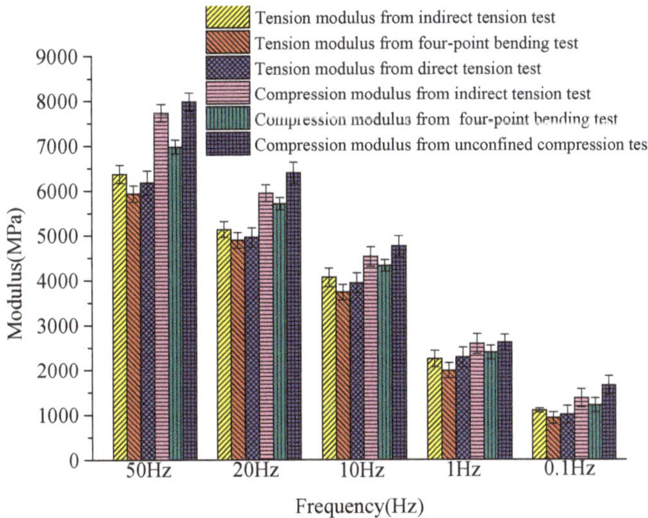

Figure 5. Comparison of six kinds of dynamic loading state.

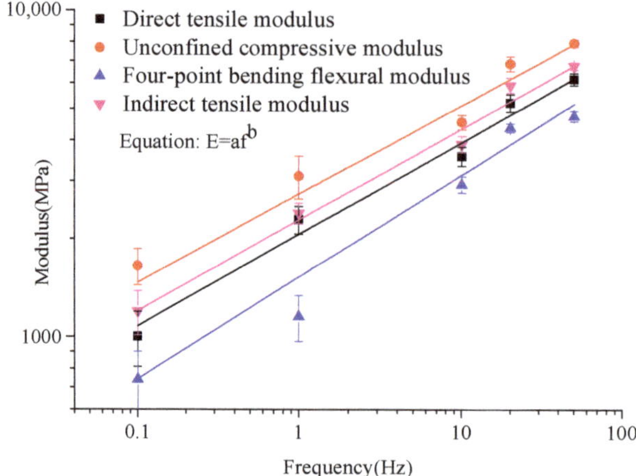

Figure 6. Fitting curve of different dynamic resilient moduli under different loading frequencies.

Table 5. Fitting parameters of different dynamic resilient moduli under different loading frequencies.

Dynamic Modulus	*a*	*b*	Correlation Coefficient
Direct tensile	2062.2475	0.2811	0.85
Unconfined compressive	2745.2190	0.2711	0.84
Flexural tensile	1529.8759	0.3122	0.80
Indirect tensile	2287.3226	0.2787	0.85

Compression resilient modulus divided by tension resilient modulus from the four-point bending test was similar under the five various loading frequencies, and the average value was approximately 1.20. Compression resilient modulus divided by tension resilient modulus from the indirect tensile test was similar for all loading frequencies, and the average value was approximately 1.18. Compression resilient modulus divided by tension resilient modulus from the direct tension experiment and unconfined compression experiment were similar for all different loading frequencies, and the average value was approximately 1.28. The compression resilient modulus and tension resilient modulus showed a similar relationship. The tension and compression moduli of the asphalt were notably differentiated.

The dynamic compression resilient modulus from the four-point bending test and the indirect tensile experiment was similar to the dynamic compression resilient modulus from the unconfined compression test (Figure 5). The dynamic tension resilient modulus from the four-point bending test and the indirect tensile experiment was similar to the dynamic tension resilient modulus from the direct test. Modulus experimental results had high susceptibility to sample form, size, and forced state. Notably, the modulus revealed variability and discreteness under various forced states. Nevertheless, even at identical forced state and loading frequency, the modulus of specimens can be relatively discrete. Asphalt modulus increased exponentially with loading frequency under various stress states (Figure 6). The stress state affected parameters *a,b* in the modulus equation, and the size of coefficients *a* and *b* varied widely. Coefficient *a* represents the declivity of the fatigue curve, and the parameter ambition represents the sequence of unconfined compressive > indirect tensile > direct tensile > flexural tensile. The power function modulus equations of modulus varied greatly inside diverse stress states.

There was a modulus deviation in the different test results. It could be attributed to the difference in the stress states under different test conditions. A large number of studies have shown that the modulus of asphalt is different when using different testing methods that have diverse stress conditions. It is easy to see that the compression and tension resilient moduli obtained from the four tests were similar. To reduce or even eliminate the effect of different forced status, loading frequency, specimen form, and specimen size, it is necessary to establish a standardized model for the stiffness characteristics of asphalt inside diverse stress states, on account of modulus and loading frequency variation.

4.2. Standardization Analysis

Data standardization includes two aspects: similar trends of data processing and dimensionless processing. Making the trend of data processing the same mainly solves the problem of different data properties, as the dimensionless processing of data makes data comparable.

Standardization analysis involves mapping the data to a range from 0 to 1 to remove the unit limit of the data and convert it into a dimensionless pure value for comparison with the indicators of diverse units and magnitudes. The data standardization process does not change the physical meaning of the data, but makes it comparable under the same scale. After standardization, the modulus of each asphalt mixture for each different loading state was converted into dimensionless parameter values that could be comprehensively analyzed.

Data pre-processing is an effective method to solve the problems related to the original data, which paves the way for further processing of the original data [19]. Standardization is one of the familiar data pre-processing methods for establishing classification and regression models for most researchers [20,21]. In min–max normalization, features are normalized in the range [0, 1] using the following equation:

$$v' = \frac{v - \min A}{\max A - \min A} \tag{6}$$

$\min A$ represents the minimum values of feature A. $\max A$ represents the maximum values of feature A. The original value of data A is expressed in v. The normalized processing value of data A is expressed as v'. The maximum eigenvalue and minimum eigenvalue are mapped to 1 and 0, respectively. A standardization model for the tensile and compressive characteristics of asphalt inside diverse stress states and a standardization model for the compressive, tensile, indirect tensile, and flexural characteristics of asphalt under diverse stress states was established based on the variation of modulus and loading frequency. The loading frequency corresponding to the 60 km/h speed of the vehicle was about 10 Hz, so 0.1 and 50 Hz were chosen as the limiting conditions in the test, considering the actual state of vehicle speed. The standardization of tensile ratio and compressive ratio for the dynamic loading state under different frequencies are shown in Figures 7 and 8. The standardization of modulus ratio versus frequency ratio achieved by the specified test methods is shown in Figure 9.

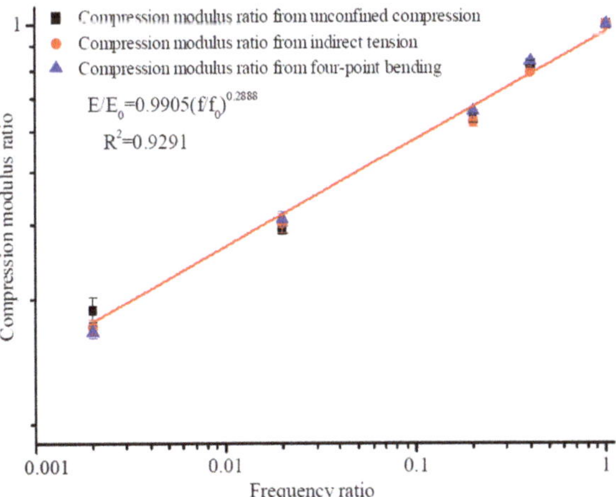

Figure 7. Standardized variation patterns for compression modulus ratio with frequency ratio.

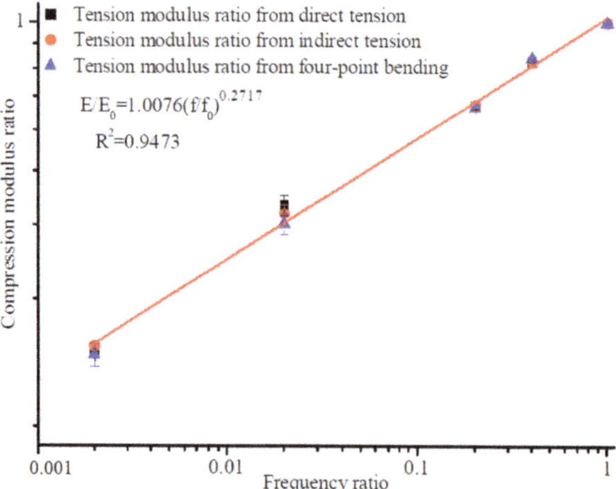

Figure 8. Standardization variation patterns for tension modulus ratio with frequency ratio.

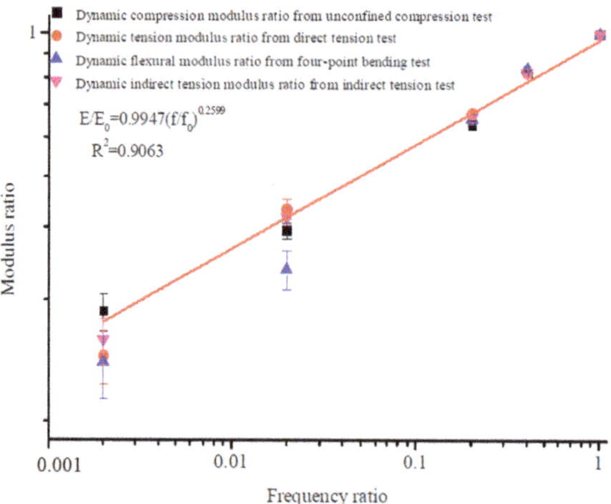

Figure 9. Standardization variation patterns for modulus ratio with frequency ratio.

Table 6. Fitting results of dynamic resilient moduli ratio by standardization under different loading frequencies ratio.

Modulus Ratio	a	b	Correlation Coefficient
Compressive dynamic modulus	0.9905	0.2888	0.93
Tensile dynamic modulus	1.0076	0.2717	0.95
Dynamic modulus	0.9947	0.2599	0.91

The fitting results obtained from Figures 7–9 inside diverse stress states showed a strong linear correlation in the modulus and frequency ratios, and the fitting correlation coefficient was extremely high. Comparisons with conventional dynamic modulus test results and modulus ratio also continuously increased with frequency ratio [6]. The difference in fitting results under diverse stress states, on account of the new approach of stiffness analysis, was extremely lessened, and it was difficult to indicate the difference in stiffness experiment results from one stress state to another. Any of the four moduli can be directly substituted into the standardization model for the flexural, compressive, tensile, and indirect tensile characteristics of asphalt mixture under diverse stress states to obtain the modulus at other frequencies.

On account of the new method of stiffness analysis, in the standardized modulus equation, the discreteness in modulus equation coefficients inside diverse stress states was greatly reduced. The values of coefficients a and b were very close to each other (Table 6), and the values of coefficient varied with the experimental conditions and stress status [21,22].

Consequences achieved from the standardization analysis modulus test revealed that modulus ratio and frequency ratio in diverse stress states were synchronously fit by the same frame of reference. Consequently, for modulus features, the standardization analysis approach is solves the conundrum that commonly plagues conventional modulus test results, and is a probable option for accurately and consistently characterizing the modulus features of asphalt inside diverse stress conditions. The united form of the standardized stiffness equations for various frequencies was obtained. The goal of standardizing modulus characteristics inside diverse stress states was accomplished, and provided a theoretical model and technical foundation for the scientific transformation of material modulus to structural modulus [23].

There is a hypothetical relationship, as follows:

$$\frac{E}{E_0} = a\left(\frac{f}{f_0}\right)^b, \tag{7}$$

where E is the dynamic modulus from the test; f is the loading frequencies; and a and b are fitting parameters. The fitting process on the logarithmic frequency scale is shown in Figures 7–9.

5. Conclusions

The aim of this paper was to develop a standardized characterization method for different moduli by using synchronous testing methods to obtain the tension and compression moduli of asphalt at different loading frequencies. Based on the closed-form solutions and laboratory measurements, the following conclusions can be drawn:

(1) The standardized model can be described as $\frac{E}{E_0} = a\left(\frac{f}{f_0}\right)^b$.

(2) The use of a standardized equation of modulus characteristics of asphalt materials under diverse stress states reduces the influence of stress status, loading frequency, sample shape, and specimen size.

(3) The standardized model takes into account the traditional tension and compression tests, indirect tension tests, and four-point bending tests in both tension and compression states. The standardized model can truly reflect the deformation resistance of asphalt pavement and eliminate the difference between tension and compression.

(4) The standardized model obtained in this paper can avoid repeated modular tests under different stress states. When the single modulus of a given frequency is substituted into the model, the modulus of any other frequency can be obtained, thus saving resources and improving the efficiency of parameter acquisition.

The preceding conclusions are limited to only one temperature. However, the resilience modulus of asphalt is significantly affected by temperature. To expand the application of this method to asphalt pavement structure design, corresponding tests for different temperatures should be performed.

Author Contributions: Conceptualization, X.F. and S.L.; methodology, S.L. and N.Z.; software, X.F. and S.L.; validation, X.F., S.L. and N.Z.; formal analysis, C.X. and Y.L.; investigation, C.X. and Y.L.; resources, S.L.; data curation, X.F., S.L. and N.Z.; Writing—Original Draft preparation, X.F., S.L. and N.Z.; Writing—Review and Editing, X.F., S.L. and N.Z.; visualization, X.F., S.L. and N.Z.; supervision, S.L.; project administration, S.L.; funding acquisition, S.L.

Funding: This work was supported by National Natural Science Foundation of China(51578081, 51608058); The Ministry of Transport Construction Projects of Science and Technology (2015318825120); Key Projects of Hunan Province-Technological Innovation Project in Industry (2016GK2096); The Inner Mongolia Autonomous

Region Traffic and Transportation Department Transportation Projects of Science and Technology (NJ-2016-35, HMJSKJ-201801) and The Hunan Province Transport Construction Projects of Science and Technology (201701).

Conflicts of Interest: The authors declare no conflict of interest.

References

1. Gao, Y.; Geng, D.; Huang, X.; Li, G. Degradation evaluation index of asphalt pavement based on mechanical performance of asphalt mixture. *Constr. Build. Mater.* **2017**, *140*, 75–81. [CrossRef]
2. Karami, M.; Nikraz, H.; Sebayang, S.; Irianti, L. Laboratory experiment on resilient modulus of BRA modified asphalt mixtures. *Int. J. Pavement Res. Technol.* **2018**, *11*, 38–46. [CrossRef]
3. Zoorob, S.E.; Suparma, L.B. Laboratory design and investigation of the properties of continuously graded Asphaltic concrete containing recycled plastics aggregate replacement (Plastiphalt). *Cem. Concr. Compos.* **2000**, *22*, 233–242. [CrossRef]
4. Loulizi, A.; Flintsch, G.; Al-Qadi, I.; Mokarem, D. Comparing resilient modulus and dynamic modulus of hot-mix asphalt as material properties for flexible pavement design. *Trans. Res. Rec.* **2006**, *1970*, 161–170. [CrossRef]
5. Shu, W.G.; You, Z.; Williams, R.C.; Li, X. Preliminary dynamic modulus criteria of HMA for Field rutting of asphalt pavements: Michigan's experience. *J. Trans. Eng.* **2011**, *137*, 37–45.
6. Lv, S.; Wang, S.; Liu, C.; Zheng, J.; Li, Y.; Peng, X. Synchronous testing method for tension and compression moduli of asphalt mixture under dynamic and static loading states. *J. Mater. Civ. Eng.* **2018**, *30*, 04018268. [CrossRef]
7. Lv, S.; Liu, C.; Chen, D.; Zheng, J.; You, Z.; You, L. Normalization of fatigue characteristics for asphalt mixtures under different stress states. *Constr. Build. Mater* **2018**, *177*, 33–42. [CrossRef]
8. Roberts, F.L.; Kandhal, P.S.; Brown, E.R.; Lee, D.Y.; Kennedy, T.W. *Hot Mix Asphalt Materials, Mixture Design, and Construction*, 2nd ed.; National Asphalt Pavement Association Education Foundation: Lanham, MD, USA, 1996.
9. Vega-Zamanillo, Á.; Calzada-Pérez, M.A.; Sánchez-Alonso, E.; Gonzalo-Orden, H. Density, adhesion and stiffness of warm mix asphalts. *Procedia Soc. Behav. Sci.* **2014**, *160*, 323–331. [CrossRef]
10. Woszuk, A.; Franus, W. Properties of the warm mix asphalt involving clinoptilolite and Na-P1 zeolite additives. *Constr. Build. Mater.* **2016**, *114*, 556–563. [CrossRef]
11. Woszukm, A.; Franus, W. A review of the application of zeolite materials in warm mix asphalt technologies. *Appl. Sci.* **2017**, *7*, 293. [CrossRef]
12. Yao, H.; Dai, Q.; You, Z.; Bick, A.; Wang, M. Modulus simulation of asphalt binder models using Molecular Dynamics (MD) method. *Constr. Build. Mater.* **2018**, *162*, 430–441.
13. Lv, S.; Liu, C.; Yao, H.; Zheng, J. Comparisons of synchronous measurement methods on various moduli of asphalt mixtures. *Constr. Build. Mater* **2018**, *158*, 1035–1045. [CrossRef]
14. M.o.T.o.t.P.s.R.o (Ministry of Transport of the People's Republic of China). *Specifications and Test Methods of Bitumen and Biminous Mixtures for Highway Engineering*; (JTG E20-2011); Renmin Communication Press: Beijing, China, 2011.
15. M.o.T.o.t.P.s.R.o (Ministry of Transport of the People's Republic of China). *Specifications for Design of Highway Asphalt Pavement*; (JTG D50-2017); Renmin Communication Press: Beijing, China, 2017.
16. Xing, C.; Tan, Y.; Liu, X.; Anupam, K.; Scarpas, T. Research on local deformation property of asphalt mixture using digital image correlation. *Constr. Build. Mater.* **2017**, *140*, 416–423. [CrossRef]
17. Specht, L.P.; Babadopulos, L.F.d.A.L.; Di Benedetto, H.; Sauzéat, C.; Soares, J.B. Application of the theory of viscoelasticity to evaluate the resilient modulus test in asphalt mixes. *Constr. Build. Mater* **2017**, *149*, 648–658. [CrossRef]
18. Lv, S.; Luo, Z.; Xie, J. Fatigue performance of aging asphalt mixtures. *Polimery* **2015**, *60*, 126–131. [CrossRef]
19. Jain, S.; Shukla, S.; Wadhvani, R. Dynamic selection of normalization techniques using data complexity measures. *Expert Syst. Appl.* **2018**, *106*. [CrossRef]
20. Datta, S.N. Min-max and max-min principles for the solution of 2 + 1 Dirac fermion in magnetic field, graphene lattice and layered diatomic materials. *Chem. Phys. Lett.* **2018**, *692*, 313–318. [CrossRef]
21. Shao, G.; Sang, N. Regularized max-min linear discriminant analysis. *Pattern Recognit.* **2017**, *66*. [CrossRef]

22. Virgil Ping, W.; Xiao, Y. Empirical correlation of indirect tension resilient modulus and complex modulus test results for asphalt concrete mixtures. *Road Mater. Pav. Des.* **2008**, *9*, 177–200. [CrossRef]
23. Lv, S.; Wang, X.; Liu, C.; Wang, S. Fatigue damage characteristics considering the difference of tensile-compression modulus for asphalt mixture. *J. Test. Eval.* **2018**, *46*, 20170114. [CrossRef]

Article

Damage Detection of Asphalt Concrete Using Piezo-Ultrasonic Wave Technology

Wen-hao Pan [1,2]**, Xu-dong Sun** [1]**, Li-mei Wu** [2]**, Kai-kai Yang** [2] **and Ning Tang** [2,*]

[1]   School of Materials Science and Engineering, Northeast University, Shenyang 110819, China;
       pwh@sjzu.edu.cn (W.-h.P.); sxd@neu.edu.cn (X.-d.S.)
[2]   School of Materials Science and Engineering, Shenyang Jianzhu University, Shenyang 110168, China;
       lmwu@sjzu.edu.cn (L.-m.W.); yangkaikai@stu.sjzu.edu.cn (K.-k.Y.)
*   Correspondence: tangning@sjzu.edu.cn; Tel.: +86-24-24690315

Received: 24 December 2018; Accepted: 29 January 2019; Published: 31 January 2019

Abstract: Asphalt concrete has been widely used in road engineering as a surface material. Meanwhile, ultrasonic testing technology has also been developed rapidly. Aiming to evaluate the feasibility of the ultrasonic wave method, the present work reports a laboratory investigation on damage detection of asphalt concrete using piezo-ultrasonic wave technology. The gradation of AC-13 was selected and prepared based on the Marshall's design. The ultrasonic wave velocities of samples were tested with different environmental conditions firstly. After that, the samples were destroyed into two types, one was drilled and the other was grooved. And the ultrasonic wave velocities of pretreated samples were tested again. Furthermore, the relationship between velocity and damaged process was evaluated based on three point bending test. The test results indicated that piezoelectric ultrasonic wave is a promising technology for damage detection of asphalt concrete with considerable benefits. The ultrasonic velocity decreases with the voidage increases. In a saturated water environment, the measured velocity of ultrasonic wave increased. In a dry environment (50 °C), the velocity the ultrasonic waves increased too. After two freeze-thaw cycles, the voidage increased and the ultrasonic velocity decreased gradually. After factitious damage, the wave must travel through or most likely around the damage, the ultrasonic velocity decreased. During the process of three point bending test, the ultrasonic velocity increased firstly and then decreased slowly until it entered into a steady phase. At last the velocity of ultrasonic wave decreased rapidly. In addition, the errors of the results under different test conditions need to be further studied.

Keywords: asphalt concrete; damage detection; ultrasonic wave; velocity

---

1. Introduction

Asphalt concrete is a widely used material for roads and pavements owing to its advantages like seamlessness, easiness of repair, high driving comfortability and lower noise. In addition to exposure to harsh environmental elements during service, traffic loads, sunlight, rain and air result in premature failure of pavement by rutting, raveling and cracking. Hence, the inevitable aging of transportation infrastructure create significant security and economic risks. This raises a new concept, preventative maintenance [1–4]. It is an important way for extending the service life of pavement in a cost effective manner. It change the passiveness to positive at the appropriate time.

For pavement maintenance, how to identify the damage and fatigue of pavement and make a cost-effective decision is a difficult task for engineers, especially appropriate time and frequency of maintenance. In the traditional process of evaluating and inspecting, the core samples are always drilled and taken to the laboratory to obtain the corresponding performance through standard test, so that the identify process is complicated and the test period is too long [5–8]. Furthermore, traditional maintenance is a passive way to repair the pavement when the pavement structure damage. It is very important to find a way to identify the maintenance time accurately and quickly. Therefore, some new non-destructive testing (NDT) techniques are applied in pavement assessment, such as ground-penetrating radar (GPR), spectral analysis of surface wave (SASW) and ultrasonic testing technology (UT) [9–13].

GPR is an electromagnetic wave method for detecting the internal structure of pavement by using high frequency pulse electromagnetic field, which can provide high resolution 2D and 3D under pavement images [14,15]. SASW has been widely used as a non-destructive testing method for underground exploration and geological experiment. The dispersion characteristics of the propagating surface wave in the pavement are measured by the phase difference between the two sensing channels. The dispersion curve is obtained by data transmission. The shape of the pavement and the shear velocity are estimated by the inversion program [16,17]. However, these two methods need to be tested in the field and analyzed in laboratory. Hence, how to realize the complete, fast and automatic recognition of the pavement damage in the field and how to guarantee the high accuracy rate, are the research topic that the relevant scholars need to solve urgently at present.

Ultrasonic testing technology is one of the most widely used nondestructive testing methods. The principle of ultrasonic testing technology is to excite elastic waves in engineering structures or materials through ultrasonic transducers. The elastic waves propagate in materials or structures with various waveforms and another transducer receives them. The characteristic parameters of elastic wave are directly related to the properties of materials or structure such as time, wave velocity, wave amplitude and wave shape.

The application of ultrasonic testing technology is in the rock and cement concrete materials [18,19] but the application in asphalt mixture is not as many [20–25]. The research is still in the exploratory stage, such as the relationship between ultrasonic velocity and airvoid, elastic modulus, fatigue life and so on. Research still needs to supplement more data to improve the research system, so as to achieve qualitative to quantitative transformation.

In this study, the gradation of AC-13 was selected and prepared based on the Marshall's design. The samples included Marshall specimens and beams. The ultrasonic wave velocities of samples were tested with different environmental conditions firstly like wet, dry and freeze-thaw. After that, the samples were destroyed into two types, one was drilled and the other was grooved. The ultrasonic wave velocities of pretreated samples were tested again. Furthermore, the relationship between velocity and damaged process was evaluated based on three point bending test.

2. Materials and Methods

2.1. Materials

Bitumen, AH-90 paving grade, was obtained from Liaohe Oilfield of Panjin City, China. The physical properties of the bitumen are given in Table 1.

Table 1. Properties of bitumen.

Property	Unit	Results	Technical Requirements	Test [26]
Penetration	dmm	88.2	80–100	T0604
Ductility	cm	127	$\geq$100	T0605
Softening point	°C	49.1	$\geq$45	T0606
Density	g/cm^3	1.03	–	T0603

2.2. Preparation of Asphalt Concrete

Gradation of AC-13 was selected and 24 samples were prepared for optimal binder percentage according to Marshall's design. The specimens were cylinders with a diameter of 101.6 mm and a height of 63.5 mm. The passing rate of AC-13 gradation for each sieve is given in Table 2. Depending on the results optimal binder percentage was 4.9%.

Table 2. Aggregate gradation for AC 13.

Sieve Size (mm)	16	13	9.5	4.75	2.36	1.18	0.6	0.3	0.15	0.075
Passing (%)	100	94.8	77.1	48.6	30.3	22.8	16.2	11.4	7.6	6.1

2.3. Experimental Methods

- Piezo-ultrasonic velocity test

The ultrasonic velocity test was carried out on TICO tester (Tectus Group, Zurich, Swiss). The testing voltage of ultrasonic wave was 250 V, the testing temperature was 15 °C and the frequency of ultrasonic wave was 30 kHz. Five test areas were selected for each specimen surface and the test results were averaged. In order to avoid the discreteness of collected data, a minimum of ten repeated measurements under the same experimental conditions (statistical sample size) would be necessary. Furthermore, five samples were prepared for each test.

- Computed Tomography

The scan images of asphalt concrete for voidage calculation were analyzed by an X-ray 3D microscope (nanoVoxel-2700, Sanying Precision Instruments Co. Ltd., Tianjin, China).

- Factitious damage

There were two types of factitious damage as shown in Figure 1. One is that a cylindrical through-hole of 5 mm diameter was drilled in the Marshall specimen.

Figure 1. The factitious damage of specimens.

The other is that some grooves with different width were carved on the surface of asphalt beam but the depth of grooves was same, 15 mm. So there are five types of asphalt beam with different grooves as following: groove of 2 mm (G-2 mm), groove of 4 mm (G-4 mm), two grooves of 2 mm (DG-2 mm), vertical groove of 2 mm (VG-2 mm) and half double groove of 2mm (HG-2 mm).

- Drying and water-saturated test

The drying test is that all the samples were placed in an oven at 50 °C for 24 h. The water-saturated test is that all the samples were immersed in water for 24 h. The ultrasonic velocity of samples were measured directly after the environmental pretreatment.

- Freeze-thaw cycle

There were two cycles. All the samples were full immersed in water for half an hour first and then kept vacuum saturation state for 15 min. After pretreatment, the samples were frozen in the refrigerator at −18 °C for 16 h and then took them out and placed in a bath box at 60 °C for melting 6 h [26].

- Three point bending test

The three point bending test was carried out on a material testing machine (WDW-T200, Tianchen, Jinan, China). The distance between the two fulcrums was 200 mm and the loading point was located at the center of the specimen in length direction. The test was carried out at a dead loading rate of 0.2 kN/s. At the same time, the ultrasonic wave detector was applied to monitor the travel velocity in the specimen in real time until the specimen was damaged.

- Healing

After heating, the bitumen was coated on the cross section of the asphalt beam and then the two cross section were quickly bonded together. At last the healed asphalt beams were placed in an oven at 50 °C for one hour.

3. Results and Discussion

3.1. Ultrasonic Test

As a key factor, air voids has been paid much attention due to it is closely related to the road performance. The pavement with larger voidage has better structure depth, anti-skid performance and friction coefficient but the performance of anti-seepage, anti-freezing and anti-moisture is poor. Therefore, to establish the relationship between voidage and piezoelectric ultrasonic velocity is the first prerequisite for using piezoelectric ultrasonic technology. Based on the same gradation, different compaction temperature is used to control the voidage of asphalt concrete. The relationship between voidage and ultrasonic velocity are shown in Figure 2.

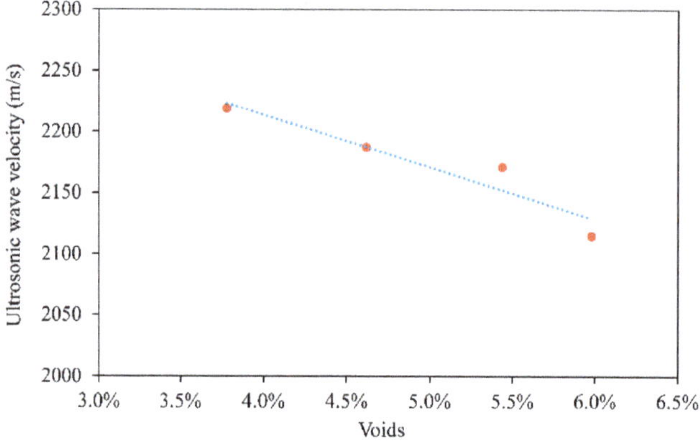

Figure 2. Relationship between ultrasonic wave velocity and voidage of the Marshall samples.

In Figure 2, the data points and trend line show the relationship between the ultrasonic wave velocity and the voidage of the Marshall samples. It is well known that the lower the compaction temperature, the higher the voidage. The voidage was 3.78%, 4.62%, 5.44% and 5.98%, respectively. Furthermore, the ultrasonic velocity decreased with the voidage increased. Obviously, there are a lot of voids in the asphalt mixture and the wave travels much faster in the air than in the aggregate, the higher the porosity, the smaller the velocity. In addition, the dispersion of the longitudinal wave was larger due to the high voidage, resulting in long travel time of ultrasonic wave and low wave velocity.

Usually the voidage of AC-13 asphalt concrete is controlled at 4–6%. In the following tests, the voidage of all Marshall specimens were controlled at 4.5% by compaction temperature. The ultrasonic velocities of Marshall specimens with different environmental conditions show in Figure 3.

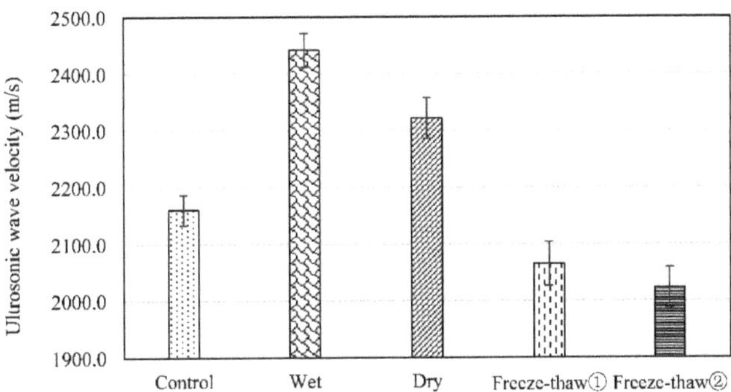

Figure 3. Ultrasonic velocity of Marshall specimens with different environmental conditions.

In a saturated water environment, the average velocity of specimens was 2442 m/s. The reason is that the water in the pores increases and the pores are basically filled with water. In addition, the velocity of ultrasonic wave propagation in water was much faster than that in air. The increase of water caused the measured wave velocity to become bigger.

In a dry environment (50 °C), the average velocity of specimens was 2321 m/s. This data was higher than that of the control group and lower than that of the saturated environment. For bitumen, the higher the temperature is, the more obvious the viscoelasticity is. In addition, the dry temperature is higher than the softening point of bitumen. The ultrasonic waves face a greater viscous resistance and the interference and resistance to the propagation of ultrasonic waves are bigger.

Furthermore, the velocity decreased in turn after two freeze-thaw cycles. Bitumen shows viscoelasticity at high temperature and elasticity at low temperature and occurrence of aging in the repeated process of high temperature and low temperature. In addition, the voids of asphalt concrete is occupied by water in the saturation environment. When water freezes, it becomes larger in volume instead of smaller and there is not enough void volume to accommodate the increased volume, so the expansion stress is generated in the void, so that the voidage increased and the ultrasonic velocity decreased gradually and the asphalt mixture was damaged at last.

3.2. Damage Detection

All the tests were repeated after the Marshall specimens had been drilled. The ultrasonic data collection included five receiver positions (P1 to P5) for one generator position (P0). Obviously, the distance at which the ultrasound wave travels was different and symmetrical, except P3, due to the damage. The test results show in Figure 4.

Figure 4 shows typical traces for an ultrasonic velocity at different receiver positions. For different environment, the collected ultrasonic data at each receiver position were similar to the data before the samples were damaged. The velocity increases under saturated and dry conditions and decreases after freeze-thaw cycles. In addition, with the increase of the wave travel length, the velocity of the ultrasonic wave in the dry environment increases rapidly, the maximum data is 4027 m/s.

For different receiver, the wave moved from generator P0 to receiver P4, P2 and P1, the velocity increased (the travel lengths also increase). At receiver P1, the velocity of all samples further increased. On the symmetrical side with the damage, similarly the plot of the average measured velocities showed a dip in the measured velocity caused by the presence of damage. Note that as the velocity of receiver P3 was slower than that of P2 and the trends of average velocity were expected. This is clearly because the wave must travel through or most likely around the damage.

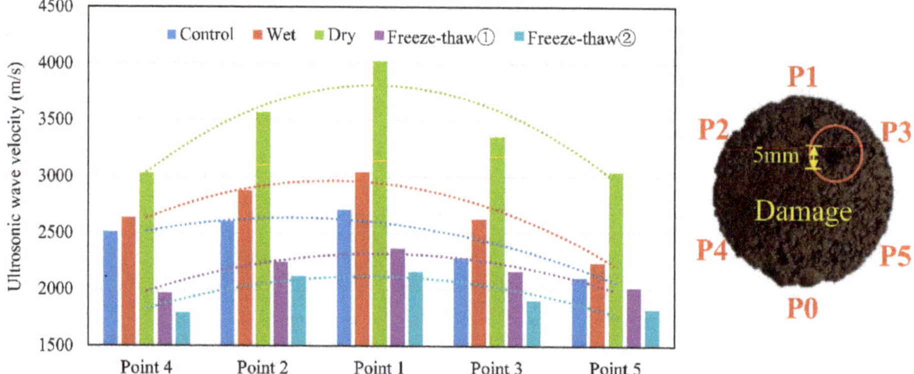

Figure 4. Ultrasonic velocity with different environmental conditions after factitious damage.

But the contact between side surface of the cylinder specimen and the sensor probe is not a complete contact but a linear contact. It can result in minor inaccuracies. To further study the influence of cracks on the velocity of ultrasonic waves, four types of grooves are carved on the different asphalt beam specimens. Figure 5 shows the test results of velocity with simulated cracks.

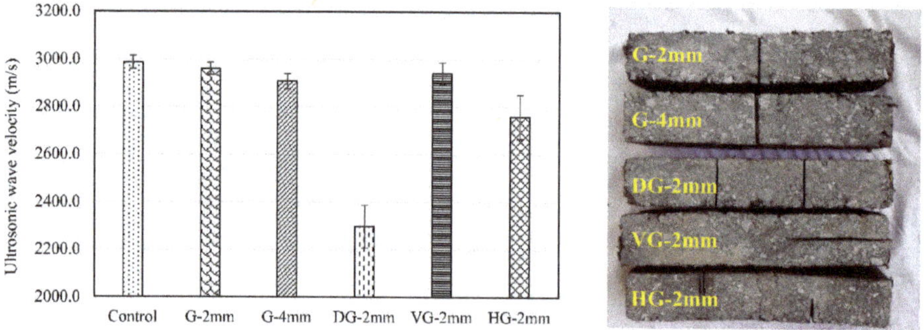

Figure 5. The velocity of ultrasonic waves with five types of grooves.

In Figure 5, for cracks width, ultrasonic velocity slightly decreased when 1mm grooves appear on the surface of asphalt concrete beam specimens. But when the crack width increases to 2 mm, the ultrasonic velocity decrease more. Note that the effect of fine cracks on ultrasonic velocity is slight and the ultrasonic velocity decreased with the increased of crack width.

Furthermore, with the presence of two grooves, the ultrasonic velocity decreased obviously and the deviation of the test results was 90.74 m/s, which was very large. When an ultrasonic wave passed through a crack, only part energy was transmitted and the air inside the crack attenuated the ultrasonic wave. When the ultrasonic wave passed through the next crack, the velocity decreased even more. In addition, the anisotropic property of asphalt concrete makes the instability of the results. Because even if the grooves were in the same position, the aggregate and bitumen content of different bitumen concrete beams were also different at the grooves.

Moreover, the two grooves have been halved. The test produced the velocity of ultrasonic wave ranging from one groove to two grooves. But the deviation of results was 93.21 m/s, which was very large, too. It revealed that there are significant positive correlations among simulated crack length, simulated crack position and ultrasonic velocity.

At last, asphalt pavement damage has not only transverse crack but also longitudinal crack. One groove was carve on the surface of asphalt beam as simulated vertical crack. The result shows the ultrasonic velocity decreased slightly. Hence, when the travel direction of ultrasonic wave was parallel to the crack direction, the change of ultrasonic velocity was slight.

3.3. Three Point Bending Test

Three-point bending test is one of the methods to evaluate the fatigue and damage of asphalt concrete. The velocity of ultrasonic wave was measured when asphalt beam specimens were applied a load until failure. Figure 6 shows the change trend of ultrasonic velocity during the whole process of damage. In Figure 6, according to the slope of curve, the damage process of asphalt beam can essentially be broken down into four phases: loading, crack growth, balance and damage.

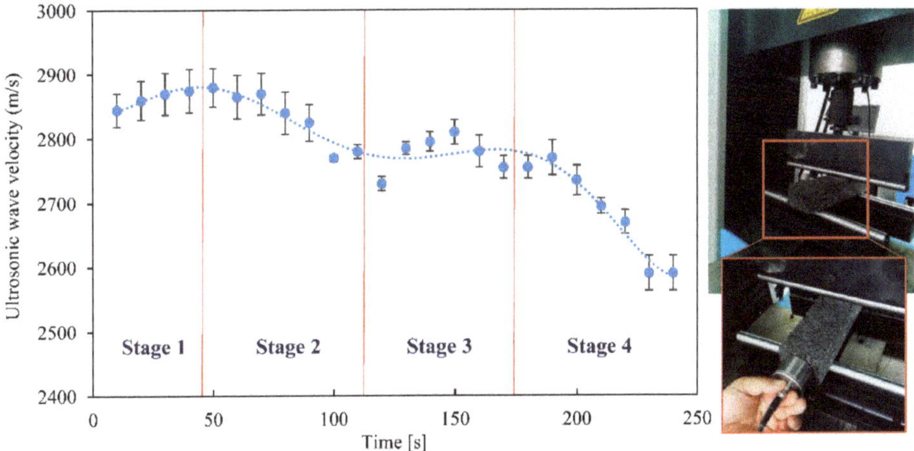

Figure 6. The change trend of ultrasonic velocity during the whole process of damage.

In the stage 1, the ultrasonic velocity increased firstly. In the first 50 s, the ultrasonic velocity increased slightly, because the flexural-tensile stress was applied to the asphalt beam, the travel lengths of ultrasonic wave increased. In addition, asphalt mixture is a typical heterogeneous material. Although the interface conditions between aggregate and bitumen is good, the mechanical properties of these two materials are different, so the deformation of asphalt mixture is very complex. Moreover the voids in the asphalt mixture occurred about 4.5% of the volume and these voids were crowded during the loading process.

In the stage 2, with the time increase (50 s to 100 s), the velocity of ultrasonic wave decreased slowly due to presence of some micro-cracks in the asphalt concrete beam. The growth rate of micro-cracks was higher than the compacted rate of the voids in asphalt concrete beam.

In the stage 3, after 100 s, the velocity of the ultrasound wave barely changed and it had entered into a steady phase. This is clearly because the displacement of the aggregate and bitumen is caused by the deformation of asphalt beam at the same time. And the growth rate of micro-cracks was close to the compacted rate of the voids.

In the stage 4, the velocity of ultrasonic wave decreased rapidly after 170 s. The micro-crack width increased visibly and cracks appeared on both sides of the larger aggregate at the lower edge of the asphalt beam. The crack grew to the critical state of penetrating and then the asphalt beam was damaged.

In addition, the velocity of ultrasonic wave in the width travel direction of the asphalt beam was measured before and after the three-point bending test. Moreover, the damaged asphalt beam was coated with new bitumen and heated up for healing. All the results were showed in Figure 6.

In Figure 7, with the increase of received length, the velocity of ultrasonic wave decreased one by one in any case. In addition, the damaged asphalt beam cannot receive ultrasonic wave at the receiving point P4 but the ultrasonic wave can be detected again after healing. It indicates that ultrasonic testing technology can be used to detect the cracking state of asphalt concrete effectively.

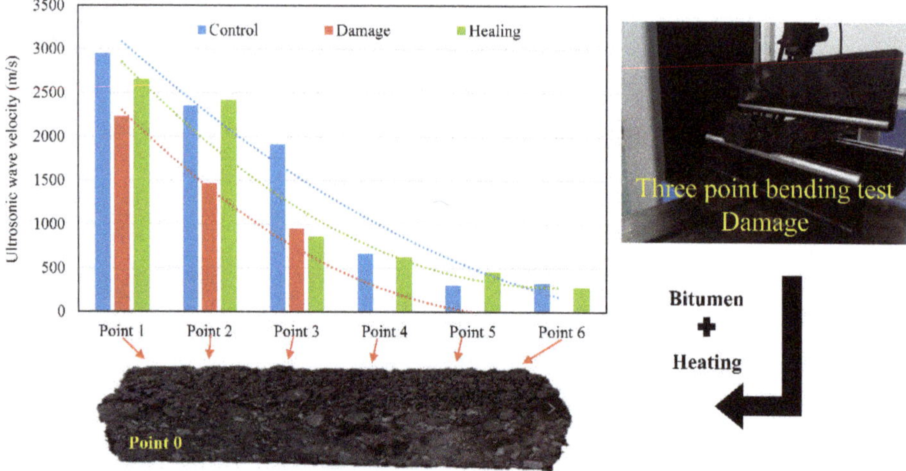

Figure 7. The velocity of ultrasonic wave in the width travel direction of the asphalt beam.

3.4. Impact Analysis

Asphalt mixture is a typical heterogeneous material, so different measured positions, different results. In order to analyze the errors, the Marshall specimens were cut into a cube with a length of 5 cm on the side. This size is close to the size of the piezoelectric ultrasonic probe. The CT scan images are shown in Figure 8. Through the calculation of image processing software (nanoVoxel-2700 system), the voidage of three specimens is 4.08% (A), 4.32% (B) and 4.56% (C), respectively.

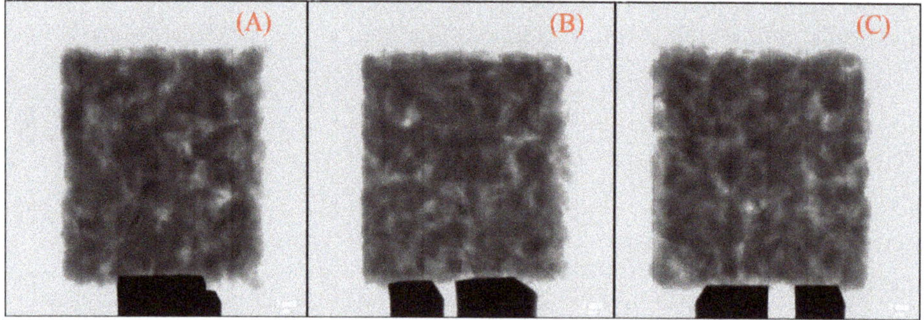

Figure 8. The CT images of cube samples with different voidage: (**A**) 4.08%, (**B**) 4.32%, (**C**) 4.56%.

The ultrasonic velocity of three specimens was 2275 m/s, 2221 m/s and 2189 m/s, respectively. The relationship between velocity and voidage remains unchanged but the value was lower than that of the Marshall specimen. Hence, voidage, temperature sensitivity, viscoelasticity, travel length of ultrasonic wave in asphalt concrete and the size of specimens are main factors affecting the ultrasonic velocity.

Compared with other studies, the ultrasonic velocity is lower than 3000 m/s in ambient [21–25]. Based on the impact analysis, it reveals that the ultrasonic velocity is the result of the combined effect of temperature, environment condition and service time. Furthermore, it is still contingency because of the reflection, attenuation, resonance and changes of velocity that occurred in an object when the propagation of ultrasonic wave. Hence the research requires a variety of supplementary data. The results of this paper are a useful supplement.

4. Conclusions

This study has indicated that piezoelectric ultrasonic wave is a promising technology for damage detection of asphalt concrete with considerable benefits. The test results demonstrated that the ultrasonic velocity decreased with the voidage increases. Furthermore, the increase of humidity and temperature lead to an increase of ultrasonic velocity. Oppositely, after two freeze-thaw cycles, the voidage increased and the ultrasonic velocity decreased gradually.

For factitious cylinder damage, the velocity of ultrasonic wave decreased at damaged position. For factitious groove damage, ultrasonic velocity decreased slightly with the increased of crack width but decreased obviously with the increased of crack quantity. In addition, the change of ultrasonic velocity was slight when the travel direction of ultrasonic wave was parallel to the crack direction.

For the relationship between ultrasonic velocity and the damaged process of asphalt beam, the damaged process can essentially be broken down into four phases: loading, crack growth, balance and damage. Correspondingly, the ultrasonic velocity increased firstly and then decreased slowly until it entered into a steady phase. At last the velocity of ultrasonic wave decreased rapidly.

In addition, voidage, temperature sensitivity, viscoelasticity, travel length of ultrasonic wave in asphalt concrete and the size of specimens are main factors. The errors of the results under different test conditions need to be further studied.

Author Contributions: Conceptualization, W.-h.P. and N.T.; Data curation, W.-h.P., L.-m. W., K.-k.Y. and N.T.; Funding acquisition, L.-m.W. and N.T.; Investigation, W.-h.P., K.-k.Y. and N.T.; Methodology, X.-d.S. and N.T.; Software, W.-h.P. and N.T.; Writing, W.-h.P., X.-d.S., L.-m.W., K.-k.Y. and N.T.

Funding: This research was funded by [National Natural Science Foundation of China] grant number [51508344], [China Postdoctoral Science Foundation] grant number [2016M591458], [Liaoning BaiQianWan Talents Program] grant number [2019] and [Support Plan for innovative talents of Liaoning Province] grant number [201954].

Acknowledgments: Thanks for Science Program and Discipline Content Education Project of Shenyang Jianzhu University. Special thanks Hong Kong Scholar Program and China Scholarship Council for sponsoring a technical visit to Politecnico di Milano.

Conflicts of Interest: The authors declare no conflict of interest.

References

1. Peshkin, D.G.; Hoerner, T.E.; Zimmerman, K.A. Optimal timing of pavement preventive maintenance treatment applications. *J. Transp. Res. Board* **2004**, *523*, 10–15.
2. Giustozzi, F.; Crispino, M.; Flintsch, G.W. Multi-attribute life cycle assessment of preventive maintenance treatments on road pavements for achieving environmental sustainability. *Int. J. Life Cycle Assess.* **2012**, *17*, 409–419. [CrossRef]
3. Haider, S.; Dwaikat, M. Estimating optimum timing for preventive maintenance treatment to mitigate pavement roughness. *J. Transp. Res. Board* **2011**, *2235*, 43–53. [CrossRef]
4. Chou, J.S.; Le, T.S. Reliability-based performance simulation for optimized pavement maintenance. *Reliab. Eng. Syst. Saf.* **2011**, *96*, 1402–1410. [CrossRef]
5. Maierhofer, C.; Arndt, R.; Röllig, M.; Rieck, C.; Walther, A.; Scheel, H.; Hillemeier, B. Application of impulse-thermography for non-destructive assessment of concrete structures. *Cem. Concr. Compos.* **2006**, *28*, 393–401. [CrossRef]
6. Winston, R.J.; Al-Rubaei, A.M.; Blecken, G.T.; Viklander, M.; Hunt, W.F. Maintenance measures for preservation and recovery of permeable pavement surface infiltration rate–The effects of street sweeping, vacuum cleaning, high pressure washing, and milling. *J. Environ. Manag.* **2016**, *169*, 132–144. [CrossRef] [PubMed]
7. Shah, Y.U.; Jain, S.S.; Parida, M. Evaluation of prioritization methods for effective pavement maintenance of urban roads. *Int. J. Pavement Eng.* **2014**, *15*, 238–250. [CrossRef]

8. Meneses, S.; Ferreira, A. Pavement maintenance programming considering two objectives: Maintenance costs and user costs. *Int. J. Pavement Eng.* **2013**, *14*, 206–221. [CrossRef]

9. Usamentiaga, R.; Venegas, P.; Guerediaga, J.; Vega, L.; Molleda, J.; Bulnes, F. Infrared thermography for temperature measurement and non-destructive testing. *Sensors* **2014**, *14*, 12305–12348. [CrossRef]

10. Hoła, J.; Bień, J.; Schabowicz, K. Non-destructive and semi-destructive diagnostics of concrete structures in assessment of their durability. *Bull. Pol. Acad. Sci. Tech. Sci.* **2015**, *63*, 87–96. [CrossRef]

11. Dumoulin, J.; Ibos, L.; Marchetti, M.; Mazioud, A. Detection of non-emergent defects in asphalt pavement samples by long pulse and pulse phase infrared thermography. *Eur. J. Environ. Civ. Eng.* **2011**, *15*, 557–574. [CrossRef]

12. Fauchard, C.; Beaucamp, B.; Laguerre, L. Non-destructive assessment of hot mix asphalt compaction/density with a step-frequency radar: Case study on a newly paved road. *Near Surf. Geophys.* **2015**, *13*, 289–297. [CrossRef]

13. Hunaidi, O. Evolution-based genetic algorithms for analysis of non-destructive surface wave tests on pavements. *NDT E International* **1998**, *31*, 273–280. [CrossRef]

14. Liu, H.; Sato, M. In situ measurement of pavement thickness and dielectric permittivity by GPR using an antenna array. *NDT E Int.* **2014**, *64*, 65–71. [CrossRef]

15. Mai, T.C.; Razafindratsima, S.; Sbartaï, Z.M.; Demontoux, F.; Bos, F. Non-destructive evaluation of moisture content of wood material at GPR frequency. *Construct. Build. Mater.* **2015**, *77*, 213–217. [CrossRef]

16. Kumar, J.; Rakaraddi, P.G. SASW evaluation of asphaltic and cement concrete pavements using different heights of fall for a spherical mass. *Int. J. Pavement Eng.* **2013**, *14*, 354–363. [CrossRef]

17. Hazra, S.; Kumar, J. SASW testing of asphaltic pavement by dropping steel balls. *Int. J. Geotech. Eng.* **2014**, *8*, 34–45. [CrossRef]

18. Rao, S.K.; Sravana, P.; Rao, T.C. Experimental studies in Ultrasonic Pulse Velocity of roller compacted concrete pavement containing fly ash and M-sand. *Int. J. Pavement Res. Technol.* **2016**, *9*, 289–301. [CrossRef]

19. Mandal, T.; Tinjum, J.M.; Edil, T.B. Non-destructive testing of cementitiously stabilized materials using ultrasonic pulse velocity test. *Transp. Geotech.* **2016**, *6*, 97–107. [CrossRef]

20. Hoegh, K.; Khazanovich, L.; Maser, K.; Tran, N. Evaluation of ultrasonic technique for detecting delamination in asphalt pavements. *J. Transp. Res. Board* **2012**, *2306*, 105–110. [CrossRef]

21. Van Velsor, J.K.; Premkumar, L.; Chehab, G.; Rose, J.L. Measuring the Complex Modulus of Asphalt Concrete Using Ultrasonic Testing. *J. Eng. Sci. Technol. Rev.* **2011**, *4*, 160–168. [CrossRef]

22. Mounier, D.; Di Benedetto, H.; Sauzéat, C. Determination of bituminous mixtures linear properties using ultrasonic wave propagation. *Construct. Build. Mater.* **2012**, *36*, 638–647. [CrossRef]

23. Cheng, Y.; Zhang, P.; Jiao, Y. Damage Simulation and Ultrasonic Detection of Asphalt Mixture under the Coupling Effects of Water-Temperature-Radiation. *Adv. Mater. Sci. Eng.* **2013**, *2013*, 1–9. [CrossRef]

24. Tigdemir, M.; Kalyoncuoglu, S.F.; Kalyoncuoglu, U.Y. Application of ultrasonic method in asphalt concrete testing for fatigue life estimation. *NDT E Int.* **2004**, *37*, 597–602. [CrossRef]

25. Norambuena, J.; Castro, D. Dynamic modulus of asphalt mixture by ultrasonic direct test. *NDT E Int.* **2010**, *43*, 629–634. [CrossRef]

26. Ministry of Transport of China. *Standard Test Methods of Bitumen and Bituminous Mixtures for Highway Engineering: JTG E20-2011*; Ministry of Transport of China: Beijing, China, 2011.

Article

Field Investigation of Clay Balls in Full-Depth Asphalt Pavement

Weiguang Zhang [1,*], Jusang Lee [2,*], Hyung Jun Ahn [3], Qiqi Le [1], Meng Wu [1], Haoran Zhu [4] and Jing Zhang [1]

[1] School of Transportation, Southeast University, Nanjing 211189, China
[2] Indiana Department of Transportation, West Lafayette, IN 47906, USA
[3] Virginia Department of Transportation, Richmond, VA 23219, USA
[4] JSTI Group, Nanjing 210019, China
* Correspondence: wgzhang@seu.edu.cn (W.Z.); jlee@indot.in.gov (J.L.); Tel.: +86-139-129-56989 (W.Z.)

Received: 7 August 2019; Accepted: 4 September 2019; Published: 6 September 2019

Abstract: Clay ball is a pavement surface defect which refers to a clump in which clay or dirt is mixed with hot asphalt mixture. Clay ball is typically caused by a combination of aggregate contamination of clay or soil, high aggregate moisture, and low production temperature at the asphalt plant. It usually appears a few weeks or months after paving under traffic load, after being liquefied and knocked from the pavement surface. Clay balls can be the source of potholing, raveling, and other issues such as moisture infiltration and reduced ride quality. This paper presents an investigation of the clay balls on US-31 one winter after construction in Hamilton County, Indiana. In order to understand the pavement condition, their severity was measured using both visual observation and infrared image collection system. In addition, a clay ball amount, its distribution pattern, and cores condition were evaluated. A precipitation effect on clay ball formation was investigated for finding a cause of the clay balls. The investigation found that infrared image collection system was appropriate in detecting the clay balls. The clay balls were elliptic in shape with 2.5 cm to 10 cm in diameter, and the maximum clay ball depth was almost penetrating the entire surface course. It was also found that the asphalt paving on the raining days or right after raining could increase the potential of clay balls. Monitoring of aggregate moisture during construction on or after raining days should be able to reduce the risk of clay balls.

Keywords: clay ball; asphalt pavement; pattern and density; infrared image collection system; field core

1. Introduction

Clay ball, also called dust ball or dust cake, is a pavement surface defect in both asphalt pavement and Portland Cement Concrete pavement. It refers to a clump in which clay or dirt is mixed with hot asphalt mixture. Clay ball may be well coated by asphalt during mixing and compaction process, while no asphalt coats on the exposed surface are found when the ball is partially sliced [1]. A clay ball is small in size and can either appear near pavement surface or inches below pavement surface [2]. A clay ball usually appears a few weeks or months after paving, but some may not appear until a full winter season of freeze-thaw cycles has occurred [3,4].

A clay ball could initiate from aggregate contamination of clay or soil before arriving on the plant site or aggregate pollution from the soil below stockpile. The clay or dirt mixes into the hot asphalt mixture with the coarse or fine aggregates at a drum and creates clumps. A clay ball could also initiate from a combination of moisture in the aggregate and lower temperatures during mixing, then fines can "cake" on the mixing paddles in the drum. In this case, lower temperatures may not be adequate to cook out the moisture in the aggregate. The fines collect with the moisture and stick to these paddles.

These clumps will be placed and compacted in the field road section together with uniform hot mix asphalt. At first, clay balls may or may not be evident. Over time, voids may be generated by fine particles absorbing water and expanding when frozen to cause a void at the surface (or very near the surface), or if traffic loadings break or crack the thin mortar-skin above the clump to expose the clay ball. The clay ball floats up to the pavement surface during traffic load since they weigh less than the aggregates and other particles in the surrounding mixture.

Several other activities could increase the possibility of clay ball initiation. For instance, central asphalt mix plants are usually more susceptible to producing clay balls, simply because they are temporarily placed on right-of-way near the project site which may have clay or loose soil underneath the stockpiles [3]. Moreover, the loader operator must apply the right amount of down pressure on the loader blade of the front-end loader. Too much pressure or too sharp an angle will cause the front blade to scrape the soil beneath the stockpile, thereby picking up the mud or clay, which will introduce soil particles into the mixture. Thus, the stockpile management should be carefully conducted to avoid the clay ball during field construction.

Scullion and Harris [2] analyzed the components of clay ball based on field cores taken from cement treated base layer. An X-ray diffraction (XRD) analysis was performed, and results show that the clay balls were predominantly smectite, mica, kaolinite, quartz, and calcite. Some of the clay balls have high silicon and aluminum concentrations typically of aluminosilicate minerals.

The existence of clay ball could deteriorate pavement. For instance, the areas with clay ball may not be stable with a source of water over time. With water, these clay balls are liquefied, eroded, or knocked from the surface by traffic load and eventually are the source of potholing and raveling [1,5–7]. Additional damages that could be caused by clay balls including moisture infiltration and reduced ride quality. It is believed that with proper repair, clay balls do not cause a long-term performance issue, such as structure related distresses [8,9].

In 2012, a number of clay balls were observed after the first winter on US-31 in Hamilton County, Indiana. The clay ball was not widely seen in Indiana previously, and it is not sure how severe they are and if the pavement performance could be negatively affected accordingly. Thus, Indiana Department of Transportation (INDOT) conducted a comprehensive evaluation with extensive forensic testing to determine the cause of such distress type. The pattern and density of these clay balls were checked as well. The "pattern" here indicates how the clay balls were located along pavement longitudinal direction (parallel to traffic), and distribution comparison among driving lane, shoulder and ramp. An infrared imaging system was also used to detect clay ball at traffic speed to see if this technology matched well with visual observation. Cores were taken from the clay ball areas and the size (diameter and depth) of each clay ball was measured. In addition, mix design, traffic, and climatic data were also collected to check if they were correlated with clay ball pattern or density as suggested elsewhere [10–12].

The objectives of this study were to (a) evaluate the accuracy of infrared image collection system in detecting position of clay balls; (b) determine the pattern and density distribution of clay balls in the field, and (c) determine the potential factors that could be correlated with clay ball pattern.

2. Materials and Methods

A high-speed infrared image collection system was used to detect the location of exposed clay balls. Distance measuring instrument (DMI) system and the Global Positioning System (GPS) were installed to obtain the accurate spatial location of each clay ball. The detection results were further confirmed by visual observation performed by workers who were responsible for clay ball repair.

The pattern of the clay ball was analyzed. In specific, a clay ball distribution along the longitudinal direction (parallel to traffic) was plotted to characterize its pattern, such as if they were distributed evenly or concentrated within specific areas, or if they were spread with fixed distance. The distribution of clay ball within the driving lane, passing lane, shoulder, and ramp was also compared.

Cores with visually observed clay balls were taken from pavement surface course, which was collected to check their shape, diameter, and depth by slicing the core in the middle of popout. Cores without visually observed clay ball were also obtained to verify if there were any existing clay balls not exposed to pavement surface. Such inspection was necessary since the existed clay ball not exposed to the surface could be the potential weak point. In the laboratory, such cores were sliced into multiple pieces to examine the existence of clay ball.

One winter after the clay ball was repaired, the researchers re-visited the clay ball areas to check if there were any newly developed clay balls or if the previously repairing of clay balls worked properly.

The construction information, such as mix design and truck coverage paving length, as well as climatic information, was collected to determine which parameter(s) could be potentially correlated with clay ball commencement and distribution pattern.

2.1. Pavement Condition

The $19.6 million interchange improvements and roadway reconstruction at the intersection of US-31 (two lanes in both directions) and SR 38 in Hamilton County, Indiana, were built in summer 2012. The amount of hot mix asphalt (HMA) was 63,818 tons for mainline, shoulders and ramps. The Average Daily Truck Traffic (AADT) was recorded as 24,814 in 2011.

In spring 2013, the interchange pavement exhibited 633 popouts from the surface course in driving lanes, shoulder, and ramps. These popouts are small size holes with fine particles surrounding the holes. Referring to the repair method used in another project constructed in 2001, the clay balls in this project were repaired by cleaning and sealing with hot poured asphalt sealant individually. Figure 1a shows a clay ball after opened by a jack hammer and cleaned by a vacuum. Figure 1b is a field core from clay ball area after slicing in the middle of the hole. Over time, these clay balls which appeared to be agglomerations of fine material disintegrate from weather and leave a hole in the mix after being washed out.

(a)   (b)

Figure 1. Field observation of clay ball. (**a**) Opened and cleaned clay ball; (**b**) sliced core from clay ball area.

2.2. Materials and Construction

The pavement type was full-depth HMA pavement with design surface thickness of 5.1 cm and total HMA thickness of 30 cm. The asphalt treated base was constructed between subgrade and asphalt layers. Two types of asphalt mix were used in the project. In the main lanes and ramps, a 9.5 mm nominal maximum aggregate size (NMAS) mix with PG76-22 binder was used for equivalent single axel load (ESAL) category 4. In the shoulder, a 9.5 mm NMAS mix was used with PG 64-22 binder corresponds to ESAL category 1.

The job mix formula (JMF) had been used for other projects with no problem. The aggregate source had been used for more than forty years without any issue. Table 1 summarizes aggregate type and percentage of each aggregate source. An aggregate drum mix dryer with a maximum capacity of 600 tons per hour, equipped with a natural gas fired aggregate dryer burner with a maximum rated power of 200 million British thermal units per hour (58.6 million watts) was used for HMA production. In general, the measured plant temperatures (159 °C to 176 °C) during construction met the INDOT specifications [13]. The paving lasted fourteen days, started on 12 October 2012 and ended on 25 October 2012.

Table 1. Summary of raw aggregate and recycled asphalt pavement (RAP).

Aggregate Type	Main Lane, %	Shoulder, %
Dolomite	24.0	30.0
Blast furnace slag	20.8	-
Dolomite sand	28.0	-
Stone sand	10.0	24.0
RAP	16.0	45.0
Baghouse fines	1.2	1.0
Aggregate total	100.0	100.0

Samples were collected and tested for a Quality Assurance (QA) purpose as shown in Table 2. INDOT QA factors for HMA production include binder content, air voids, and voids in mineral aggregate (VMA). Overall HMA QA pay factors were 1.02 and 1.035 for main lane and shoulder mix, respectively, which indicates that the HMA exceeded average quality (pay factor equals to 1.0). Another INDOT HMA construction QA factor, the percentage of density, was also listed in Table 2. Overlay, pay factor for the density, were higher than 1.0 except one lot with main lane construction (with 0.94 pay factor).

Table 2. Summary of Quality Assurance (QA) test results.

Parameter	Main Lane	Shoulder
Binder, %	6.85	5.65
Air voids, %	4.56	4.35
VMA, %	15.58	16.3
Density, %	92.57	93.25

2.3. Field Core

Two types of cores were taken: cores with clay balls and cores without clay ball from good and bad sections based on visual observation. Popouts were only observed at pavement surface, and thus the coring depth was 2 inches, which covered the entire surface course depth. These cores were drilled on 12 August 2013 and were delivered back to the laboratory for detailed inspection. Figure 2 presents a schematic plot that shows the project layout with core location and construction sequence. As shown, the cores were taken from paving sections with low frequencies of clay balls, such as core sample locations 4, 6, 8, and 13 to 16, as well as from paving sections with high frequencies of clay balls, such as core sample locations 5, 7, 9, 10, and 11. Cores were taken from driving lane, passing lane, shoulder, and ramp. Note that a couple of samples were taken within each core location, and a total of 633 cores were obtained.

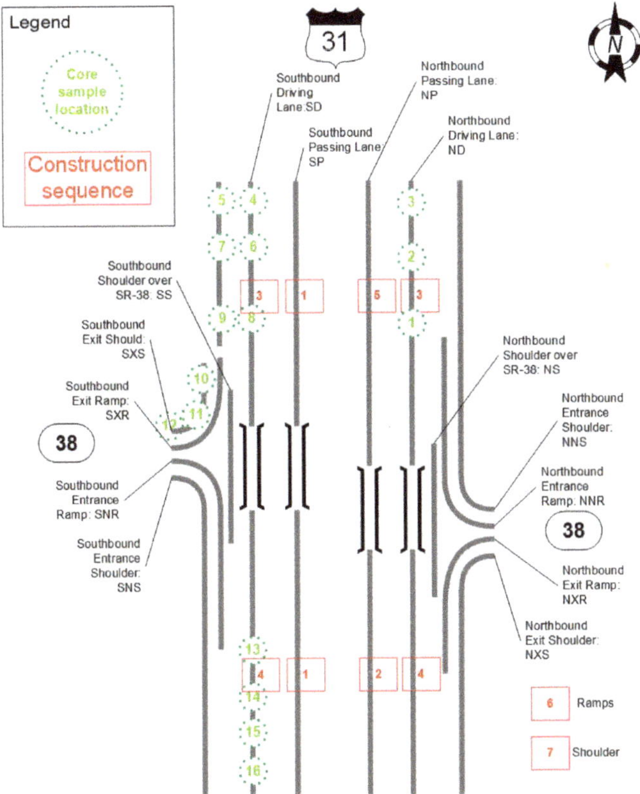

Figure 2. Schematic of US-31 and SR-38 with core location and construction sequence.

2.4. Infrared Image Collection System

The detection of clay balls can be achieved by measuring the density or texture of asphalt pavement. Pavement sections with observed clay balls are usually characterized with extreme high in place air voids or very non-uniform texture distribution [14,15]. In this study, in order to rapidly detect clay balls, pavement surface images were collected and analyzed using the INDOT pavement infrared image collection system integrated with an infrared (IR) camera, a gray scale high-speed camera and a right of way camera (iPhone 4S) as shown in Figure 3. The system consists of a van, three cameras that were fixed to a bar, and a computer to acquire image data. As shown, the IR camera was placed at the very top of the bar and took photos that were perpendicular to the pavement in vertical direction. The gray scale high-speed camera was set in the middle. The right of way camera was put close to pavement surface which was used to capture clay ball. A distance measuring instrument (DMI) system was used to measure the distance the van drove. The IR camera was triggered at a fixed distance based on DMI data. The right of way camera embedded an iPhone 4S cell phone to take photos and used GPS to determine location. Both cameras covered the width of one entire traffic lane. The operation speed ranged between 5 and 10 mph. The image collection was conducted from 9:30 to 15:30 on 17 September 2013. Two replicates were conducted in the morning and in the afternoon, respectively to avoid any potential test error.

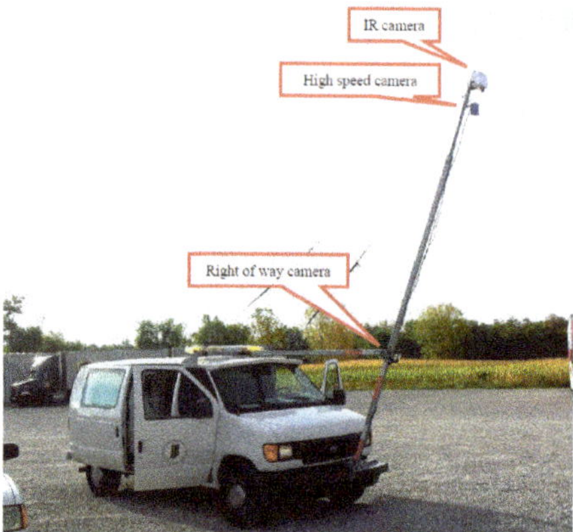

Figure 3. Infrared image collection system.

3. Results and Discussion

3.1. Infrared Image Collection System

Figure 4a,b shows the images taken from the IR camera and the right of way camera, respectively. Note that the two cameras were taking photos simultaneously. As shown, all the five popout locations in Figure 4b were successfully captured by the IR camera in Figure 4a. The different colors in the IR image indicate temperature differentiation. The clay balls were identified in these areas with high temperature differentiate (i.e., higher temperature in the middle with lower temperature around). In general, a total of 633 holes (346 holes on the mainline and shoulders and 287 holes from ramps) were cleaned and filled with asphalt sealant, among which 564 cores were successfully captured by the IR camera. Among all the clay balls (69 in total) that were not successfully captured by the IR, the majority (50 in total) were missed due to the very small size of clay balls (i.e., with diameter less than 1 cm). It is noted that the angle of the two photos are not exactly the same: The IR camera took photos vertically perpendicular to pavement surface, whereas the right of way camera took photos with an angle to pavement surface.

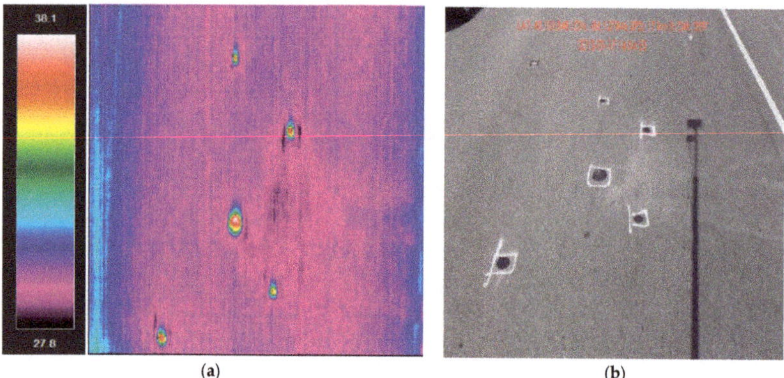

(a)   (b)

Figure 4. Clay ball images from (**a**) IR camera and (**b**) right way of camera.

3.2. Distribution of Clay Ball

A few small voids caused by clay balls can be ignored as a byproducts of the heavy construction process, but many significant voids in the pavement surface warrants investigation, such as the case in this study. Table 3 shows the amount of clay balls based on each paving sequence. The construction date is also presented. As seen, the amount of clay ball varies greatly from one day to another. The highest clay ball every 61 m was 2.82 from shoulder area (paving sequence 7), whereas the lowest clay ball very 61 m was 0.49 from passing lane area (paving sequence 1).

Table 3. Summary of project information.

Paving Sequence	Date	Length, m	Number of Clay Balls	Clay Ball Every 61 m
1	10/12/2012	2631	21	0.49
2	10/13/2012	1564	14	0.55
3	10/15/2012	2134	79	2.26
4	10/16/2012	3136	76	1.48
5	10/17/2012	762	10	0.80
6	10/21/2012	3039	53	1.05
7	10/24/2012	6935	321	2.82

Figure 5 presents the clay ball distribution in the longitudinal direction (parallel to the direction of vehicle traffic) for both driving lane and ramp. The majority of the popouts were found in shoulder, ramp, and various locations in the driving lanes, shoulder, and ramp but few in the passing lanes. Additionally, most popouts were from the pavement location between wheel paths, whereas very few of these pop-out locations were witnessed within the wheel paths.

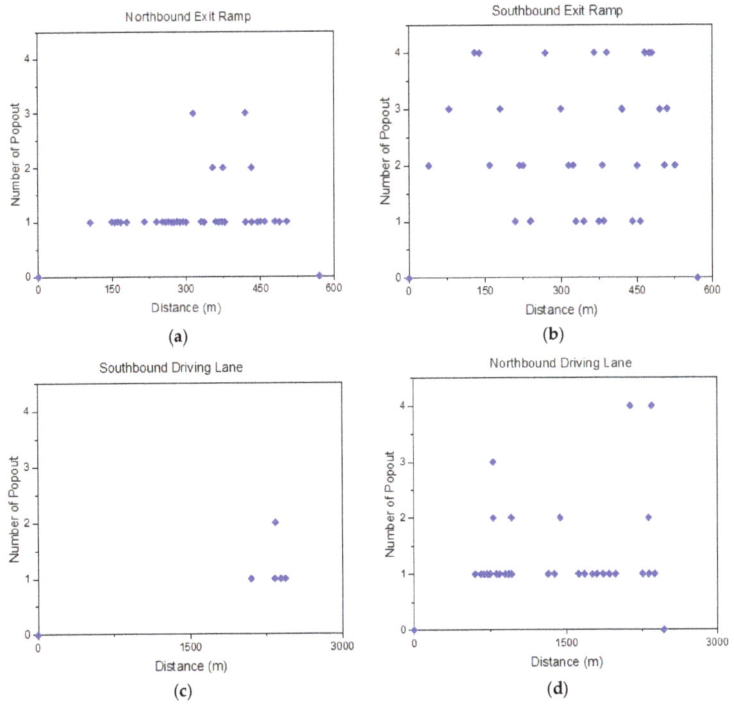

Figure 5. Clay ball distribution within (**a**) southbound exit ramp, (**b**) northbound exit ramp, (**c**) southbound driving lane, and (**d**) northbound driving lane.

As shown in Figure 5, the number of popout within southbound exit ramp is almost evenly distributed with around 61 m to 73 m as the gap distance. It should be noted that the typical laydown lengths of 20-ton surface mix in a truck with 165 lbs/yd^2 rate were 55 mt for main lane (3.7 m width) and 66.4 m for shoulder (3 m width). Thus, it is possible that the non-uniform mix between two trucks could be a reason that caused clay ball in this area.

However, such trend was not observed in northbound exit ramp or driving lane. In those areas, the clay balls were distributed either as close as several meters and as far as one hundred meters. The fact that the "pop-out" locations do not seem to be consistent throughout the mix and the fact that these areas seem to be confined to what would amount to a typical truckload of material at a time based on length and concentration of the pop-outs and the randomness of these locations should rule out the possibility of the paving operation being an issue.

3.3. Size and Shape of Clay Ball

The depth, size, and shape are major parameters to characterize a clay ball [16–20] and were measured. The six-inch core samples taken from popout locations were cut to reveal a rectangular cross-section. All the sliced cores were taken between or away from the wheel-path to avoid the effect of trafficking compaction. In order to have a side view of clay ball, cores were cut in the middle of a clay ball opening hole using a diamond coated wire saw. The typical circular saw used to cut the asphalt core had a 3.3 mm thick saw bit. Furthermore, the lateral movement of the saw bit in cutting operation may have caused a wider sample thickness loss. To overcome this limitation, a diamond-impregnated wire saw was utilized. This saw is a customized product designed to make a series of programmed cuts for the sample cores. The diameter of the diamond-impregnated wire was only 0.2 mm, with very small lateral movements due to fixed pulleys in the saw system. The samples were dry-cut to keep the clay ball inside intact.

Figure 6 shows two clay ball examples that were squeezed, and their openings were collapsed. The left figure was pictured upside down, and the right figure shows sliced core in the middle of the hole. The diameter of each clay ball was measured and was ranged from 1 to 4 inches in diameter. Considering the ellipse sphere shape of the clay ball, several measurements were conducted and averaged. As seen, the hole on pavement surface is small in scale but the clay ball size is much bigger than the hole. The depth of the clay ball almost punctures the entire surface course thickness of 2 inches. It is also noted that good amounts of fines remained in the voids.

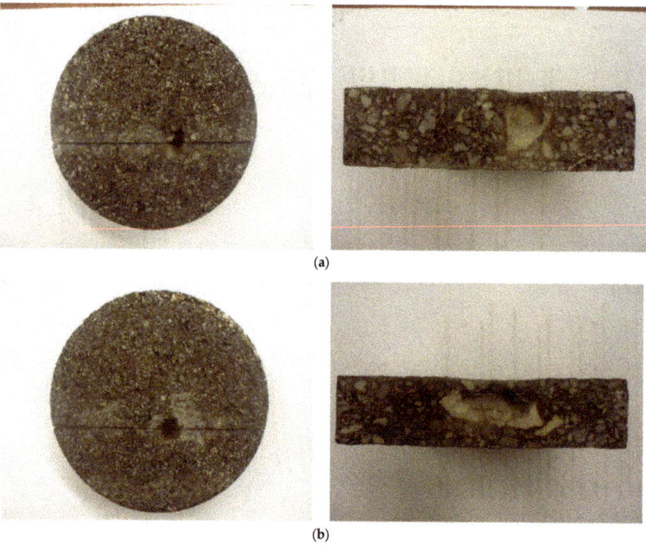

(a)

(b)

Figure 6. Clay balls with (**a**) one inch in diameter and (**b**) two inches in diameter.

3.4. Cores Without Visually Observed Popouts

As aforementioned, a couple of cores were taken from pavement areas without clay balls to verify if there was any clay ball not exposed to the pavement surface. In the laboratory, each six-inch core was sliced into several pieces. Figure 7a,b shows an example of sliced core from top and side, respectively. As seen in Figure 7b, the samples are very clean and no clay ball was found. Additionally, there was no aggregate size larger than 1 inch in diameter, the critical size of clay ball for the popout. All the rest cores were sliced and no clay balls were observed as well, even for the cores taken from bad sections (with the large frequency of clay balls). This indicates that in the project evaluated all the clay balls were exposed to the pavement surface.

(a)   (b)

Figure 7. Cut surface of core sample; (**a**) top-down view and (**b**) inside of core.

3.5. Relationship between Moisture and Clay Ball

During paving, there were seven raining days, and no construction was conducted on those days. Figure 8 shows precipitation distribution within each month in 2012, as well as the average monthly precipitation based on years from 1981 to 2010. It is shown in the figure that precipitations in October 2012 were much high compared to its companion averaged from 1981 to 2010.

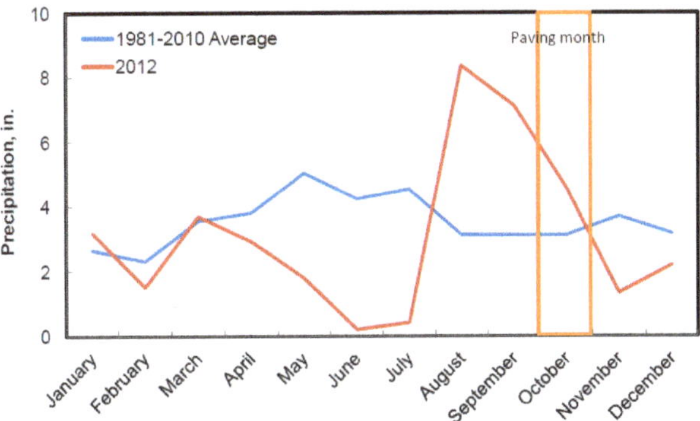

Figure 8. Comparison of monthly precipitations.

Figure 9 presents the precipitation during the days the road was paved with the paving sequences 1 to 7. As shown, the more precipitation could result in more moisture in aggregate stockpiles, and the wet aggregates may cause the material to become sticky and bind together, which increases the possibility of clay balls in a drum mixer. In general, paving in a day after the rain showed higher clay ball densities than others. For instance, paving sequences 3 and 7 were constructed one day after raining and showed the highest clay ball density. It is noted that typically aggregate moisture was checked before construction, while raining during paving days could greatly increase aggregate moisture and increase the possibility of clay balls. Thus, check for aggregate moisture during construction, especially on or after raining could be necessary.

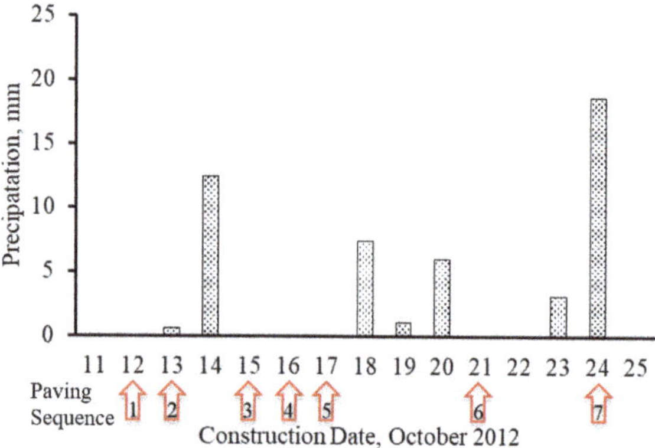

Figure 9. Comparison of daily precipitations to paving dates.

3.6. Revisit of Pavement Sections with Clay Ball

After two winters, the pavement section was revisited. It was found that the number of clay balls did not increase, which means that no new clay balls were developed. The patching worked very well without loss of materials. This indicates that clay balls should be a one-time distress and may not affect the long-term field performance of pavement as long as it is well repaired.

4. Conclusions

This study evaluated the clay balls that were observed on US-31 in Hamilton County, Indiana. The detection of clay ball using infrared image collection system was evaluated. The clay ball amount and distribution pattern were also checked. In addition, field cores were taken and the depth and diameter of clay balls were measured. The cores without visually observed popouts were also collected to see if any clay balls were not detected. Precipitation information during construction was also collected to check the potential relationship between field moisture and clay ball density. Based on the analysis, the following conclusions can be drawn:

- Based on engineering judgement, the infrared image collection system was found to be accurate enough (detection accuracy of 564 out of 633) in pinning the clay ball location.
- Regarding cores with clay balls, the clay ball diameter ranged from 2.5 cm to 10 cm, and the maximum clay ball depth is almost penetrating the entire course's surface. Most clay balls were elliptic in shape.
- The asphalt paving on raining days or right after raining could increase the potential of clay balls. It is recommended to check the aggregate moisture right after or on raining days and take the necessary steps such as extended mixing time to reduce the risk of clay balls.

Author Contributions: Data curation, H.J.A.; formal analysis, Q.L.; investigation, W.Z.; project administration, J.L.; resources, H.J.A.; software, M.W. and J.Z.; supervision, J.L.; validation, M.W. and J.Z.; visualization, H.Z.; writing—original draft, W.Z.; writing—review and editing, J.L. and W.Z.

Funding: The authors gratefully acknowledge the financial support of National Key R&D Program of China (grant number 2016YFE0108200). The authors also want to thank the sponsorship by the CCCC First Highway Consultants CO., LTD. (with project number 8521002546) and the Qilu Transportation Group (with grant number 2018B51) for the data analysis conducted in this study.

Acknowledgments: The authors would like to thank the Joint Transportation Research Program and the Indiana Department of Transportation for its sponsorship of this project.

Conflicts of Interest: The authors declare no conflict of interest.

References

1. Hoover, J.M. Surface improvement and dust palliation unpaved secondary roads and streets. In *Iowa Highway Research Board Project HR-151*; Iowa State Highway Commission: Wapello, IA, USA, 1973.
2. Scullion, T.; Harris, P. Forensic evaluation of three failed cement-treated based pavements. *Transp. Res. Rec.* **1998**, *1611*, 10–18. [CrossRef]
3. R&T Update Concrete Pavement Research & Technology. Clay Ball Prevention and Repair: Stockpile Management is Key. Available online: http://1204075.sites.myregisteredsite.com/downloads/RT/RT5.04.pdf (accessed on 9 August 2019).
4. Miller, J.S.; Bellinger, W.Y. *Distress Identification Manual for the Long-Term Pavement Performance Program*; FHWA-HRT-13-092; Federal Highway Administration: McLean, VA, USA, 2014.
5. Ryu, S.; Kim, T.; Kim, Y. Feature-based pothole detection in two-dimensional damage. *Transp. Res. Record J. Transp. Res. Board* **2015**, *2528*, 9–17. [CrossRef]
6. Zhang, W.; Shen, S.; Wu, S. Comparison of the relative long-term field performance among various Warm Mix Asphalt (WMA) Pavements. *Transp. Res. Rec. J. Transp. Res. Board* **2018**, *2672*, 200–210. [CrossRef]
7. Ding, X.; Chen, L.; Ma, T.; Ma, H.; Gu, L.; Chen, T.; Ma, Y. Laboratory investigation of the recycled asphalt concrete with stable crumb rubber asphalt binder. *Constr. Build. Mater.* **2019**, *203*, 552–557. [CrossRef]
8. Ding, X.; Ma, T.; Huang, X. Discrete-element contour-filling modeling method for micro-and macro-mechanical analysis of aggregate skeleton of asphalt mixture. *J. Transp. Eng. Part B Pavements* **2019**, *145*, 04018056. [CrossRef]
9. Zhang, Y.; Ma, T.; Ling, M.; Huang, X. Mechanistic sieve size classification of aggregate gradation by characterizing load carrying capacity of inner structures. *J. Eng. Mech.* **2019**, *145*, 04019069. [CrossRef]
10. Dong, Q.; Huang, B.; Jia, X. Long-term cost-effectiveness of asphalt pavement pothole patching methods. Transp. *Res. Rec. J. Transp. Res. Board* **2014**, *2431*, 49–56. [CrossRef]
11. Koch, C.; Jog, G.M.; Brilakis, I. Automated pothole distress assessment using asphalt pavement video data. *J. Comput. Civ. Eng.* **2013**, *27*, 370–378. [CrossRef]
12. Ma, T.; Wang, H.; He, L.; Zhao, Y.; Huang, X.; Chen, J. Property characterization of asphalt binders and mixtures modified by different crumb rubbers. *J. Mater. Civ. Eng.* **2017**, *29*, 04017036. [CrossRef]
13. Indiana Department of Transportation. Section 400-Asphalt Pavement. In *Standard Specifications*; Indiana Department of Transportation: Albany, IN, USA, 2016.
14. Praticò, F.G. Metrics for management of asphalt plant sustainability. *J. Constr. Eng. Manag.* **2017**, *143*, 04016116 [CrossRef]
15. Megali, G.; Cacciola, M.; Ammendola, R.; Moro, A.; Praticò, F.G.; Morabito, F.C. Assessing reliability and potentiality of nonnuclear portable devices for asphalt mixture density measurement. *J. Mater. Civ. Eng.* **2010**, *22*, 0000088. [CrossRef]
16. Roberts, F.L.; Kandhal, P.S.; Brown, E.R.; Lee, D.Y.; Kennedy, T.W. (Eds.) *Hot Mix Asphalt Materials, Mixture Design and Construction*; National Asphalt Paving Association Education Foundation: Lanham, MD, USA, 1996.
17. Jo, Y.; Ryu, S.K.; Kim, Y.R. Pothole detection based on the features of intensity and motion. *Transp. Res. Rec. J. Transp. Res. Board* **2016**, *2595*, 18–28. [CrossRef]
18. Zhang, W.; Shen, S.; Wu, S.; Chen, X.; Xue, J.; Mohammad, L.N. Effects of in-place volumetric properties on field rutting and cracking performance of asphalt pavement. *J. Mater. Civ. Eng.* **2019**, *31*, 04019150. [CrossRef]
19. Ma, T.; Zhang, D.; Zhang, Y.; Wang, S.; Huang, X. Huang. Simulation of wheel tracking test for asphalt mixture using discrete element modelling. *Road Mater. Pavement Des.* **2018**, *19*, 367–384. [CrossRef]
20. He, R.; Huang, X.; Zhang, J.; Geng, Y.; Guo, H. Preparation and evaluation of exhaust-purifying cement concrete employing titanium dioxide. *Materials* **2019**, *12*, 2182. [CrossRef] [PubMed]

Article

Experimental Study on Wet Skid Resistance of Asphalt Pavements in Icy Conditions

Boxiang Yan [1], Huanhuan Mao [1,*], Sai Zhong [1,2,*], Pengfei Zhang [1] and Xiaoshan Zhang [1]

[1] State Key Laboratory of Silicate Materials for Architectures, Wuhan University of Technology, Wuhan 430070, China; yanboxiang@whut.edu.cn (B.Y.); pengfeizhang@whut.edu.cn (P.Z.); zxs971206@163.com (X.Z.)

[2] School of Information and Mathematics, Yangtze University, Jingzhou 434000, China

* Correspondence: mhhyue@whut.edu.cn (H.M.); lanlv@whut.edu.cn (S.Z.); Tel.: +86-8765-1462 (H.M.); +86-1582-7151-239 (S.Z.)

Received: 16 March 2019; Accepted: 9 April 2019; Published: 12 April 2019

Abstract: In this research, the durability of skid resistance during the ice melting process with temperature increasing from −5 °C to 10 °C was characterized by means of a British Pendulum Skid Tester. Four types of pavement surfaces were prepared and tested. The difference between two antiskid layers prepared with bitumen emulsion was the aggregate. The detailed angularity and form 2D index of fine aggregates used for antiskid surfaces, characterized by means of the Aggregate Image Measure System (AIMS) with micro image analysis methods, were then correlated with British Pendulum Number (BPN) values. Results indicate that skid resistance has the lowest value during the ice-melting process. The investigated antiskid layers can increase the surface friction during icy seasons. In icy conditions, the skid resistance behavior first worsens until reaches the lowest value, and then increases gradually with increasing temperature. Results from ice-melting conditions on four investigated pavement surfaces give the same temperature range where there will be lowest skid resistance. That temperature range is from 3 °C to 5 °C. A thicker ice layer will result in a lower skid resistance property and smaller "lowest BPN".

Keywords: asphalt pavement; skid resistance; durability; antiskid surface; image analysis

1. Introduction

Asphalt pavements, considered as flexible pavements, are widely used in both road and airfield engineering. It was reported that more than 90% of highways were constructed by asphalt materials [1,2]. A durable and sustainable designed asphalt pavement would contribute a lot to economic, social and environmental aspects, as well as enhancing driving safety.

At all stages of pavement service life, the surface should have some sort of roughness to ensure sufficient friction between traffic wheels and the pavement surface [3,4]. Skid resistance is a measure of resistance of pavement surface to sliding or skidding of the vehicle [5–7]. The texture of pavement surface is of prime importance in providing skid resistance. It is a common fact that the lower the skid resistance, the higher the percentage of traffic accidents, especially in the wet and winter seasons.

Typically, the Locked Wheel Skid Trailer (LWST) and British Pendulum Tester (BP Tester) are the most common and accepted methods of measuring pavement skid resistance. LWST, widely used in the United States, is standardized in ASTM E 274 specifications and measures the sliding force of a locked tire on a wetting pavement surface. The BP Tester is one of the most widely used devices to determine low speed related skid resistance in the laboratory. The value obtained is called the British Pendulum Number (BPN).

However, one of the limitations of the current BP Tester is that it lacks any correction for temperatures. Previous research had concluded that BPN values are dependent on the pavement temperature [8]. Temperature influences friction properties because it changes the physical properties of tire rubber and asphalt pavement surfaces, which are both viscoelastic materials. Bazlamit reported that skid resistance of asphalt pavement decreases during seasons with warmer temperature and increases during seasons with colder temperature. Such phenomenon is related to the stiffness properties of asphalt materials and rubber that used during skid resistance tests [9].

Prof. Liang used a BP Tester to characterize the temperature effect on the measured frictional properties of a hot-mix asphalt mixture surface [10]. He concluded that an increase of temperature would result in a decrease of friction, and the effect of an HMA pavement, rubber slider and water temperatures influenced the measured frictional properties significantly. Based on the data of skid resistance of a wetted pavement, Bazlamit concluded that an approximately linear relationship can be finalized between skid resistance and temperature [8]. Table 1 summarizes the temperature ranges that have been carefully studied in literatures. Obviously, most of the related work was focused on ambient temperature or pavement surface temperature higher than 10 °C.

Table 1. Temperature effect on the friction behavior from literatures.

Reference	Temperature Conditions	Findings
Burchett JL, 1980 [11]	From 16.8 °C to 19 °C	• Correlations between skid resistance and temperature were not good;
Elkin BL, 1980 [12]	Seasonal changes	• Skid resistance was highest in the spring, dropped off noticeably during the summer, and began to recover in late fall;
Subhi MB, 2005 [8]	At 0, 10, 20, 30 and 40 °C	• Skid resistance decreases with increased temperature; • Linear relationship between skid resistance and temperature;
Bianchini A, 2011 [13]	From 10 °C to 35 °C	• It is possible to define a reference air temperature to which friction measured at any other air temperature value can be adjusted;
Baran ED, 2011 [14]	From 10 °C to 60 °C	• Temperature correction relationship for skid resistance measurements were proposed;
Robert YL, 2012 [10]	From 4.4 °C to 60 °C	• Increase of temperature resulted in decrease of friction;
Scarpas A, 2013 [15]	From 0 °C to 60 °C	• Effect of Pavement temperature and ambient temperature was modeled; • Higher PT and AT would result in lower friction;
Hadiwardoyo SP, 2013 [16]	From 30 °C to 55 °C	• Skid number values decreased with increasing temperature;
Wang DW, 2014 [17]	Surface temperature: 15–45 °C Water temperature: 10–30 °C Air temperature: 15–30 °C Tire temperature: 20–30 °C	• All four of the acquired temperatures are negatively related to the friction coefficient

However, cold winter conditions will influence the micro-texture of aggregates and stiffness of tire rubber itself. Hence, research on temperature effect in colder temperature range is important as well, typically in icy conditions. In a cold winter, a better understanding of skid resistance behavior will lead to safer traffic by ensuring logical deicing techniques and skid maintenance in asphalt pavements to ensure better skid resistance [18].

Accordingly, this paper reports on the durability of skid resistance in very low temperature condition, especially in an ice-melting condition lower than 0 °C, has been developed and is discussed herein. BPN values under ice melting conditions were measured to examine the skid performance of antiskid surfaces under winter road condition. The angularity and surface texture index of the used fine aggregates were characterized by mean of Aggregate Image Measure System (AIMS) with micro image analysis methods. The detailed angularity and surface texture were then correlated with BPN values under ice-melting conditions. Conclusions from this research can be used as guidance for pavement maintenance in the winter time.

2. Materials and Methods

Asphalt mixture gradation of AC-16 (asphalt concrete, Panjin north asphalt co. LTD, Panjin, China) and SMA-16 (stone-matrix asphalt mixture, Panjin north asphalt co. LTD, Panjin, China) were first prepared as a traditional asphalt pavement surface layer. The void content is 4.2% for AC-16 and 4% for SMA = 16. Their skid resistance at ice-melting conditions were first tested. Another two antiskid surfaces were prepared by applying bitumen emulsion and fine aggregates onto AC and SMA surfaces. The construction steps of the antiskid surface are as follows:

Firstly, beams with 380 ± 5 mm length, 50 ± 6 mm height and 63 ± 6 mm width were cut from mixture slabs. Secondly, bitumen emulsion was uniformly sprayed onto the specimen surface with a ratio of 0.7 kg/m^2. Table 2 presents the properties of used bitumen emulsion. Thirdly, clean and fine cubic aggregates were applied to the bitumen emulsion and let them set down into bitumen film with a ratio of 2.2 kg/m^2. Two types of aggregates, basalt and dolerite, were used. Table 3 concludes the characteristics of the used fine aggregates. Both aggregates are alkali aggregates, so that bitumen emulsion with approximately pH of 2.0 was used.

Table 2. Characteristics of bitumen emulsion for antiskid surface layer.

	Properties	Values	Specifications
	Appearance	Dark brown liquid	–
	Density [g/cm^3]	0.99	ASTM D6973-16
	Acidity/Alkalinity	Approx. 2.0 pH	–
	Viscosity at 25 °C [mPa·s]	36.7	ASTM D4402-15
Recovered binder by distillation	Weight [%]	47.5	ASTM D6997-12
	Penetration at 25 °C	62	ASTM D5-13
	Softening point	50	ASTM D36-14e1

Table 3. Characteristics of fine aggregates for antiskid surface layer.

Aggregate Types	Properties	Values		Specifications
Basalt	Apparent specific gravity	2.876		ASTM C128-15
	Aggregates size		0.3–0.6	
Dolerite	Apparent specific gravity	2.912		ASTM C128-15
	Aggregates size		0.6–1.18	

After bitumen emulsion is cured, loose aggregates were swept away and finally the antiskid layer was then obtained on the cut surfaces. Figure 1 compares the visualized features of the investigated specimens, with antiskid layer or ice layer on top. Three specimens were prepared and test for every type of antiskid surface.

<table>
<tr><td>Surface layer</td><td>Surface layer covered with Antiskid-B</td><td>Surface layer covered with Antiskid-D</td><td>Antiskid-B with ice layer</td></tr>
</table>

Figure 1. Visual feature of the investigated surfaces.

2.1. British Pendulum Skid Test

A BP Tester was employed to measure the surface frictional properties during ice melting process in accordance with ASTM E303-93. The values measured, BPN, represent the frictional properties obtained with the apparatus. A BP Tester is a low-speed device, about 10 km/h, that measures the skid resistance related to surface micro-texture rather than macro-texture. Antiskid layers in this research used bitumen emulsion-based chip seal technology, which therefore lets the micro-texture of surface domains cause the pavement friction. Therefore, BP Tester can be used to correlate BPN during ice melting process. Table 4 shows BPN values of the investigated surfaces without ice at room temperature of 28 °C. The average BPN value of the SMA surface is 74.8, which is slightly higher than that of the AC surface. The two applied antiskid surfaces can provide better antiskid resistance, while Antiskid-B has higher increment than that of Antiskid-D.

Table 4. British Pendulum Number (BPN) values of the investigated surfaces without ice.

Surface	BPNs					Average BPN	Increment
AC-16	75	70	73	70	70	71.6	–
AC16 plus Antiskid-B	80	80	81	81	80	80.4	13%
AC16 plus Antiskid-D	72	72	75	72	75	73.2	2.2%
SMA-16	76	75	73	75	75	74.8	–
SMA16 plus Antiskid-B	85	80	80	85	80	82	9.6%
SMA16 plus Antiskid-D	75	75	76	76	76	75.6	1.1%

2.2. Aggregates Surface Characteristics Analyses

Skid resistance is a function of different factors including the micro-texture and macro-texture of the pavement surface. Micro-texture is highly dependent on the surface characteristics of aggregates, such as angularity and surface texture [19,20]. The Aggregate Image Measure System (AIMS) is an image analysis tool, which can investigate the shape properties of both coarse and fine aggregate including angularity, sphericity and surface texture index of coarse aggregates, 2D form of fine aggregates.

Basalt aggregate with nominal size of 0.3 to 0.6 mm, and dolerite aggregate with nominal size of 0.6 to 1.18mm, were used to prepare the antiskid surface layer. AIMS has been employed to characterize the angularity and 2D form of these two types of fine aggregate. Figure 2 presents the visualized fine aggregates and their corresponding image during AIMS analysis.

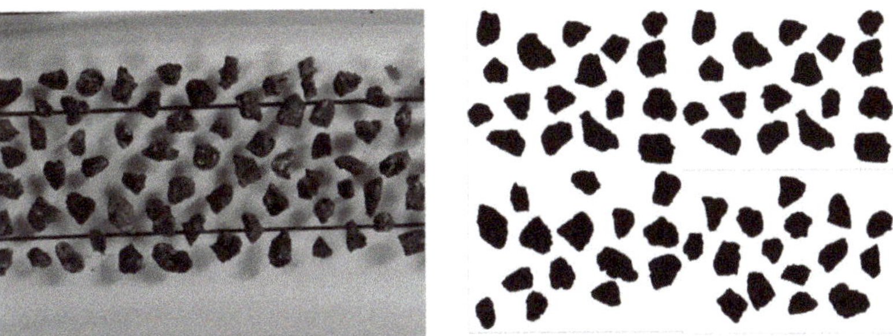

Figure 2. Visualized fine aggregates (left) and Aggregate Image Measure System (AIMS) image (right).

3. Results and Discussion

The ice melting process has three stages, initial stage, melting stage and end stage. At the initial stage, the pavement surface is completely covered with ice, while the melting stage has both ice and water. At the end stage, the pavement is then covered with cold water and no ice present at all. Ice thickness of 1mm and 2 mm was prepared under following steps. Firstly, the beams were immersed in water for 2 h at room temperature, and then placed in the refrigerating chamber at −5 °C for 5 h with plastic bag packed. Secondly, the top side of every beam was carefully fenced and a certain amount of water was then stored to create an ice layer with the expected thickness. The first step was needed to first fill voids in the beam and hence prevent the water from leakage when it was poured onto the beam surface. This process is explained with Figure 3.

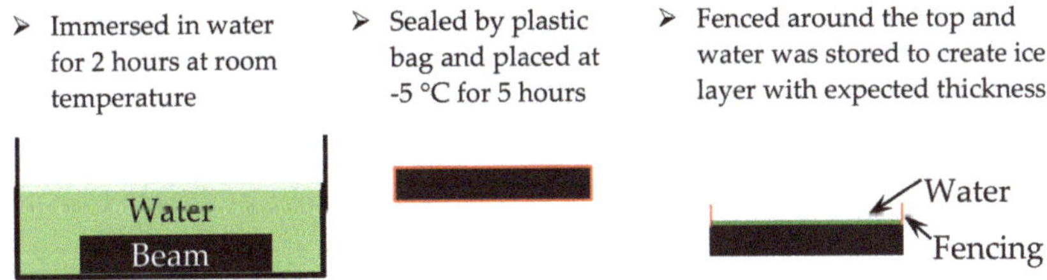

Figure 3. Process of prepare ice layer with certain thickness on beams.

The ice-melting process was characterized by putting specimens under BP Tester at room temperature condition of 28 °C. The BPN values and surface temperatures were then continuously recorded every three minutes. In this research, the 1mm thick ice was completely melted at 7 °C, and 7.2 °C for the 2 mm thick ice layer. The surface temperature of specimens in this research was defined as the temperature of the entire specimen's surface during the ice melting process, which means it was either temperature of the ice surface or temperature of the melted water. Therefore, the initial BPN value represents the skid resistance property of the pavement surface when it is fully covered with ice. The following BPN values illustrate the skid resistance behaves when the surface is influenced by ice and water. At the end, BPN values illustrate the surface friction characteristics under the water condition.

Figure 4 compares the BPN values of AC-16 and SMA-16 under the ice-melting condition. Research data clearly illustrates that the SMA-16 surface has a higher BPN value and better skid resistance that of AC-16 surface, no matter what temperature condition and ice accumulation characteristic of the surface has.

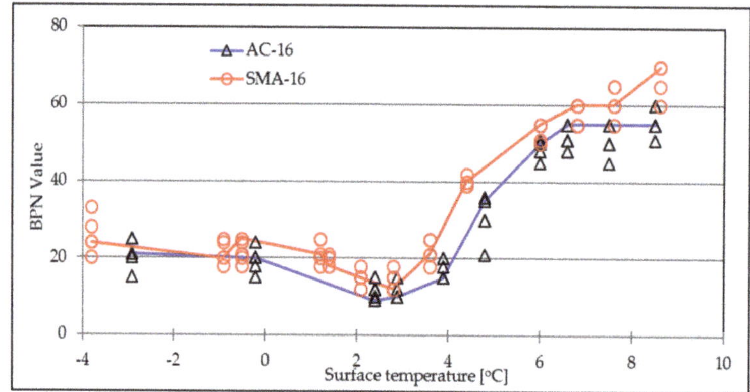

Figure 4. BPN values of traditional surface layers during ice-melting process.

The relationship between BPN and skid resistance properties on pavements is concluded in Figure 5 [21]. It can be seen that minimum BPN of 42 must be achieved to prevent traffic from skidding accidents. This means that the investigated SMA-16 and AC-16 surface cannot provide a sufficient friction property when it is covered with ice and in the beginning of ice-melting stage. When the surface temperature increases to higher than 6 °C, both surfaces can provide BPN values higher than 42.

BPN Value	Skid resistance
27 and below	Very low
27–32	Low
32–37	Medium
37–42	Good
42+	Excellent

Figure 5. Relationship between BPN and skid resistance on pavements.

3.1. Surface Temperature Dependencies

Figure 6 indicates the ice effect on BPN values of the asphalt surface, as well as temperature influence. One millimeter ice will dramatically decrease the skid resistance of the asphalt pavement surface, the BPN values drops from 70–80 to as low as around 10. Surface temperature has the same influence contribution on both AC surface with and without the ice layer. In the range of 0 °C to 6 °C, BPN values decrease slightly and then tend to increase when the surface temperature increases.

Figure 7 presents the relationship between surface temperature and BPN values during the ice-melting process. Skid resistance at the pavement surface decreases at the very beginning of ice melting process. Initial skid resistance is relatively low until the thick ice film is worn off the top of surface by a rubber slider from the BP Tester, resulting in increasing values of BPN. Antiskid-B and Antiskid-D show the same results of BPN trends.

The ice-melting process results in a concave trend between friction and surface temperature (melting time as well). There is a temperature at which the minimum friction value at pavement surface is achieved during the ice-melting process. Table 5 concludes the lowest BPN values that occurred during ice melting at their corresponding surface temperatures. Firstly, it can clearly be seen that the investigated antiskid layers can significantly increase the surface friction properties. The lowest BPN values of AC and SMA surfaces have been enhanced from 11.6 and 14.4 to a value of around 20, respectively, when the antiskid surface has been applied. Secondly, the corresponding temperatures have been increased slightly as well, from 2.5 °C to 5 °C.

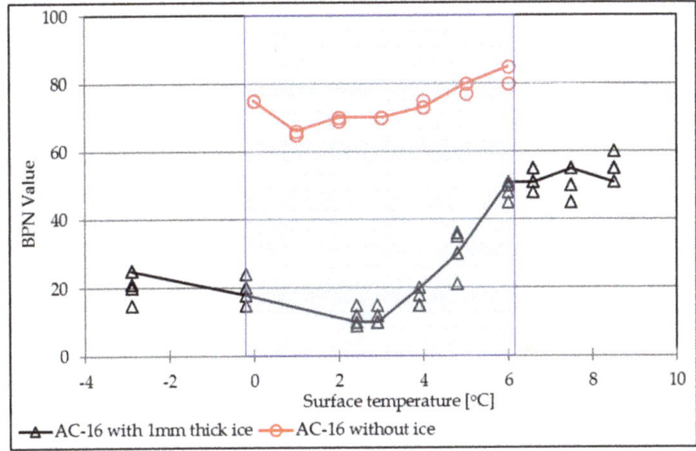

Figure 6. Ice effect on the BPN values.

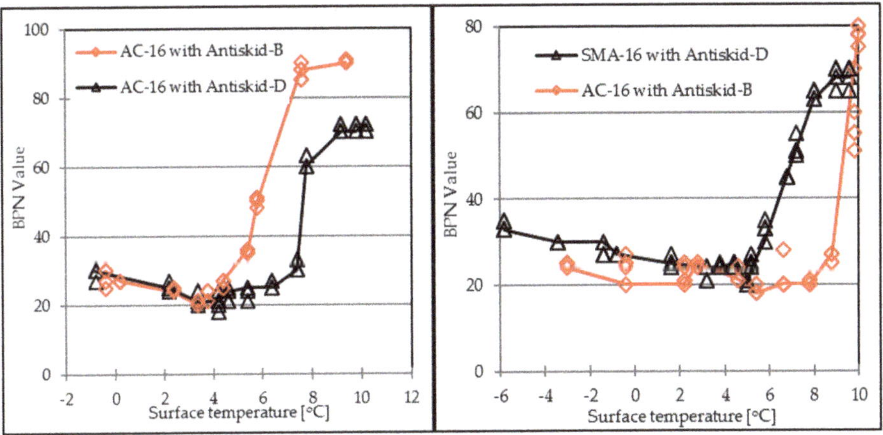

Figure 7. Relationship between surface temperature and BPN value.

Table 5. Lowest temperature and lowest BPN values.

Surface Type	Lowest BPN	Surface Temperature Where Lowest BPN Exists [°C]
AC16	11.6	2.4
AC16 plus Antiskid-B	20	3.4
AC16 plus Antiskid-D	21	4.2
SMA16	14.4	2.8
SMA16 plus Antiskid-B	18	5.4
SMA16 plus Antiskid-D	20	5

Figure 8 explains the resulting concave trend between friction and surface temperature by means of macrotexture. In the initial stage, the surface is fully covered with an ice layer. During the melting stage, water will flow from the peak to the bottom between aggregates, resulting in lower average depth between peak points and bottom points. Therefore, the lowest BPN values will be achieved in this stage. At the end stage, ice has been completely melted and water has been flowed away and evaporated as well. A lower temperature condition will result in aggregates with higher surface micro-roughness and a rubber slider of increased stiffness [22]. So the real antiskid surface has been achieved and the maximum friction value can be achieved. However, the temperature dependency in this research was concluded with an icy surface. Ice and cold water have non-negligible influence. Conducting skid resistance analysis in a cold temperature chamber could provide further valuable recommendations.

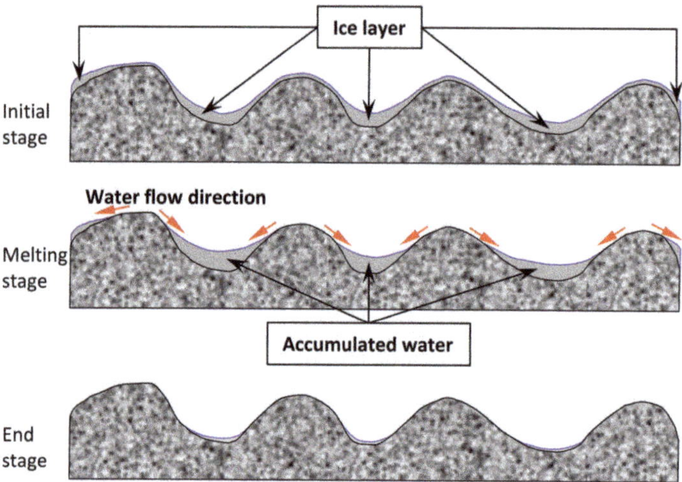

Figure 8. Graphic explanation on the macrotexture changes during ice melting process.

3.2. Aggregate and Gradation Dependencies

Figure 9 compares the BPN trends during the ice-melting process between different aggregate types. The ice thickness is 1 mm. The graph shows the maximum BPN values of four investigated antiskid surfaces are in agreement with the results presented in Table 4.

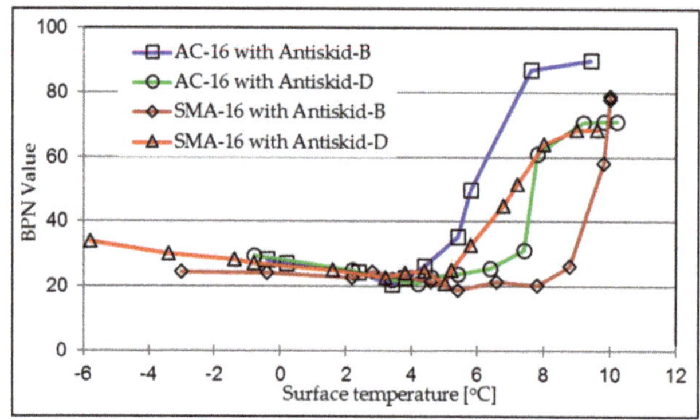

Figure 9. BPN values of investigated antiskid layers.

Antiskid-B on both AC and SMA surfaces presents the highest BPN values at the end stage, but with a far different trend during the ice-melting process. On the AC structure, friction property gets gradually enhanced when the surface temperature is higher than 4 °C, while it keeps decreasing until 8 °C and then increases significantly to the maximum value on the SMA surface. Antiskid-D presents the same BPN varying trend, as well as the maximum value at the final stage, both on AC and SMA structures.

Figure 9 illustrates that when the same surface construction method was used, basalt aggregate would provide better skid resistance than that from dolerite aggregates. In the ice-melting condition, the four curves illustrate that the surface friction will dramatically increase to the maximum BPN value within less than 6 min, at the later period in the melting stage.

Angularity from AIMS analysis describes variation at the particle boundary that influences the overall shape. Higher angularity values illustrate more angular shape. A perfect circle has a small limited value of angularity, which is close to zero [23]. The definition in AIMS analysis provides a relative scale from 0 to 10,000 for angularity measurement. Figure 10 compares the angularities between basalt aggregate and dolerite aggregate.

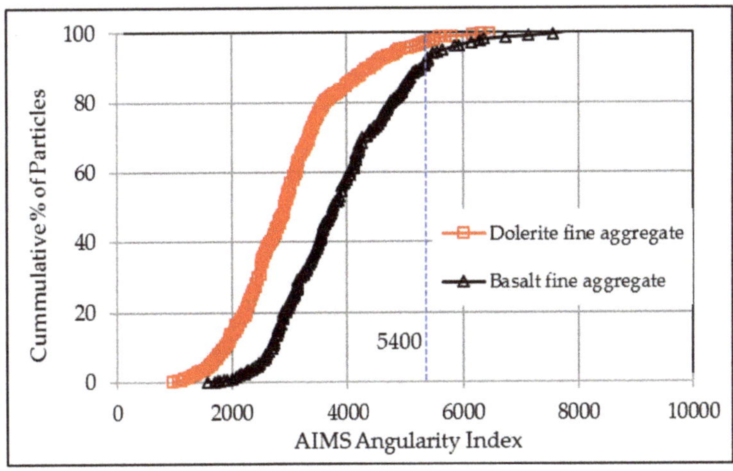

Figure 10. Angularity distributions of the fine aggregates used.

Firstly, according to the definition of AIMS angularity, aggregates with values higher than 5400 fall into the extreme category. Figure 10 illustrates that both basalt and dolerite aggregates presented very small angularities, with less than 10% of aggregates with angularity values higher than 5400. Secondly, it can clearly be seen that basalt fine aggregate had higher angularity values than that of dolerite aggregate. According to literatures [24,25], higher angularity will result in better aggregate interlocking and higher friction coefficient, which also consistent with the BPN results discussed in Figure 9.

AIMS also defined Form2D and described the relative form from 2-dimensional images of fine aggregates, as Figure 2 shows. Form2D index is explained by Equation (1). It has a relative scale from 0 to 20, while 0 means a perfect circle.

$$\text{Form2D} = \sum_{\theta=0}^{\theta=360-\Delta\theta} \left[\frac{R_{\theta+\Delta\theta} - R_\theta}{R_\theta} \right], \tag{1}$$

where: R_θ is the radius of a particle at an angle of θ, $\Delta\theta$ is the incremental difference in the angle.

Figure 11 shows that basalt aggregate and dolerite aggregate have similar distributions of form2D index. This means that basalt aggregates have quite same sphericity as dolerite aggregates have, with only small difference at ranges that form2D values are smaller than 8.

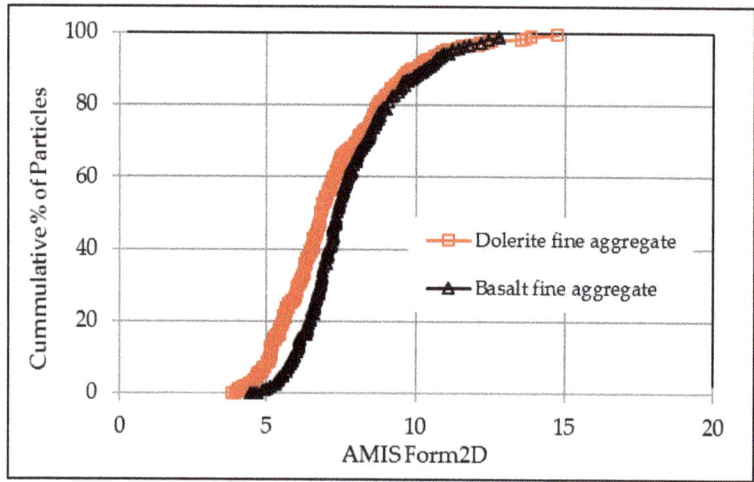

Figure 11. Form2D distributions of the used fine aggregates.

3.3. Ice Thickness Dependencies

Fujimoto conducted a quantitative evaluation of the relation between ice thickness on a road surface and the skid resistance property. He confirmed that the skid resistance decreases with increases in the ice film thickness [26]. This research investigates the ice thickness dependency of surface friction by comparing the BPN values during the ice-melting process between 1mm thick and 2 mm thick ice conditions. The research results are presented in Figure 12. It illustrates that thicker ice layer will result in a lower skid resistance property and smaller "lowest BPN". The lowest BPN for the 1 mm condition is 20, while it is 18 for the 2 mm condition. Very obviously that thicker ice layer needs more time to fully melt, hence the longer melting stage.

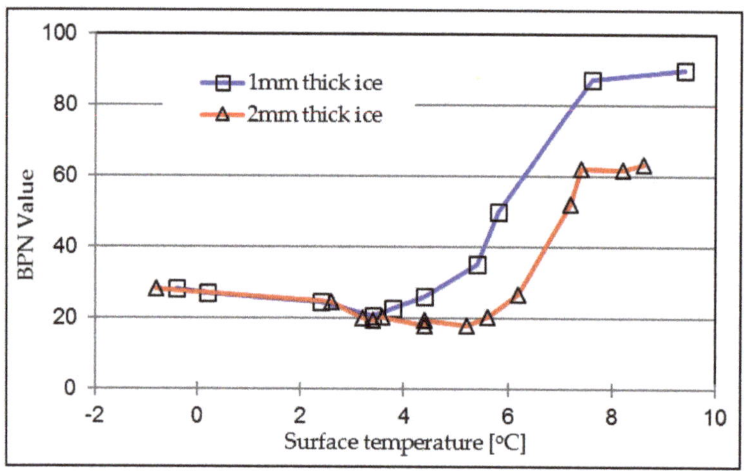

Figure 12. Influence of ice thickness on BPN values on asphalt concrete (AC) with antiskid-B.

4. Conclusions

The durability of skid resistance during the ice-melting process at temperature from −5 °C to 10 °C was characterized by means of a BP Tester. SMA-16, AC-16 and two types of antiskid surface layer, with 1mm thick and 2 mm thick ice layers, were studied. Based on the research results, the following conclusions can be drawn:

1. Investigated surfaces have a concave trend between friction and surface condition. Lowest BPN value occurs during the ice-melting stage, and then increases gradually with increasing temperature. Melted ice will result in water flow from peak points to gorges between aggregates, leading to lower average depth. Hence the smallest macrotexture and then lowest friction property will be presented.

2. The investigated antiskid layers can significantly increase the surface friction prope rties during icy seasons. The lowest BPN values of AC and SMA surfaces during the ice-melting process have been enhanced from 11.6 and 14.4 to a value of around 20, respectively, when an antiskid surface has been applied. Aggregate image analysis indicated that basalt fine aggregate had higher angularity values than that of dolerite aggregate, resulting in a better friction coefficient. But the lowest BPN values of the two studied antiskid layers are the same.

3. A thicker ice layer will result in a lower skid resistance property and smaller "lowest BPN". The lowest BPN for the 1mm condition is 20, while it is 18 for the 2 mm condition.

Author Contributions: Conceptualization, B.Y. and H.M.; methodology, B.Y. and S.Z; software, H.M. and P.Z.; validation, B.Y., S.Z. and X.Z.; formal analysis, B.Y. and X.Z.; investigation, S.Z.; resources, H.M. and P.Z.; data curation, B.Y. and H.M.; writing—original draft preparation, B.Y. and S.Z.; writing—review and editing, S.Z. and H.M.; visualization, S.Z. and H.M.; supervision, S.Z.

Funding: This research was funded by National Natural Science Foundation of China (No. U1733121) and the 111 Project (No. B18038).

Acknowledgments: This research was funded by National Natural Science Foundation of China and 111 Project, and the authors gratefully acknowledge the financial support.

Conflicts of Interest: The authors declare no conflict of interest.

References

1. Pan, P.; Wu, S.; Xiao, Y.; Liu, G. A review on hydronic asphalt pavement for energy harvesting and snow melting. *Renew. Sustain. Energy Rev.* **2015**, *48*, 624–634. [CrossRef]

2. Cui, P.; Wu, S.; Xiao, Y.; Wan, M.; Cui, P. Inhibiting effect of layered double hydroxides on the emissions of volatile organic compounds from bituminous materials. *J. Clean. Prod.* **2015**, *108 (Pt A)*, 987–991. [CrossRef]

3. Herndon, D.A.; Xiao, F.; Amirkhanian, S.; Wang, H. Investigation of Los Angeles value and alternate aggregate gradations in OGFC mixtures. *Constr. Build. Mater.* **2016**, *110*, 278–285. [CrossRef]

4. Guo, Y.; Ma, B.; Zhi, Z.; Tan, H.; Liu, M.; Jian, S.; Guo, Y. Effect of polyacrylic acid emulsion on fluidity of cement paste. *Coll. Surfaces A Physicochem. Eng. Asp.* **2017**, *535*, 139–148. [CrossRef]

5. Asi, I.M. Evaluating skid resistance of different asphalt concrete mixes. *Build. Environ.* **2007**, *42*, 325–329. [CrossRef]

6. Xiao, Y.; van de Ven, M.F.C.; Molenaar, A.; Wu, S.; Verwaal, W. Surface texture of antiskid surface layers used on runways. In Proceedings of the TRB 90th Annual Meeting, Washington, DC, USA, 23–27 January 2011; p. 11.

7. Chen, Z.; Wu, S.; Xiao, Y.; Zeng, W.; Yi, M.; Wan, J. Effect of hydration and silicone resin on Basic Oxygen Furnace slag and its asphalt mixture. *J. Clean. Prod.* **2016**, *112 (Pt 1)*, 392–400. [CrossRef]

8. Bazlamit, S.M.; Reza, F. Changes in asphalt pavement friction components and adjustment of skid number for temperature. *J. Transp. Eng.* **2005**, *131*, 470–476. [CrossRef]

9. Kogbara, R.B.; Masad, E.A.; Kassem, E.; Scarpas, A.; Anupam, K. A state-of-the-art review of parameters influencing measurement and modeling of skid resistance of asphalt pavements. *Constr. Build. Mater.* **2016**, *114*, 602–617. [CrossRef]

10. Khasawneh, M.A.; Liang, R.Y. Temperature effect on frictional properties of HMA at different polishing stages. *Jordan J. Civ. Eng.* **2012**, *6*, 39–53.

11. Burchett, J.L.; Rizenbergs, R.L. Seasonal variations in the skid resistance of pavements in Kentucky. *Transp. Res. Rec.* **1980**, *788*, 6–14.

12. Elkin, B.L.; Kercher, K.J.; Gulen, S. Seasonal variation in skid resistance of bituminous surfaces in Indiana. *Transp. Res. Rec. J. Transp. Res. Board* **1980**, *777*, 50–58.

13. Bianchini, A.; Heitzman, M.; Maghsoodloo, S. Evaluation of temperature influence on friction measurements. *J. Transp. Eng.* **2011**, *137*, 640–647. [CrossRef]

14. Baran, E. Temperature influence on skid resistance measurement. In Proceedings of the 3rd International Surface Friction Conference, Safer Road Surfaces—Saving Lives, Gold Coast, Australia, 15–18 May 2011; p. 18.

15. Anupam, K.; Srirangam, S.K.; Scarpas, A.; Kasbergen, C. Influence of temperature on tire -pavement friction: Analyses. *Transp. Res. Rec. J. Transp. Res. Board* **2013**, *2369*, 114–124. [CrossRef]

16. Hadiwardoyo, S.P.; Sinaga, E.S.; Fikri, H. The influence of Buton asphalt additive on skid resistance based on penetration index and temperature. *Constr. Build. Mater.* **2013**, *42*, 5–10. [CrossRef]

17. Wang, D.; Schacht, A.; Schmidt, S.; Oeser, M.; Steinauer, B. Influence of the Testing Temperatures on Skid Resistance Measurements on Roads. Available online: https://trid.trb.org/view/1316575 (accessed on 12 April 2019).

18. Ma, T.; Geng, L.; Ding, X.; Zhang, D.; Huang, X. Experimental study of deicing asphalt mixture with anti-icing additives. *Constr. Build. Mater.* **2016**, *127*, 653–662. [CrossRef]

19. Xiao, Y.; Wang, F.; Cui, P.; Lei, L.; Lin, J.; Yi, M. Evaluation of fine aggregate morphology by image method and its effect on skid-resistance of micro-surfacing. *Materials* **2018**, *11*, 920. [CrossRef] [PubMed]

20. Cui, P.; Xiao, Y.; Yan, B.; Li, M.; Wu, S. Morphological characteristics of aggregates and their influence on the performance of asphalt mixture. *Constr. Build. Mater.* **2018**, *186*, 303–312. [CrossRef]

21. Institute MoTHR. *JTG H20-2007 China Highway Performance Assessment Standards*; China Communications Publishing House: Beijing, China, 2007.

22. Smith, R.H. *Analyzing Friction in the Design of Rubber Products and Their Paired Surfaces*; CRC Press: Boca Raton, FL, USA, 2008.

23. Kuang, D.; Wang, X.; Jiao, Y.; Zhang, B.; Liu, Y.; Chen, H. Influence of angularity and roughness of coarse aggregates on asphalt mixture performance. *Constr. Build. Mater.* **2019**, *200*, 681–686. [CrossRef]

24. Chen, S.; Yang, X.; You, Z.; Wang, M. Innovation of aggregate angularity characterization using gradient approach based upon the traditional and modified Sobel operation. *Constr. Build. Mater.* **2016**, *120*, 442–449. [CrossRef]

25. Wang, D.; Chen, X.; Xie, X.; Stanjek, H.; Oeser, M.; Steinauer, B. A study of the laboratory polishing behavior of granite as road surfacing aggregate. *Constr. Build. Mater.* **2015**, *89*, 25–35. [CrossRef]

26. Fujimoto, A.; Kiriishi, M.; Tokunaga, R.; Takahashi, N.; Ishida, T. Method for predicting change in skid resistance on icy road surface by use of deicing salt. In Proceedings of the 94th TRB Annual Meeting, Washington, DC, USA, 11–15 January 2015; p. 14.

Article

Optimizing Gradation Design for Ultra-Thin Wearing Course Asphalt

Wentian Cui, Kuanghuai Wu, Xu Cai, Haizhu Tang and Wenke Huang *

School of Civil Engineering, Guangzhou University, Guangzhou 510006, China; wentiancui1@163.com (W.C.); wukuanghuai@163.com (K.W.); cx_caixu@163.com (X.C.); 13553932632@163.com (H.T.)
* Correspondence: h.wenke@gzhu.edu.cn; Tel.: +86-135-1275-7010

Received: 2 December 2019; Accepted: 30 December 2019; Published: 2 January 2020

Abstract: In recent years, ultra-thin wearing course asphalt mixture has been widely used in the reconstruction of old road surfaces and the functional layer of new road surfaces due to its good road performance. To improve the rutting resistance of ultra-thin wearing course asphalt mixture, this research presents an Ultra-thin Wearing Course-10 (UTWC-10) asphalt mixture with good high-temperature stability and skid resistance based on the Taylor system standard mesh specifications. The Course Aggregate Void Filling (CAVF) method is used to design the UTWC-10 asphalt mixture, which is compared with two other traditional ultra-thin wearing course asphalt mixtures on the basis of different laboratory performance tests. The high-temperature rutting test data shows that the rutting dynamic stability (DS) index of the UTWC-10 asphalt mixture is much higher than that of traditional wearing course asphalt mixtures, as it has better high-temperature stability. Moreover, anti-sliding performance attenuation tests are conducted by a coarse aggregate polishing machine. The wear test results show that the skid resistance of the UTWC-10 asphalt mixture is promising. The anti-sliding performance attenuation test can effectively reflect the skid resistance attenuation trend of asphalt pavement at the long-term vehicle load. It is verified that the designed UTWC-10 asphalt mixture shows excellent high-temperature rutting resistance and skid resistance, as well as better low temperature crack resistance and water stability than the traditional wearing course asphalt mixtures.

Keywords: functional pavement layer; ultra-thin wearing course asphalt mixture; high-temperature stability; pavement performance test

---

1. Introduction

As the economy continues to grow and urbanization deepens, the traffic load is also rising year by year [1,2]. Most of the roads will suffer from defects such as cracking, loosening, and deformation after 3–5 years of service, which will sharply weaken important pavement performance factors like water damage resistance, high-temperature stability, and skid resistance. These defects will not only shorten the service life of roads but will also bring severe threats to comfort and safety [3]. To reduce the occurrence of these pavement defects, researchers have started to study preventive pavement maintenance technologies [4]. The ultra-thin wearing course asphalt mixture can enhance the whole flatness and skid resistance of roads, reduce pavement noise, improve driving comfort and safety, prolong the time until surface cracking, fix local potholes, and other defects on pavement occur, as well as enhance the service life of roads [5–7]. Therefore, it has been widely used in the pavement rehabilitation for the extension of the service life of roads.

As an overlay, the ultra-thin wearing course is the uppermost layer in pavement structures. In case of insufficient rutting resistance of ultra-thin wearing course, defects such as waves, displacements, and ruts will occur after a repetitive traffic load, thus affecting the slip resistance and driving comfort of pavement and threatening the safety of driving. However, few studies of ultra-thin wearing course were focused on the high-temperature resistance to permanent deformation [8–12].

At present, coarse aggregate is mostly concentrated at 4.75–9.5 mm, when the gradation design of ultra-thin wearing course asphalt mixture is designed to be open- or semi-graded. However, because the distance between the 4.75 mm mesh hole and 9.5 mm mesh hole is slightly larger, the uniformity of the coarse aggregate cannot be controlled. This will reduce the contact points between coarse aggregates and affect the spatial structure of the overall framework of the asphalt mixture. As a result, the embedded extrusion force between coarse aggregates will be reduced, the dynamic stability (DS) of the ultra-thin wearing course will be low, and rutting resistance will be insufficient [13,14]. Therefore, 8 mm (two meshes) and 6.7 mm (three meshes) mesh holes are added between 4.75 mm (four meshes) and 9.5 mm according to the Taylor system standard mesh specifications [15].

This research presents a new gradation design method to improve the high-temperature performance of the Ultra-thin Wearing Course-10 (UTWC-10) asphalt mixture based on the Course Aggregate Void Filling (CAVF) designed method and the mechanical performances were evaluated with two other commonly used open-graded friction courses OGFC-7 and Novachip-B asphalt mixture.

2. Materials and Methods

2.1. Raw Materials

2.1.1. Aggregates

Diabase, consisting of 5–10 mm macadam, 3–5 mm macadam, and 0–3 mm stone chips was used as the aggregate of asphalt mixture. Table 1 summarizes the basic performance test results. Tests were conducted according to the Chinese specifications [16].

Table 1. Properties of Aggregates.

Aggregate	Test Items	Standard Results	Test Results		
			5–10	3–5	0–3
Coarse aggregate	Crushed value (%)	≤26	4.9	5.3	-
	Abrasion value (%)	≤28	6.1	6.1	-
	Polished value PSV	≥42	49	-	-
	Needle shape (%)	≤15	3.2	-	-
	Water absorption rate (%)	≤2.0	0.87	0.96	-
	Gross bulk density (g/cm^3)	-	2.90	2.89	-
	Apparent density (g/cm^3)	>2.6	2.93	2.92	-
	Adhesion to asphalt (level)	≥5	5	-	-
Fine aggregate	Apparent density (g/cm^3)	≥2.5	-	-	2.93
	Sand equivalent (%)	≥60	-	-	87
	Methylene blue value (g/kg)	≤25	-	-	14

2.1.2. Asphalt Binder

Here, high viscosity asphalt was used as asphalt binder. Table 2 shows the test results obtained for basic rheological properties. Tests were conducted according to the Chinese specifications [17].

2.1.3. Mineral Filler

The filler used was alkaline limestone mineral filler. The impurities in the mineral filler were removed. Table 3 summarizes the obtained performance test results. Tests were conducted according to the Chinese specifications [16].

Table 2. Properties of asphalt binder.

Test Items	Unit	Standard Results	Test Results
Penetration index (25 °C)	0.1 mm	40–80	43.1
Softening Point $T_{R\&B}$	°C	≥70	81.6
Ductility (5 °C)	cm	≥20	29
Relative density of asphalt	g/cm^3	-	1.023
Rotational viscosity 135 °C	Pa·s	-	2.958
Flexible recovery 25 °C	%	≥85	92
Rotating film oven test (163 °C)			
Quality loss	%	≤0.6	0.09
Penetration ratio	%	≥65	79
Ductility 5 °C	cm	≥15	20

Table 3. Properties of mineral filler.

Test Items		Unit	Standard Results	Test Results
Apparent relative density		g/cm^3	≥2.5	2.864
Hydrophilic coefficient		—	<1	0.6
Plasticity index		—	<4	1
Water content		%	≥1	0.08
Particle size range	<0.6 mm	%	100	100
	<0.15 mm	%	90–100	99.1
	<0.075 mm	%	70–100	97.7

2.2. Methods and Preparation

2.2.1. Asphalt Mixture Gradation Design

In the design of the target mix proportion of ultra-thin wearing course UTWC-10, the skid resistance and high-temperature stability performance of asphalt mixture pavement need to be considered. The gradation curve can be more reasonably controlled by adding 6.7 mm mesh holes and 8 mm mesh holes between 4.75 mm mesh holes and 9.5 mm mesh holes, in accordance with the Taylor system standard mesh specifications. Based on the porosity, the Course Aggregate Void Filling (CAVF) method was used to design three asphalt mixtures of ultra-thin wearing course with different porosities, and UTWC-10 was compared with OGFC-7 and Novachip-B. UTWC-10 was used as an example to illustrate the application of the CAVF method in the gradation design of asphalt mixture of ultra-thin wearing course, and that in Novachip-B and OGFC-7 was similar. The gradation design of fine aggregate was designed by the Fowler series as shown in Equation (1). The design process is shown in Figure 1 and durability test process is summarized in Figure 2. According to the specifications [16], the gross bulk density ρ_b of the coarse aggregate was 2.90 g/cm^3, and the bulk density ρ of the compacted state was 1.72 g/cm^3. Therefore, it can be calculated from Equation (2) that the VCA$_{DRC}$ was 40.7%. The amount of mineral filler (q_p), the target void ratio (V_v) the asphalt content (q_a) were 5%, 12%, and 5%, respectively. The amounts of coarse and fine aggregate were calculated using Equations (3) and (4) [18]:

$$P = \left(\frac{d}{D}\right)^n \tag{1}$$

$$VCA_{DRC} = \left(1 - \frac{\rho}{\rho_b}\right) \times 100 \tag{2}$$

$$q_c + q_f + q_p = 100 \tag{3}$$

$$\frac{q_c \times (VCA_{DRC} - Vv)}{100 \times \rho} = \frac{q_f}{\rho_{af}} + \frac{q_p}{\rho_f} + \frac{q_a}{\rho_a} \tag{4}$$

where, P is the percentage of aggregate passing through the mesh size (%); d is the mesh size in mm; D is the maximum particle size of the aggregate in mm; n is the index, $0.3 \leq n \leq 0.5$; ρ_b is the gross bulk density of the synthetic coarse aggregate in g/cm^3; ρ is the density of accumulation under rammed state in g/cm^3; ρ_{af} is the apparent relative density of fine aggregate in g/cm^3; ρ_a is the relative density of asphalt in g/cm^3; and ρ_f is the relative density of mineral filler.

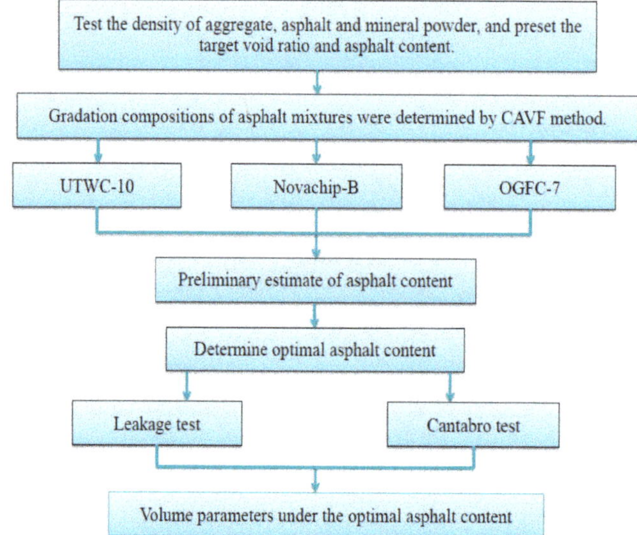

Figure 1. Design process of three asphalt mixtures.

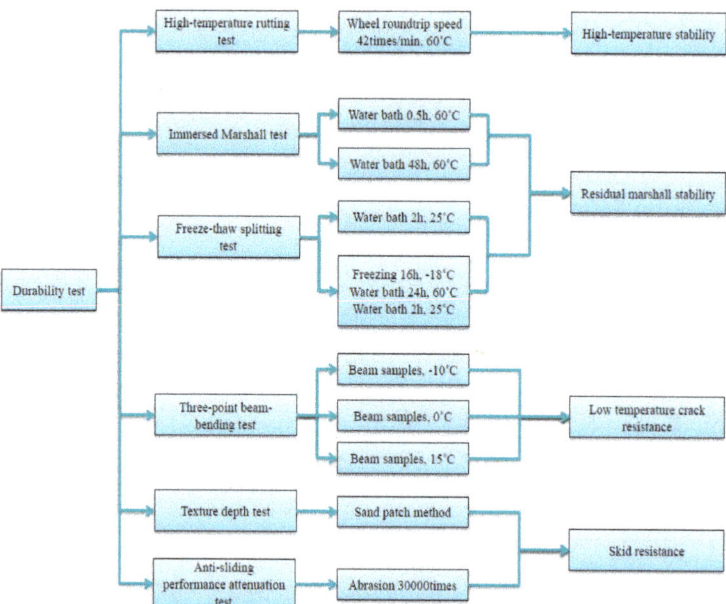

Figure 2. Flow chart of the durability test of three asphalt mixtures.

Then, ρ_{af} = 2.93 g/cm^3, ρ_a = 1.023 g/cm^3, and ρ_f = 2.864 g/cm^3 were substituted in Equation (4). The quantity of coarse aggregate (q_c) was found to be 72.2%, and the quantity of fine aggregate (q_f) was 22.8%. The gradation of coarse and fine aggregates was fitted to a curve. The mesh hole passing rate and gradation curves of UTWC-10, Novachip-B, and OGFC-7 were designed by the CAVF method. The gradation compositions of the three ultra-thin wearing courses are shown in Table 4, and the gradation curves are shown in Figure 3.

Table 4. The gradation compositions of the three asphalt mixtures.

Gradation Type	Passing Rate (%) under Different Mesh Apertures (mm)										
	13.2	9.5	8	6.7	4.75	2.36	1.18	0.6	0.3	0.15	0.075
Novachip-B	100	97.6	-	-	32.5	25.3	17.6	13.9	9.8	7.5	5.6
OGFC-7	100	100	-	98.3	86.2	20.3	16.8	15.4	12.3	10.0	6.4
UTWC-10	100	98.8	96.5	87.7	34.0	27.8	23.3	16	11.4	7.7	5

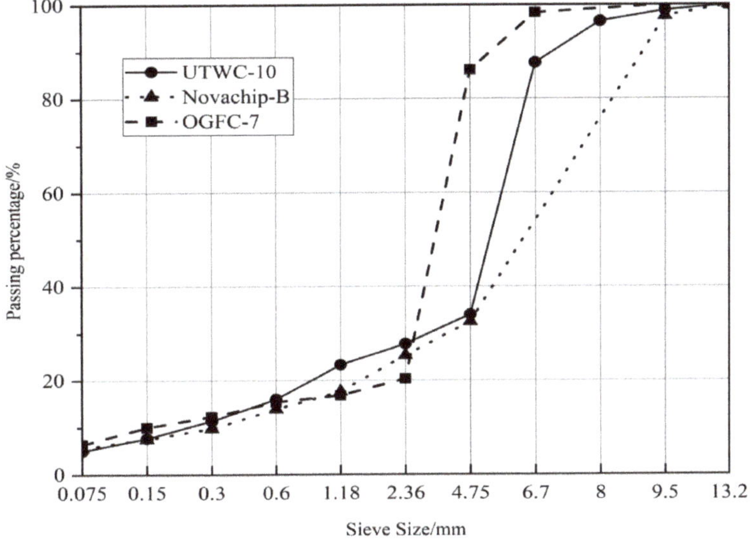

Figure 3. The gradation curves of the three asphalt mixtures.

2.2.2. Determination of the Optimal Asphalt Content in Asphalt Mixture of the Ultra-Thin Wearing Course

The optimal asphalt content of UTWC-10 and Novachip-B were determined using the asphalt film thickness and void ratio, where the optimal thickness was 10 μm [19] and the target porosity was 12%. The optimal asphalt content of OGFC-7 was determined to be 4.5% through a leakage test and scattering test.

The molded five groups' asphalt content of asphalt mixture are 4.4%, 4.7%, 5.0%, 5.3% and 5.6% respectively, with 0.3% spacing distance between each group. Then, the optimum asphalt content is confirmed to be 5.0% for Novachip-B and UTWC-10 according to bulk density, void fraction, VMA, VFA, asphalt membrane thickness test data, and technical requirements of corresponding test.

The optimal asphalt content of OGFC-7 was determined through leakage test and scattering test. The five groups asphalt mixture with asphalt content to be 3.9%, 4.2%, 4.5%, 4.8%, and 5.1% shall be stirred. Marshall compaction instrument is used to mold the test specimen. It compacts 25 times on both sides. The formed specimen is put in the Los Angles abrasion machine to turn 300 circles at a speed of 33 r/min. Then, the ratio of lost weight and the original weight of the specimen is the flying loss. When the asphalt content of the open graded friction course OGFC-7 is between 4.4% and 4.6%, the change rate in leakage analysis loss and flying loss varies greatly. And since both the two losses of the asphalt mixture at this time meet the specification requirement, 4.5% is the optimum asphalt content of OGFC-7.

The volume parameters of the three asphalt mixtures under the optimum asphalt content are shown in Table 5, and the detailed volume parameters of the three asphalt mixtures can be seen in Tables A1–A5 in Appendix A.

Table 5. The volume parameters of thea three asphalt mixtures.

Asphalt Mixture Type	Optimal Asphalt Content (%)	Void Ratio (%)	VMA (%)	VFA (%)	Stability (kN)	Flow Value (mm)
UTWC-10	5.0	12.1	21.9	44.7	8.64	27.4
Novachip-B	5.0	12.3	22.4	43.5	8.43	32.1
OGFC-7	4.5	20.3	28.3	28.3	6.80	38.5

2.3. Methods and Tests

2.3.1. High-Temperature Rutting Test

Rutting dynamic stability (DS) index can be used to check the asphalt texture stability at a high temperature. The larger the DS value, the better asphalt mixture performance in resistance to deformation and high temperature. Rutting modeling machine of asphalt mixture is used to mold the 300 mm × 300 mm × 50 mm rutting plate. The wheel tracking tests instrument is shown in Figure 4. Then, it is put for curing for 48 h. Rut test can be conducted after 5 h at temperature 60 °C. Wheel pressure and roundtrip speed of 0.7 MPa and 42 times/min were applied in tests. The wheel driving direction is consistent with the rolling direction in specimen molding. The deformation of asphalt mixture is 45 min, and 60 min is recorded separately. Total round trip times divides the gap of specimen deformation in 60 min and 45 min to gain the value of DS. DS can be calculated using Equation (5):

$$DS = \frac{(t_2 - t_1) \times 42}{d_2 - d_1} \times c_1 \times c_2 \tag{5}$$

where d_1 and d_2 are tracking depths at 45 and 60 min, t_1 and t_2 are 45 and 60 min, respectively; c_1 and c_2 are correction factors.

Figure 4. Wheel tracking tests instrument.

2.3.2. Three-point Beam-bending Test

The low temperature crack resistance of asphalt mixture was evaluated by the small beam specimen bend test at low temperature. The formed rutting plate is cut to 250 mm × 30 mm × 35 mm trabecular specimen. Then, all specimens will be put in the incubator for about 6 h to enable the interior of specimen to reach the given temperature. The test temperature is −10 °C, 0 °C, and 15 °C. The specimen reaching the temperature shall be taken out to be put in the two-point support frame. The universal testing machine will load by means of mid-point loading at a speed of 50mm/min. The maximum bending strength and strain at failure were calculated and employed as evaluation indices for asphalt mixture low temperature crack resistance. Calculations were performed using Equations (6)–(8):

$$R = \frac{3LP}{2bh^2} \tag{6}$$

$$\varepsilon = \frac{6hd}{L^2} \tag{7}$$

$$S = \frac{R}{\varepsilon} \tag{8}$$

where ε is maximum bending strain at failure; P is breaking load (N); R is damage strength (MPa); S is stiffness modulus (MPa); h is cross-section sample height (mm); L sample length (mm); d is mid-span deflection at sample breaking point (mm); and b is test specimen width across middle section (mm).

2.3.3. Immersed Marshall Test

The immersed Marshall method is quite simple and highly practical. The standard and formed Marshall test specimens are divided into two groups. One group is cured in 60 °C thermostatic water tank for 30 min, then goes for the Marshall stability test. The other group is cured for 48 h in 60 °C thermostatic water tank, then goes for the stability test. The ratio of the two stability values is residual stability. The closer the residual stability is to 1, the better asphalt mixture water stability. Calculations were performed based on Equation (9):

$$MS = \frac{S_2}{S_1} \tag{9}$$

where MS is test specimen residual stability (%); S_1 is stability after immersing test specimen in water for 30 min (kN); and S_2 is stability after immersing test specimen in water for 48 h (kN).

2.3.4. Freeze-Thaw Splitting Test

The Marshall test method is adopted to mold two test specimen groups. One group is immersed in 25 °C water bath for 2 h, then its splitting strength (R_1) is tested. The other group is immersed for vacuum treatment and water saturation for 15 min at 98 kPa, then the vacuum valve is switched on to cure for 30 min in water bath in ordinary pressure. 10 mL water shall be poured after the vacuum and water-saturated test specimen is sealed with a plastic bag. Next, it is tightened, sealed, and put in the −18 °C incubator to cure for 16 h. After that, the specimen saved in low temperature is taken out from the incubator and put in the 60 °C thermostatic water tank to cure for 24 h. At last, the test specimen is taken out from the hot water bath and put in 25 °C water. As can be seen from Figure 5, specimens were then removed to perform splitting tests according to the Chinese specifications [17]. After 2 h of constant temperature curing, its splitting strength (R_2) is tested. The ratio of R_2 and R_1 is splitting tensile strength ratio (TSR). Calculations were performed using Equations (10) and (11):

$$R = 0.006287\frac{P}{h} \tag{10}$$

$$TSR = \frac{R_2}{R_1} \tag{11}$$

where TSR is freeze thaw splitting strength ratio; R is splitting tensile strength (MPa); R_1 and R_2 are average splitting tensile strength without and after freeze-thaw cycle (MPa), respectively; P is single specimen test load (N); and h is single specimen height (mm).

Figure 5. Freeze-thaw splitting test.

2.3.5. Texture Depth Test

The sand patch method is commonly used in texture depth test, which is simple and widely applied. Firstly, 0.15 mm to 3 mm dry and clean sands shall be prepared. Then, they are filled into a 25 mL sand measuring cylinder, knock on the cylinder and bulldoze the cylinder mouth. After the surface of the rutting plate specimen is cleaned, fine sands are poured into the cylinder slowly. Then, the sands are spread outward to a circle with a push plate as much as possible and filled in the interspace of the test specimen. Lastly, the diameter of the circle in the two vertical directions can be measured and its average value shall be gained. The calculation method is shown in Equation (12):

$$TD = \frac{1000V}{\pi D^2/4} = \frac{31831}{D^2} \tag{12}$$

where TD is pavement texture depth (mm); V is sand volume (cm^3); and D is paved sand average diameter (mm).

2.3.6. British Pendulum Number (BPN) Test

The BPN test makes use of pendulum type friction coefficient measuring instrument (BPN tester) to get the BPN in bituminous pavement and cement concrete pavement in order to evaluate the anti-slide performance of the pavement in wet environments. The film in the bottom of the pendulum bob stands for the wheel. The pendulum bob falls from a certain height and the film in the bottom will rub for a certain distance on the pavement before swinging back. The height of swinging back is the pendulum of this section. In this article, the test specimen of the rut plate standards for tested pavement.

2.3.7. Anti-Sliding Performance Attenuation Test

In order to simulate the actual paving thickness of ultra-thin wearing course, 1.5 cm thick rutting plate is molded according to the method described in Section 2.3.1. The rutting plate is cut to 30 mm × 1.5 mm × 8 mm slices and put in the steps of standard test film. Then, glue sands are made in the test of polishing value of coarse aggregate and filled in the interspace of the test film (Figure 6). After the test specimen demolds, it shall be put in the road wheel of the polishing machine. The rubber load and function times of the polishing machine shall be adjusted to test the BPN variance before and after abrasion. The value of BPN can be obtained by using pendulum type friction coefficient measuring instrument (BPN tester). The anti-slide performance attenuation acts as the main index to evaluate the anti-slide performance of asphalt mixture.

(a)   (b)

Figure 6. Wear test system: (**a**) coarse aggregate polishing machine; (**b**) ultra-thin wearing course specimen.

3. Results and Discussion

3.1. High-Temperature Rutting Resistance

The dynamic stability of the three asphalt mixtures meets the specification of being not less than 3000 times/mm [17]. For the UTWC-10 asphalt mixture shown in Table 6, the high-temperature stability was far more than that of OGFC-7, and the DS was 77.2% more than that of OGFC-7. Compared with Novachip-B, the high-temperature stability of UTWC-10 was strongly improved, and the DS was 36.9% higher than that of Novachip-B. The reason for the improvement in the high-temperature stability of the ultra-thin wearing course is that when UTWC-10 was optimized for gradation design, 8 mm (two meshes) and 6.7 mm (three meshes) mesh holes were added between 4.75 mm (four meshes) and 9.5 mm, according to the Taylor system standard mesh specifications. Coarse aggregates with a particle size of 4.75–6.7mm accounted for about 50% of the gradation. This increased the contact points between coarse aggregates, which made the spatial structure of the overall framework more reasonable and improved the embedded extrusion forces between coarse aggregates [20,21].

Table 6. The high-temperature rutting test results of the three asphalt mixtures.

Gradation	45 min Deformation (mm)	60 min Deformation (mm)	Deformation Difference (mm)	DS (times/mm)	Mean Value of DS (times/mm)
UTWC-10	1.26	1.38	0.12	5250.00	5568
	1.30	1.41	0.11	5727.27	
	1.28	1.39	0.11	5727.27	
Novachip-B	1.47	1.63	0.16	3937.5	4025
	1.51	1.66	0.15	4200.00	
	1.49	1.65	0.16	3937.5	
OGFC-7	2.10	2.30	0.20	3150.00	3150
	2.13	2.33	0.21	2985.78	
	2.15	2.34	0.19	3314.22	

3.2. Low Temperature Crack Resistance

As can be seen from Figure 7, the stress and strain changes of the UTWC-10, Novachip-B, and OGFC-7 asphalt mixtures were similar. With temperatures from −10 °C to 15 °C, the bending strain of UTWC-10 increased by 34.5%, that of Novachip-B increased by 32.8%, and that of OGFC-7 increased by 12.9%. The bending strain of UTWC-10 increased the most with the increase in temperature. As the temperature rose, the ductility of the asphalt increased, improving the strain of the asphalt mixture. The bending strength of the three asphalt mixtures tended to be low with an increasing temperature, with a strength drop of 13.5% for UTWC-10, 14.3% for Novachip-B, and 28.5% for OGFC-7. The bending strength of UTWC-10 was least affected by temperature, because the rising temperature gradually increased the embedded squeeze forces within the framework of the UTWC-10 asphalt mixture. The bending stiffness modulus is the ratio of the bending strength and bending strain, and it is an important index for the evaluation of the low-temperature crack resistance of the asphalt mixture. The smaller the bending stiffness is, the better the elastoplasticity of the asphalt mixture is at the same damage strength. As the temperature increased, the bending stiffness modulus of UTWC-10 decreased by 35.7%, that of Novachip-B decreased by 35.4%, and that of OGFC-7 decreased by 36.7%. Because UTWC-10 had the fastest reduction in the bending stiffness modulus and the smallest bending stiffness modulus, it was shown to have the best crack resistance at low temperatures.

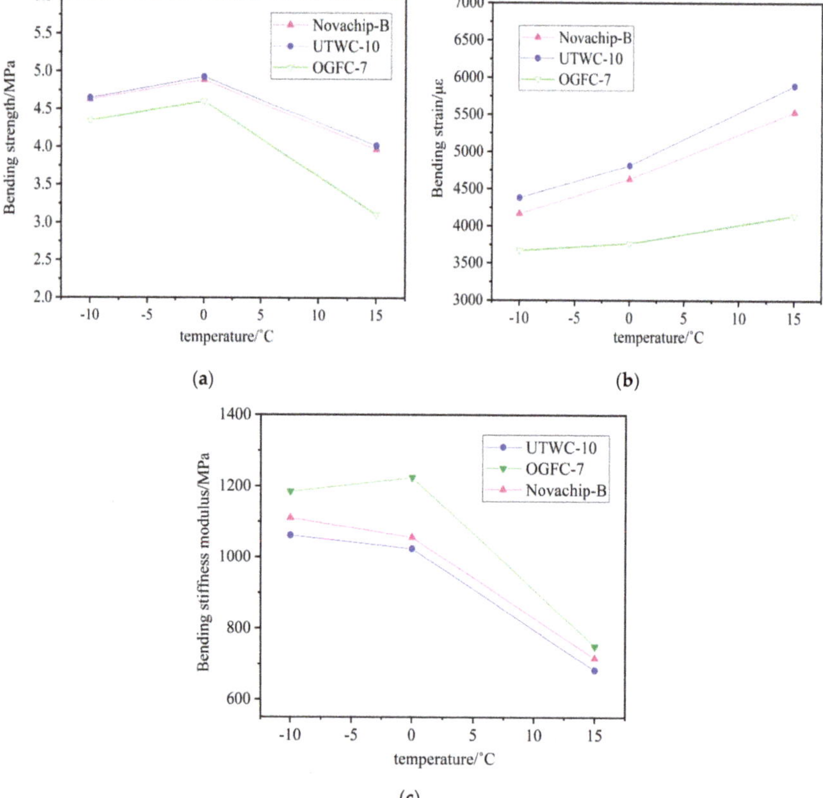

Figure 7. The various performance indexes of the three asphalt mixtures: (**a**) relationship between the bending strain and temperature; (**b**) relationship between the bending strength and temperature; and (**c**) relationship between the bending stiffness modulus and temperature.

3.3. Water Stability

3.3.1. Immersed Marshall Test Results

Figure 8 shows that the residual stability of the three asphalt mixtures meets the requirements of the specifications [17]. The residual stability of the three asphalt mixtures is above 90%, and the residual stability of UTWC-10 was 94.3%, that of Novachip-B was 92.9%, and that of OGFC-7 was 90.8%. Due to the high strength of the high viscosity asphalt used, the damage effect of the water immersion on the specimen is greatly reduced, so the residual stability of the OGFC-7 asphalt mixture with large void ratio was also good. The residual stability of utwc-10 is high and has good water stability.

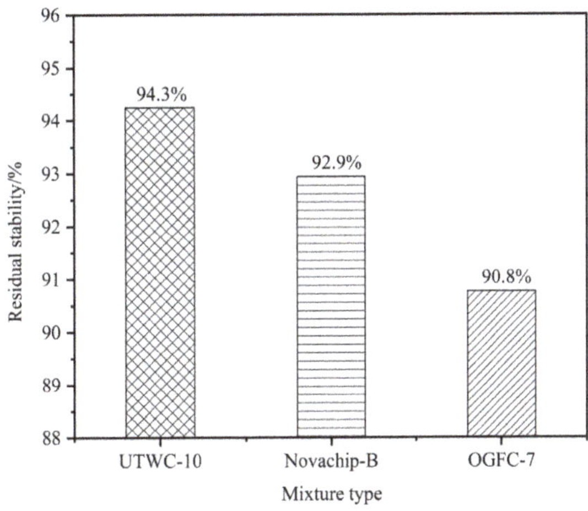

Figure 8. The average residual stability of the three asphalt mixtures.

3.3.2. Freeze-Thaw Splitting Test Results

It can be seen from Table 7 that the TSR value of UTWC-10 was 92.3%, the TSR value of Novachip-B was 90.4%, and the TSR value of OGFC-7 was 86.4%. Because the TSR value of UTWC-10 was the largest, the water stability performance of the UTWC-10 asphalt mixture was the best. The water porosity of the asphalt mixture will increase at low temperatures and produce tensile stress, causing micro cracks inside materials and thus attenuating the mechanical properties of the asphalt mixture. In the drainage material test of the OGFC-7 asphalt mixture, water was able to fill in most of the voids to reduce the free space of water after heaving. Therefore, the mechanical properties of heaved materials attenuated the fastest. UTWC-10 has a good framework strength and large density, and the heaving force of water has little effect on the mechanical properties of materials, so its freeze–thaw splitting strength ratio was shown to be the highest.

Table 7. Freeze-thaw splitting test results of three asphalt mixtures.

Asphalt Mixture Type	Original Splitting Strength (R_1)/MPa	Freeze-thaw Splitting Strength (R_2)/MPa	Strength Ratio	Standard Results
UTWC-10	0.765	0.706	92.3%	
Novachip-B	0.757	0.684	90.4%	≥80%
OGFC-7	0.543	0.469	86.4%	

3.4. Asphalt Mixture Surface Roughness

Figure 9 shows that the texture depths of the three asphalt mixtures meet the requirements of the specifications [17] greater than or equal to 0.55 mm. The texture depth of UTWC-10 was shown to be

0.75 mm, that of Novachip-B was 0.73 mm, and that of OGFC-7 was 0.92 mm. OGFC-7 was shown to have the largest texture depth. The research results show that the void ratio is proportional to the texture depth index. The larger the void ratio is, the larger the texture depth is [22,23].

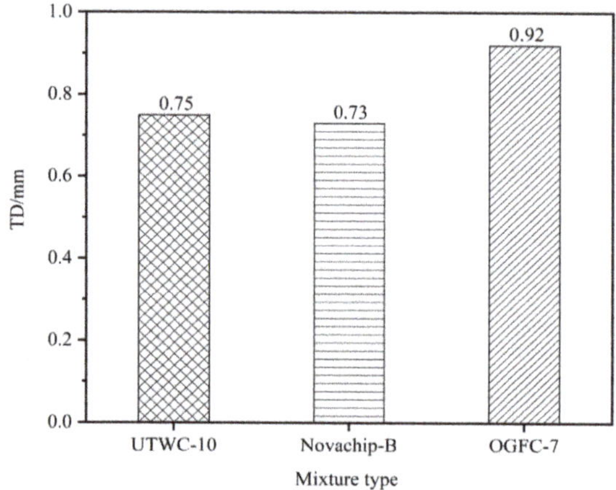

Figure 9. The texture depths of the three asphalt mixtures.

3.5. Anti-sliding Performance Attenuation

The anti-sliding performance attenuation test was completed by using a coarse aggregate polishing machine as shown in Figure 6. It can be seen from Figure 10 that as wear times rose, the pendulum BPN of the three asphalt mixtures showed a significant downward trend. The wearing results show that among the three ultra-thin wearing course, the attenuation rate of the anti-sliding performance of UTWC-10 was 15.80%, that of Novachip-B was 16.05%, and that of OGFC-7 was 18.18%. OGFC-7 was shown to have the highest anti-sliding performance attenuation rate. The attenuation rate of the anti-sliding performance of UTWC-10 and Novachip-B was similar. Besides, the BPN of Novachip-B was smaller than that of UTWC-10 and the skid resistance of UTWC-10 was the best. The anti-sliding performance attenuation test was able to effectively reflect the skid resistance attenuation trend of asphalt pavement at a long-term vehicle load.

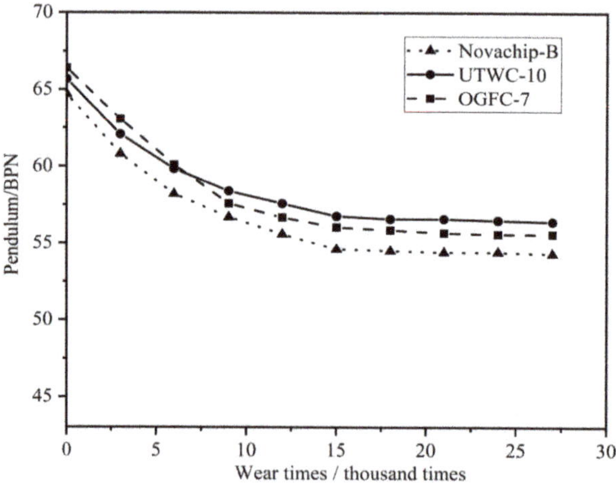

Figure 10. The anti-sliding performance attenuation curves of the three asphalt mixtures.

4. Conclusions

This study proposed a UTWC-10 (Ultra-thin Wearing Course-10) asphalt mixture with good high-temperature stability and skid resistance. In the gradation design of UTWC-10, 8 mm (two meshes) and 6.7 mm (three meshes) mesh holes were added between 4.75 mm and 9.5 mm based on the Taylor system standard mesh specifications. Based on the laboratory tests and results discussion, the following conclusions can be drawn:

(1) The DS of UTWC-10 asphalt mixture is as high as 5568 times/mm, which is much larger than Novachip-B and OGFC-7 asphalt mixture.

(2) The test results of low temperature bending beam tests, immersed Marshall tests, and freeze–thaw splitting tests proved that UTWC-10 asphalt mixture has satisfied crack resistance at low temperatures and the ability to resist water damage.

(3) The results of texture depth test and pendulum test confirmed that UTWC-10 asphalt mixture can provide good compactness and frictional resistance.

(4) The anti-sliding performance attenuation test system employed in the paper can closely simulate the skid resistance attenuation of roads at a long-term vehicle load. The test results show that the skid resistance of UTWC-10 is the best.

Author Contributions: Investigation and project administration, K.W.; writing—review and editing, W.H.; writing—original draft, W.C.; formal analysis and validation, W.C. and H.T.; supervision, X.C. All authors have read and agreed to the published version of the manuscript.

Funding: This publication is supported by National Natural Science Foundation of China (No. 51908154, No.51878193) and Science and Technology Program of Guangzhou (No. 201804010231).

Acknowledgments: The authors thank all those who contributed in the experimental part of this study.

Appendix A. Volume Parameters of the Three Asphalt Mixtures

The detailed volume parameters of the three asphalt mixtures and the leakage test results of UTWC-10 and Novachip-B were in Tables A1–A5.

Table A1. The volume parameters of UTWC-10.

Asphalt Content (%)	Void Ratio (%)	VMA (%)	VFA (%)	Stability (kN)	Flow Value (mm)	Asphalt Film Thickness (µm)
4.4	14.1	23.6	40.1	6.69	27.0	7.6
4.7	13.3	23.5	43.4	7.40	32.3	8.7
5.0	12.1	21.9	44.7	8.64	27.4	10.3
5.3	11.3	21.5	47.4	7.25	30.7	11.1
5.6	11.0	21.2	48.1	6.30	34.4	12.3

Table A2. The leakage test results of UTWC-10.

Sample Number	Mixture Quality (g)	Adhesive Quality (g)	Leakage Loss (%)	Average Leakage Rate (%)	Test Temperature (°C)
1	1000.8	0.84	0.08		
2	1000.1	0.77	0.08	0.08	185
3	1000.3	0.91	0.09		
4	1000.1	0.72	0.07		

Table A3. The volume parameters of Novachip-B.

Asphalt Content (%)	Void Ratio (%)	VMA (%)	VFA (%)	Stability (kN)	Flow Value (mm)	Asphalt Film Thickness (μm)
4.4	15.2	25.3	40.4	6.72	33.4	8.1
4.7	13.4	24.6	42.3	7.51	31.5	9.3
5.0	12.3	22.4	43.5	8.43	32.1	10.5
5.3	11.6	22.1	45.9	7.38	30.8	11.6
5.6	11.1	21.7	47.2	6.67	33.6	12.8

Table A4. The leakage test results of Novachip-B.

Sample Number	Mixture Quality (g)	Adhesive Quality (g)	Leakage Loss (%)	Average Leakage Rate (%)	Test Temperature (°C)
1	1000.1	0.71	0.07		
2	1000.4	0.93	0.09	0.09	185
3	1000.6	1.14	0.11		
4	1000.3	0.82	0.08		

Table A5. The volume parameters of OGFC-7.

Asphalt Content (%)	Leakage Loss (%)	Scattering Loss (%)	Void Ratio (%)	VMA (%)	VFA (%)	Stability (kN)	Flow Value (mm)
3.9	0.08	24.87	24.8	27.7	25.4	5.82	34.8
4.2	0.11	19.12	22	28.1	26.7	6.53	36.8
4.5	0.17	13.88	20.3	28.3	28.3	6.80	38.5
4.8	0.33	13.29	19	28.6	30.5	5.41	41.2
5.1	0.47	11.6	18.3	29.1	32.9	5.07	40.3

References

1. Xue, G. Study on design method and road performances of ultra-thin wearing course SMA-5 asphalt mixture. *Highw. Transp. Res. Dev.* **2009**, *26*, 18–21.
2. Qian, Z.; Lu, Q. Design and laboratory evaluation of small particle porous epoxy asphalt surface mixture for roadway pavements. *Constr. Build. Mater.* **2015**, *77*, 110–116. [CrossRef]
3. Zhao, D.; Kane, M.; Do, M.T. Effect of aggregate and asphalt on pavement skid resistance evolution. *Am. Soc. Civ. Eng.* **2010**, *46*, 8–18.
4. Sun, X.L.; Zhang, X.N.; Cai, X. Accelerated test-based study of long-term pavement performance of micro-surfacing. *J. Tongji Univ.* **2012**, *40*, 691–695.
5. Krummenauer, K.; Andrade, J.O. Incorporation of chromium-tanned leather residue to asphalt micro-surface layer. *Constr. Build. Mater.* **2009**, *23*, 574–581. [CrossRef]
6. Zhang, H.J.; Li, H.; Zhang, Y. Performance enhancement of porous asphalt pavement using red mud as alternative filler. *Constr. Build. Mater.* **2018**, *160*, 707–713. [CrossRef]
7. Wang, X.W.; Gu, X.Y.; Dong, Q. Evaluation of permanent deformation of multilayer porous asphalt courses using an advanced multiply-repeated load test. *Constr. Build. Mater.* **2018**, *160*, 19–29. [CrossRef]
8. Wu, K.H.; Zhang, X.N. Design of SMA-5 asphalt mixture for thin overlays. *J. South China Univ. Technol.* **2006**, *34*, 43–46.
9. Liu, C.H.; Sha, Q.L. Comparision testing research on aggregate gradation of ultra-thin layer asphalt concrete SAC-10. *China J. Highw. Transp.* **2005**, *18*, 7–13.
10. Audrius, V.; Tadas, A.; Viktoras, V.; Aleksandras, J.; Boris, F.; Ewa, Z. Asphalt wearing course optimization for road traffic noise reduction. *Constr. Build. Mater.* **2017**, *152*, 345–356.
11. Hu, M.J.; Li, L.H.; Peng, F.X. Laboratory investigation of OGFC-5 porous asphalt ultra-thin wearing course. *Constr. Build. Mater.* **2019**, *219*, 101–110. [CrossRef]
12. Wan, J.M.; Wu, S.P.; Xiao, Y.; Fang, M.J.; Song, W.; Pan, P.; Zhang, D. Enhanced ice and snow melting efficiency of steel slag based ultra-thin friction courses with steel fiber. *J. Clean. Prod.* **2019**, *236*, 117613. [CrossRef]

13. Nekkanti, H.; Putman, B.J.; Danish, B. Influence of aggregate gradation and nominal maximum aggregate size on the performance properties of OGFC mixtures. *Transp. Res. Rec.* **2019**, *1*, 240–245. [CrossRef]

14. David, A.H.; Xiao, F.P.; Serji, A. Investigation of Los Angeles value and alternate aggregate gradations in OGFC mixtures. *Constr. Build. Mater.* **2016**, *110*, 278–285.

15. ISO 565-1990. *Test Sieves-Metal Wire Cloth, Perforated Metal Plate and Electroformed Sheet-Nominal Sizes of Openings*; International Organization for Standardization: Geneva, Switzerland, 1990.

16. JTG E42-2005. *Testing Procedures of Aggregate for Highway Engineering in China*; General Administration of Quality Supervision, Inspection and Quarantine of the People's Republic of China: Beijing, China, 2005.

17. JTG E20-2011. *Standard Test Method of Bitumen and Bituminous Mixture for Highway Engineering in China*; General Administration of Quality Supervision, Inspection and Quarantine of the People's Republic of China: Beijing, China, 2011.

18. Zhang, X.N.; Wang, S.H.; Wu, K.H. CAVF method of asphalt mixture composition design. *Highway* **2001**, *12*, 17–20. (In Chinese)

19. Ghazi, G.; Aroon, S. Mixture-property-independent asphalt film thickness model. *Mater. Today Commun.* **2019**, *19*, 482–486.

20. Reza Pouranian, M.; Haddock, J.E. Determination of voids in the mineral aggregate and aggregate skeleton characteristics of asphalt mixtures using a linear-mixture packing model. *Constr. Build. Mater.* **2018**, *188*, 292–304. [CrossRef]

21. Ding, X.H.; Ma, T. Effects by property homogeneity of aggregate skeleton on creep performance of asphalt concrete. *Constr. Build. Mater.* **2018**, *171*, 205–213. [CrossRef]

22. Praticò, F.G.; Vaiana, R. A study on the relationship between mean texture depth and mean profile depth of asphalt pavements. *Constr. Build. Mater.* **2015**, *101*, 72–79. [CrossRef]

23. Plati, C.; Pomoni, M.; Stergiou, T. Development of a mean profile depth to mean texture depth shift factor for asphalt pavements. *Transp. Res. Rec.* **2017**, *1*, 156–163. [CrossRef]

 *materials*

Article

The Improvement of Moisture Resistance and Organic Compatibility of SrAl₂O₄: Eu²⁺, Dy³⁺ Persistent Phosphors Coated with Silica–Polymer Hybrid Shell

Lei Lyu [1], Yuxian Chen [1], Liting Yu [1], Rui Li [1,*], Liu Zhang [2] and Jianzhong Pei [1,*]

[1] School of Highway, Chang'an University, Xi'an 710064, China; lyulei@chd.edu.cn (L.L.); chenyuxian@chd.edu.cn (Y.C.); yuliting@chd.edu.cn (L.Y.)

[2] Shijiazhuang Municipal Design and Research Institute Co., Ltd., 35 Jianshe South Street, Shijiazhuang 050000, China; 2015121190@chd.edu.cn

*   Correspondence: lirui@chd.edu.cn (R.L.); pei@chd.edu.cn (J.P.)

Received: 28 November 2019; Accepted: 14 January 2020; Published: 16 January 2020

Abstract: The existing road surface marking with poor visibility at night results in traffic safety hazards in insufficient lighting roads. This study aims to prepare the dedicated aluminate-based persistent phosphors considering the integrated pavement environment, as the first step to achieve the durable luminescent road surface marking. SrAl₂O₄: Eu²⁺, Dy³⁺ persistent phosphors coated with silica–polymer hybrid shell were prepared by chemical precipitation and sol-gel method to improve moisture resistance and organic compatibility. The optimum silane coupling agent type and dosage, the surfactant dosage, the optimum sodium silicate dosage, and the coating reaction time in silica shell and polymer shell coating were studied based on the moisture resistance test. The silica–polymer hybrid shell coating balances the organic compatibility and thermal stability as compared to the silica or polymer shell coating in the oil absorption test and thermogravimetric analysis. Ex-Em Spectra, XRD, and SEM method were used to characterize the persistent phosphors, indicating the preparation does not destroy the persistent phosphors. The outstanding durable properties of SrAl₂O₄: Eu²⁺, Dy³⁺ persistent phosphors coated with silica–polymer hybrid shell as shown in this research is crucial for its potential application in waterborne luminescent coatings of road surface marking.

Keywords: SrAl₂O₄ persistent phosphors; silica–polymer hybrid shell; moisture resistance

1. Introduction

The road surface marking is an important transportation infrastructure to convey various official information, especially to guide traffic on the channelized traffic roads [1–3]. Room temperature solvent paint, hot melt paint, heating solvent paint, and water-based paint are the four mainstream types of road surface marking paints. Due to the slow construction speed and serious environment pollution, room temperature solvent paint and hot melt paint have been gradually phased out [4,5]. Besides, the road surface marking produced by above method has poor visibility at night, resulting in traffic safety hazards in insufficient lighting roads.

Energy-storing luminescent materials are alternative road coatings to provide enhanced visibility by self-illuminating. They convert absorbed external energy into fluorescence when excited, relying on their own lattice defects to generate energy-level transitions. The energy storing luminescent materials are classified into sulfides, aluminates and aluminum silicon salts. Aluminate-based persistent phosphors are the most widely used and studied luminescent material for its high quanta efficiency, long after-glow life, and good chemical stability [6]. The SrAl₂O₄: Eu²⁺, Dy³⁺ crystalizes in a monoclinic system and consists of rings formed by six-corner-sharing oxide aluminate tetrahedra, of which two Sr²⁺ ions (r = 126 pm in VIII coordination) are replaced by Eu²⁺ (r = 125 pm) and Dy³⁺ (r = 97 pm) respectively [7]. The persistent time of $SrAl_2O_4$: Eu^{2+}, Dy^{3+} with a strong emission centered at 520 nm (green) was found to be longer than 16 h [8]. However, the luminescent structure of aluminate-based persistent phosphors is gradually damaged at 86.5 °C due to the thermal quenching.

Many researchers have demonstrated the possibility of using aluminate-based persistent phosphors in waterborne paints for road surface markings based on laboratory and field tests [9–11]. OssN329 road in the Netherlands is the first road achieving the achieve the self-luminescence of road surface markings using aluminate-based persistent phosphors in water-based paint. However, the road surface markings lost the luminous function in less than 14 days were caused by the rainfall [12]. Literatures also suggest that poor moisture resistance and organic compatibility is the challenge for the aluminate-based persistent phosphor to be widely used in waterborne paints of road surface marking [13–15]. Furthermore, little research focused on the thermal stability of the aluminate-based persistent phosphors used in the road surface markings. The highest temperature of pavement surface is over 60 °C, leading to the potential damage to the aluminate-based persistent phosphors [16]. There has also not been much experience of developing the dedicated durable aluminate-based persistent phosphors considering the integrated pavement environment.

The surface coating modification of aluminate-based persistent phosphors is a widely used method to enhance the moisture resistance and the compatibility with organic solvents of waterborne paints. The surface coating modification includes organic coating and inorganic coating. Inorganic coating refers to coating silica, metal oxides, or metal halides on the surface of persistent phosphors by sol-gel method, liquid phase precipitation method and high-temperature solid-state method. Sodium silicate, ethyl orthosilicate and alumina are widely used inorganic coating agents, which has been proved to improve the moisture resistance of coated luminescent materials [17–19]. In organic coating, the organic modifier covers the surface of persistent phosphors by forming chemical bonds or electrostatic adsorption to improve the organic compatibility [20]. However, single inorganic or organic single shell coating could not balance the moisture resistance and organic compatibility of aluminate-based persistent phosphors.

In this paper, $SrAl_2O_4$: Eu^{2+}, Dy^{3+} persistent phosphors were subject to the silica shell coating, polymer shell coating, and the silica–polymer hybrid shell coating. Coating properties were evaluated based on the moisture resistance, organic compatibility, thermal stability, luminosity, and microstructure characterization to explore the optimum composition as shown in Figure 1. The objective of this study was to prepare the dedicated aluminate-based persistent phosphors considering the integrated pavement environment, as the first step to achieve the durable luminescent road surface marking. The simultaneous use of inorganic and organic coating combines their advantages on the moisture resistance and organic compatibility to achieve the balanced design. The moisture resistance and thermal stability of $SrAl_2O_4$: Eu^{2+}, Dy^{3+} persistent phosphors were comprehensively evaluated based on the integrated pavement environment for the first time.

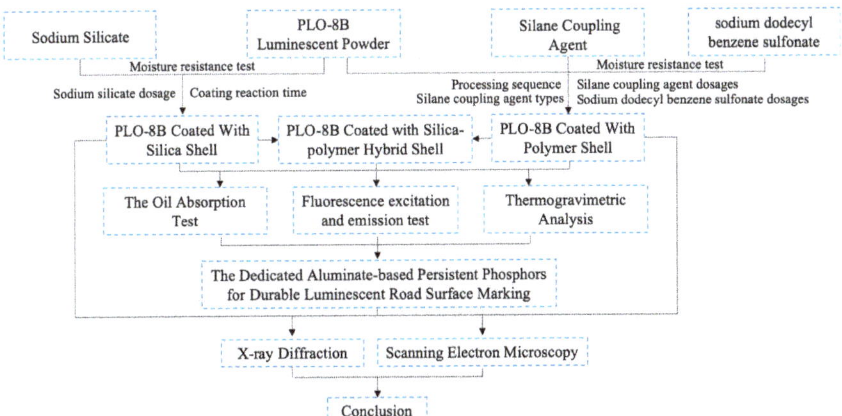

Figure 1. Flow chart of test method.

2. Experiments

2.1. Raw Materials

PLO-8B luminescent powder ($SrAl_2O_4$: Eu^{2+}, Dy^{3+} persistent phosphors) was purchased from Luming Technology Group. The silane coupling agent, anhydrous ethanol, acrylic monomer (AA), methyl methacrylate (MMA), ethylene glycol, and sodium silicate were produced by Xi'an Chemical Reagent Factory.

2.2. Preparation

The PLO-8B luminescent powder coated with silica shell, polymer shell and silica–polymer hybrid shell were prepared to evaluate the effect of compositions on coating properties as shown in Figure 2.

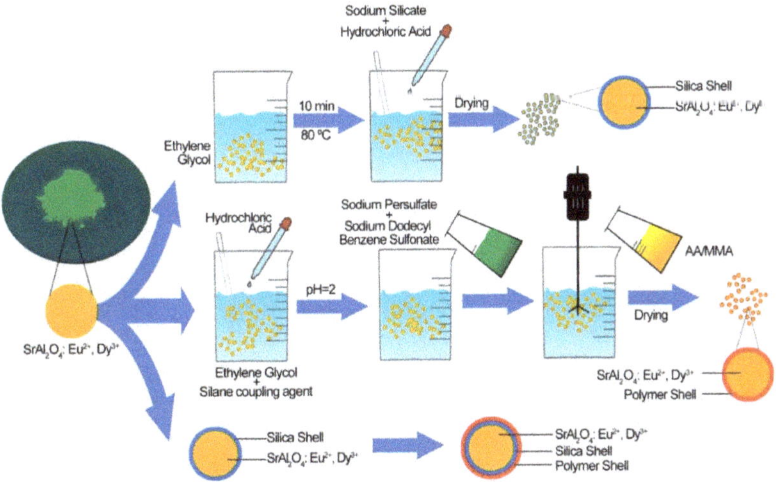

Figure 2. The preparation process of PLO-8B luminescent powder coated with silica shell, polymer shell and silica–polymer hybrid shell.

2.2.1. Preparation of Persistent Phosphors Coated with Silica Shell

The PLO-8B luminescent powder was coated with sodium silicate by heterogeneous precipitation method in ethyl alcohol. In the acidic solution, sodium silicate gradually deposits into Si(OH)$_4$ colloid as shown in Reaction 1, and bonding with the hydroxyl group having high activation energy to form a dense protective shell.

$$Na_2SiO_3 + 2H^+ + (X-1)H_2O \rightarrow SiO_2 \cdot XH_2O + 2Na^+ \tag{1}$$

In this preparation, ethylene glycol and PLO-8B luminescent powder was dispersed for 10 min by the parallel feed process. The sodium silicate was poured to solution pH = 9 (80 °C) and stirring quickly. Then the mixture was washed, filtered and dried to obtain PLO-8B luminescent powder coated with silica shell. The initial material dosages of the coating with silica shell is shown in Table 1.

Table 1. Dosages of the coating with silica shell.

Reagent	Weight
PLO-8B luminescent powder (PLP)	5 g
Ethylene glycol	50 g
Sodium silicate	4 g

2.2.2. Preparation of Persistent Phosphors Coated with Polymer Shell

Literatures suggested the pretreatment with transparent and well-compatibility silane coupling agent is necessary for persistent phosphors to enhance the modification effect of the organic coating [21]. Due to the alkoxy functional groups of the silane coupling agent produce highly active hydroxyl functional groups after hydrolysis reaction, the first shell was formed on the surface of persistent phosphors. After the pretreatment, the persistent phosphors with a silicon dioxide shell was then coated with the polymer shell from the polymerization reactions of acrylic acid monomer (AA) and methyl methacrylate (MMA) by sol-gel method to achieve the graft modification of the polymer shell as shown in Figure 3 [21].

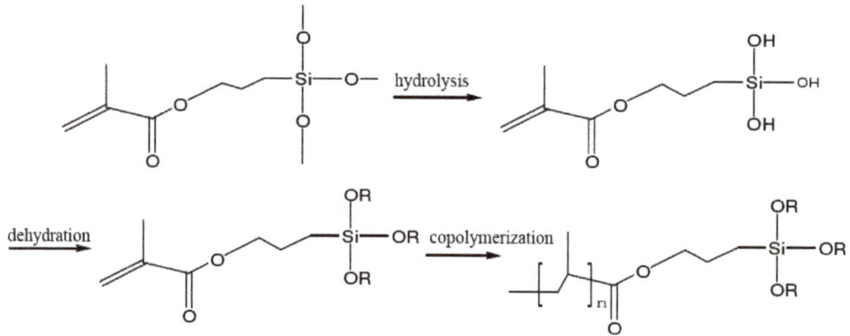

Figure 3. The reaction process of persistent phosphors coated with polymer shell.

Specifically, PLO-8B luminescent powder, silane coupling agent and absolute ethanol were mixed in the container. The pH of the solution was adjusted to 2 by adding hydrochloric acid and placed at 80 °C for 4 h. Then, the fully dissolved solution of sodium persulfate and sodium dodecyl benzene sulfonate was added. The mixture of AA/MMA was slowly poured in low speed stirring. After 5 min stationary and removing the upper white suspension, the persistent phosphors coated with polymer shell was obtained by washing with absolute ethanol, filtering and drying. The initial material dosages of the coating with polymer shell is shown in Table 2.

Table 2. Dosages of the coating with polymer shell.

Reagent	Weight
PLO-8B luminescent powder (PLP)	10 g
Silane coupling agent	1 g
Anhydrous ethanol	70 g
Acrylic acid monomer (AA)	0.5 g
Methyl methacrylate (MMA)	0.5 g
Sodium persulfate	1 g
Sodium dodecyl benzene sulfonate	0.15 g
Water	30 g

2.2.3. Preparation of Persistent Phosphors Coated with Silica-Polymer Hybrid Shell

Based the former tests, sodium silicate and acrylic acid monomer/methyl methacrylate was used as the inorganic modifier and organic modifier, respectively. The coating process is the same as above. The initial material dosages of the silica–polymer hybrid shell coating is shown in Table 3 [22].

Table 3. Material dosages of the coating of silica–polymer hybrid shell.

Step	Reagent	Weight
	PLO-8B luminescent powder (PLP)	5 g
Inorganic coating	Ethylene glycol	50 g
	Sodium silicate	4 g
	Silane coupling agent	0.2 g
	Anhydrous ethanol	70 g
	Acrylic acid monomer (AA)	0.1 g
Organic coating	Methyl methacrylate (MMA)	0.1 g
	Sodium persulfate	1 g
	Sodium dodecyl benzene sulfonate	0.15 g
	Water	30 g

2.3. Characterization Method

2.3.1. The Moisture Resistance Test

Improving the moisture resistance of the persistent phosphors is the main purpose of coating, which also deeply affects its service life. Therefore, the effects of different factors in silica or polymer shell on the moisture resistance of the persistent phosphors were investigated.

$$SrAl_2O_4 + 4H_2O \rightarrow Sr^{2+} + 2OH^- + 2Al(OH)_3 \downarrow \tag{2}$$

As showed in the Reaction (2), $SrAl_2O_4$: Eu^{2+}, Dy^{3+} persistent phosphors are destroyed with water, which contributes to the degradation of their luminescent property and the increase of solution pH [23]. Due to the longest average rainfall duration in China being 5.4 hours based on the Chinese meteorological statistics [24,25], pH of the solution after 6 hours was selected to evaluate the moisture resistance. In the test, 0.5 g of luminescent powder was soaked in 40 °C water. The change of pH with time was measured using a digital pH test pen. The growth rate of solution pH (PR) soaked with PLO-8B luminescent powder coated with different shells was calculated to evaluate the improvement of moisture resistance by the following Equation:

$$PR = \frac{SP_f - SP_i}{SP_i}$$

where

SP_f: the solution pH after the samples soaked for 6 h;
SP_i: the initial solution pH soaked with samples.

2.3.2. The Oil Absorption Test

The oil absorption was evaluated according to ISO787-5-1980 to analyze the compatibility between the coated persistent phosphors and organic solvents [26]. The refined linseed oil (acid value: 5–7 mg/g) was gradually added into the 10 g luminescent powder, stirred evenly until all powder stacked together. 20–25 min is the maximum acceptable test time to guarantee the accuracy. The oil absorption is indicated by the minimum oil consumption in mass percentage. The lower minimum oil consumption in mass percentage means the better organic compatibility.

2.3.3. Thermogravimetric Analysis

The thermal stability of the persistent phosphors was evaluated by a comprehensive thermogravimetric analyzer (TGA/DSC 3+, METTLER TOLEDO, Zurich, Switzerland) in the range from room temperature to 300 °C under nitrogen atmosphere, and the heating rate was 10 °C/min.

2.3.4. Fluorescence Excitation and Emission Test

The excitation and emission spectra of the persistent phosphors were tested by FLs980 full-featured Steady/Transient Fluorescence Spectrometer (Edinburgh Instruments, Edinburgh, UK). Excitation spectrum was scanned from 250 nm to 500 nm. Emission spectrum was monitored from 350 nm to 700 nm.

2.3.5. X-Ray Diffraction

The crystal phase properties were analyzed by X-ray diffraction (XRD) using Bruker D8 ADVANCE (Bruker AXS, Karlsruhe, Germany) as shown in 2θ scan mode from 10 to 60° (2θ) using Cu Kα radiation, with a step width of 0.02° and a time interval of 0.1 s per step [27].

2.3.6. Scanning Electron Microscopy

Scanning electron microscopy (SEM) (S-4800, Hitachi, Japan) was used to analyze the morphological characteristics of persistent phosphors. The resolution increased from 3 to 10 nm with the test voltage ranging from 3 to 40 kV. The accelerating voltage was 3.0 kV [28].

3. Results and Discussions

3.1. Moisture Resistance of Silica Shell Coating

3.1.1. Effect of the Coating Reaction Time

Obviously, the compactness of silica shell increases synchronously with the prolongation of reaction time until the reactions ending, contributing to the improvement of the moisture resistance. However, fully polycondensation reaction is impossible in the coating process due to its resource intensive and time consuming (over 10 h). Thus, three samples of silica shell coating (the coating reaction time is 2, 3, and 4 h respectively) were prepared to determine the optimum time balancing the moisture resistance and preparation time. As shown in Figure 4, the solution pH soaked with PLO-8B luminescent powder coated for 3 and 4 h decreased by 25% and 27.5% compared with the uncoated samples, indicating the moisture resistance of PLO-8B luminescent powder was significantly improved. 3 h was selected as the optimum reaction time for the following tests due to the shorter coating reaction time and similar moisture resistance.

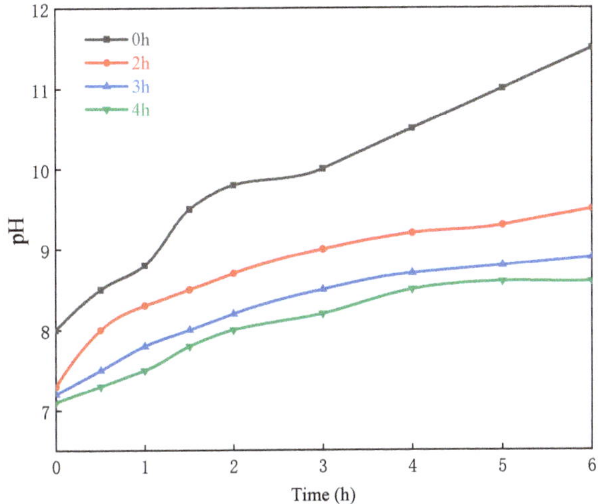

Figure 4. The change of solution pH soaked with PLO-8B luminescent powder coated with different reaction time.

3.1.2. Effect of the Sodium Silicate Dosage

$SiO_2 \cdot XH_2O$ content in Reaction 1 is determined by the sodium silicate dosage resulting in the different thicknesses of the silica shell and coating properties. Five samples of silica shell coating (the sodium silicate dosages are 2, 4, 6, 8, and 10 g respectively) were prepared to determine the optimum dosage. Figure 5 shows the increased sodium silicate dosages decreased the solution pH, indicating effective improvement of the moisture resistance. This improvement was significant as the sodium silicate dosage was less than 4 g, contributing to the 20.34% decreasing compared to the uncoated samples. Further increase of sodium silicate dosages contributed little to improving the moisture resistance of persistent phosphors. Thus, 4 g was selected as the optimum sodium silicate dosage for the following tests.

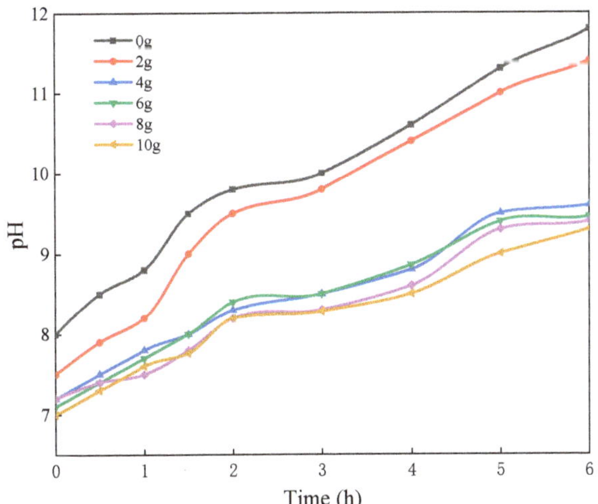

Figure 5. The change of solution pH soaked with PLO-8B luminescent powder coated with different sodium silicate dosages.

3.2. Moisture Resistance of Polymer Shell Coating

3.2.1. Effect of the Processing Sequence

The different preparation sequences of hydrochloric acid and the silane coupling agent from literatures was investigated to achieve better pretreatment. As shown in Reaction 2, the yellow $SrAl_2O_4$: Eu^{2+}, Dy^{3+} persistent phosphor is destroyed and turn white after reacting with H_2O. The degree of hydrolysis reaction was evaluated by the image software based on the value of Y(yellow) in the CMYK(cyan, magenta, yellow, and black) color mode to determine the preparation sequences. Solutions soaked with samples prepared by different sequences was shown in Figure 6. The Y values in the sample with silane coupling agent added first is obvious higher, indicating the better coating for preventing the $SrAl_2O_4$: Eu^{2+}, Dy^{3+} persistent phosphor from hydrolysis reaction.

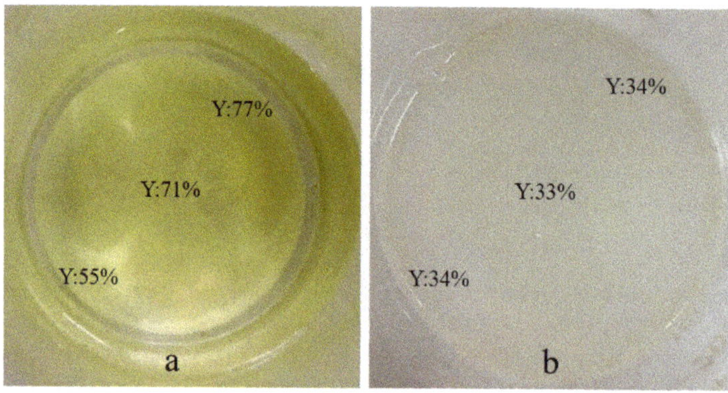

Figure 6. The Y value in hydrolysis solutions of different preparation sequences: (**a**) Add the silane coupling agent first; (**b**) Adjust pH first.

3.2.2. Effect of the Silane Coupling Agent Types

KH560 and KH570 are the mainly silane coupling agents used in the coatings industry. Two silane coupling agents were used to prepare the PLO-8B luminescent powder coated with the polymer shell respectively to determine the optimal type. The moisture resistance was evaluated as shown in Figure 7. The solution was initially acidic due to the hydrogen ions added in the preparation. The untreated PLO-8B luminescent powder hydrolyzed rapidly in 40 °C water resulting in a significant increase in solution pH and almost hydrolyzed completely after 5 h. Two silane coupling agents slowed down the increase of solution pH, indicating the moisture resistance was improved. The solution pH soaked with PLO-8B luminescent powder coated with KH560 and KH570 decreased by 25% and 48.79%, respectively. Obviously, KH570 is more suitable to polymer shell coating for PLO-8B luminescent powder than KH560.

Molecular structures of different silane coupling agents are shown in Figure 8. Siloxy groups forming the highly active silanol after the hydrolysis reaction are observed both in the KH560 and KH570, which bonds with the hydroxyl groups on the surface of persistent phosphors and undergo polycondensation by the "Si–OH" functional group. However, the methacrylic group in KH570 is unique. It is copolymerized with MMA and sodium persulfate to generate a denser polymer shell further improving the moisture resistance.

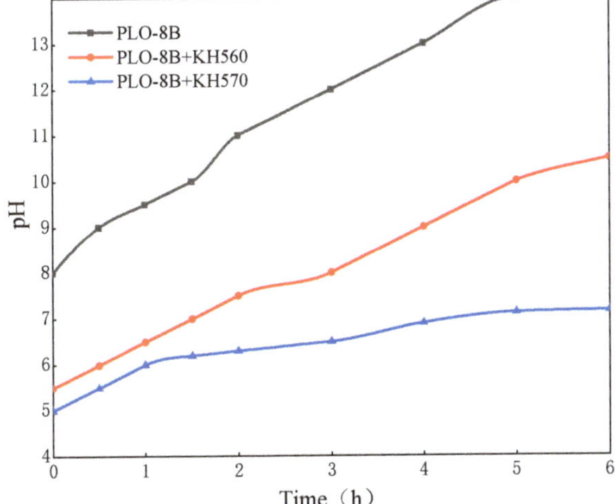

Figure 7. The change of solution pH soaked with PLO-8B luminescent powder coated with different silane coupling agents.

KH560

KH570

Figure 8. Molecular structures of different silane coupling agents.

3.2.3. Effect of Silane Coupling Agent Dosages

Three samples of polymer shell coating (silane coupling agent dosages are 10, 15, and 20 wt% respectively) were prepared to determine the optimum silane coupling agent dosage. As shown in Figure 9, the solution pH soaked with coated samples decreases clearly. However, the excessive silane coupling agent (20 wt% silane coupling agent) formed the loose and easily broken polymer shell, resulting in the sharp rise of solution pH after the polymer shell broken in 1.5 h. The solution pH soaked with PLO-8B luminescent powder coated with 10 wt% and 15 wt% silane coupling agent decreased by 47.14% compared to the uncoated sample, indicating the significant improvement of moisture resistance. Herein 10 wt% silane coupling agent formed a thinner polymer shell contributing to the less negative impact on the luminescent properties. Thus 10 wt% was selected as the optimum dosage of the silane coupling agent.

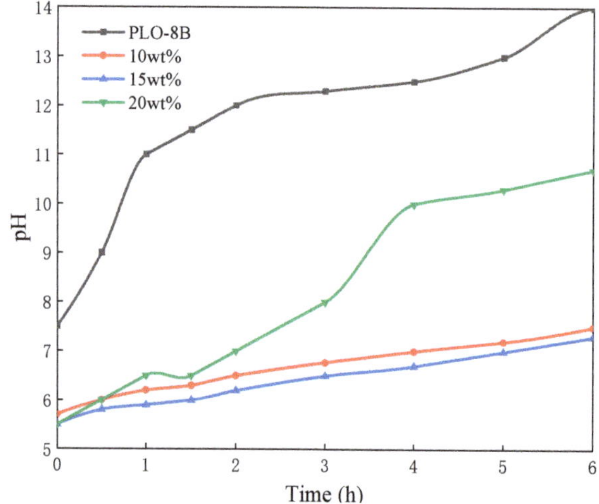

Figure 9. The change of solution pH soaked with PLO-8B luminescent powder coated with different silane coupling agent dosages.

3.2.4. Effect of Sodium Dodecyl Benzene Sulfonate Dosages

The surfactant (sodium dodecyl benzene sulfonate) dosage determines the conversion rate of the methacrylic monomer, relating to the quality of the polymer shell. Three samples of polymer shell coating (sodium dodecyl benzene sulfonate dosages are 0.1, 0.15, 0.2 wt%, respectively) were prepared to select the optimum surfactant dosage. As Figure 10 shows, the solution pH started significantly increasing from 1.5 h, indicating the polymer shell had been severely damaged. Obviously, the moisture resistance of the polymer shell coating samples with 0.1 or 0.2 wt% surfactant was much worse than 0.15 wt%. Too much surfactant (0.2 wt%) leads to the violent polymerization reaction. The newly grown polymer shell rapidly aggregates on the surface of PLO-8B luminescent powder, resulting in the loose structure of the polymer shell. On the contrary, too less surfactant (0.1 wt%) limits the formation of polymer shell. Thus, 0.15 wt% is selected as the optimum surfactant dosage.

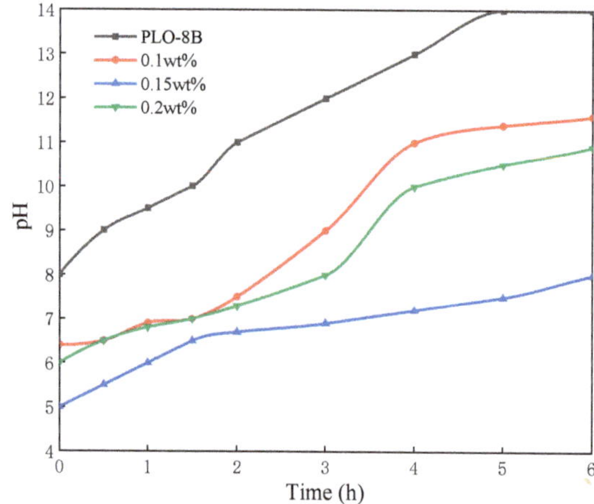

Figure 10. The change of solution pH soaked with PLO-8B luminescent powder coated with different sodium dodecyl benzene sulfonate dosages.

3.3. *Moisture Resistance of Silica-Polymer Hybrid Shell Coating*

PLO-8B luminescent powder coated with silica–polymer hybrid shell (PLO-8B-SP) was prepared to compare the moisture resistance with PLO-8B luminescent powder coated with silica shell (PLO-8B-S) and polymer shell (PLO-8B-P). As Figure 11 shows, the silica–polymer hybrid shell coating significantly slowed down the increase of solution pH, indicating the improvement in the moisture resistance. The growth rate of solution pH soaked with PLO-8B, PLO-8B-S, PLO-8B-P, and PLO-8B-SP was 75%, 18%, 32%, and 21%, respectively as shown in Figure 12. The PLO-8B-S and PLO-8B-P was prepared based the on the optimal composition from the previous tests. Obviously, all types of shell coating significantly improve the moisture resistance of PLO-8B luminescent powder. The growth rate of solution pH soaked with PLO-8B-S, PLO-8B-P, and PLO-8B-SP was decreased by 75.93%, 57.89%, and 71.72%, respectively compared to PLO-8B. The moisture resistance of PLO-8B luminescent powder coated with silica–polymer hybrid shell was significantly improved compared to polymer shell coating, indicating the silica shell coating contributes the most in the test.

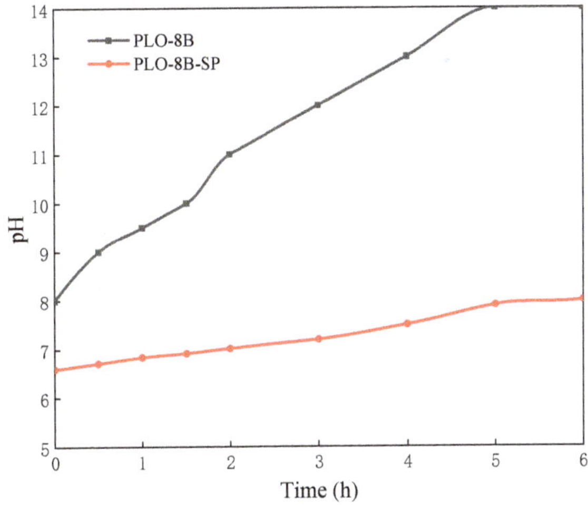

Figure 11. The change of solution pH soaked with PLO-8B luminescent powder coated with silica–polymer hybrid shell.

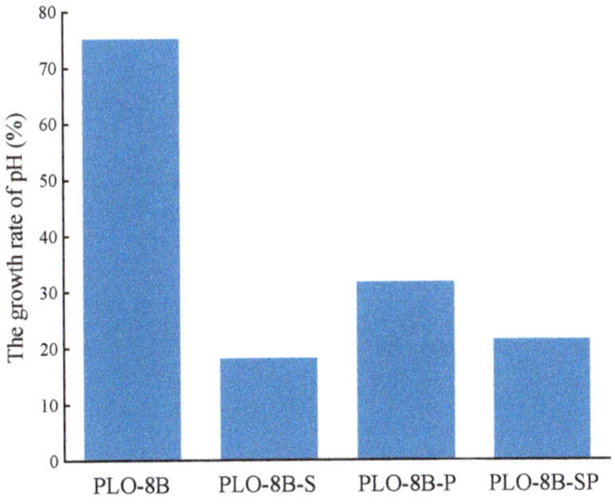

Figure 12. The growth rate of solution pH soaked with PLO-8B luminescent powder coated with silica shell, polymer shell and silica–polymer hybrid shell.

3.4. Organic Compatibility of Persistent Phosphors

The oil absorption of persistent phosphors shows the organic compatibility, indicating the compatibility with other organic solutions in waterborne coatings such as auxiliaries and binders [29–31]. Four samples were prepared to evaluate the organic compatibility based on the oil absorption test according to the ISO787-5-1980 as shown in Figure 13. The organic compatibility of PLO-8B-S, PLO-8B-P, and PLO-8B-SP was improved by 7.41%, 40.74%, and 33.33%, respectively indicating the silica shell coating has little effect on the oil absorption. Both the polymer shell coating and the silica–polymer hybrid shell coating greatly improved the organic compatibility of PLO-8B luminescent powder. Due to the decrease in silane coupling agent and AA/MMA dosage, the organic compatibility improvement of PLO-8B luminescent powder coated with silica–polymer hybrid shell is slightly weaker than that of PLO-8B luminescent powder coated with polymer shell.

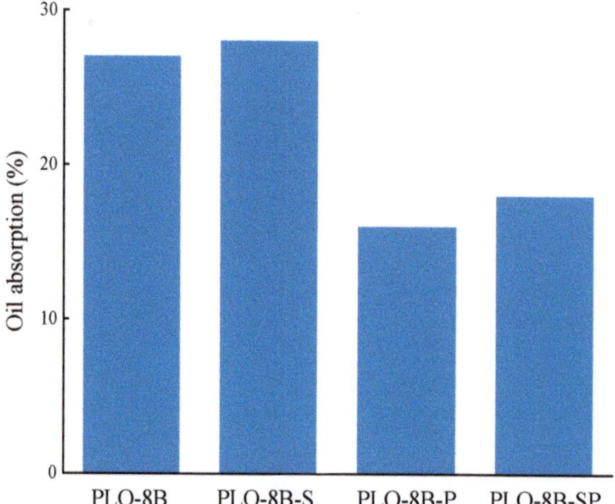

Figure 13. Oil absorption of PLO-8B luminescent powder.

3.5. Thermal Stability of Persistent Phosphors

The comprehensive thermogravimetric analyzer (TGA/DSC 3+, METTLER TOLEDO) was used to measure the TG curves of PLO-8B luminescent powder coated with silica shell, polymer shell, and silica–polymer hybrid shell as shown in Figure 14. PLO-8B-P constantly lost mass with the temperature increasing due to the volatilization of polymer shell. The mass loss of PlO-8B-S was within 0.5%, indicating the better thermal stability. The initial thermal weight loss behavior of PLO-8B-SP was consistent with PLO-8B-P, due to the same polymer shell. However, the less silane coupling agent and AA/MMA dosage formed the thinner polymer shell coating the PLO-8B luminescent powder in PLO-8B-SP. Thus, the thermal weight loss of PLO-8B-SP slowed down gradually. Obvious, the easily decomposable polymer shell coating is not suitable to prepare the dedicated aluminate-based persistent phosphors for the durable luminescent road surface marking, due to the extremely high temperatures of pavement surface in some cases. The silica and silica–polymer hybrid shell contributed to the better thermal stability of PLO-8B-S and PLO-8B-SP, indicating the ability to ensure the moisture resistance in the wider range of temperature.

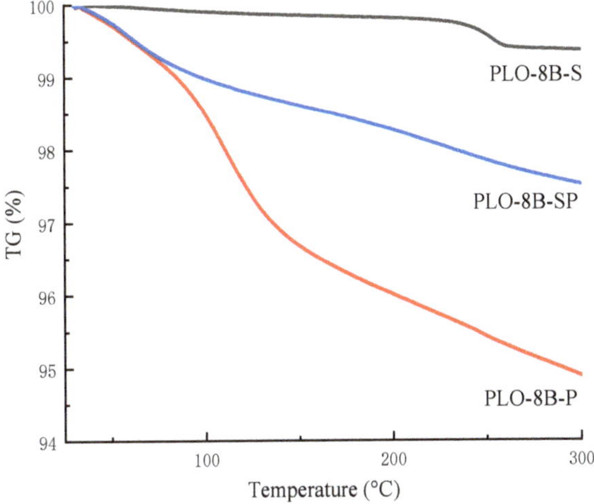

Figure 14. TG curves of PLO-8B luminescent powder coated with silica shell, polymer shell, and silica–polymer hybrid shell.

3.6. Luminous Performance of Persistent Phosphors

FLs980 full-featured Steady/Transient Fluorescence Spectrometer (Edinburgh) was used to measure the excitation and emission spectra of PLO-8B luminescent powder (PLO-8B) and PLO-8B luminescent powder coated with silica–polymer hybrid shell (PLO-8B-SP). The luminescence properties were analyzed by comparing the spectral peaks and peak areas. According to Figure 15, the peak positions of PLO-8B-SP in the excitation spectrum and the emission spectrum are consistent with the PLO-8B, while the peak decreased. Due to part of light was reflected or absorbed by the silica–polymer hybrid shell, the persistent phosphors absorbed less light energy, resulting in the attenuate of emitted light. The peak emission spectrum and excitation spectrum of the PLO-8B-SP decreased by 23.4% and 13.7%, respectively. Obviously, the silica–polymer hybrid shell coating affects the brightness but does not damage the persistent phosphors.

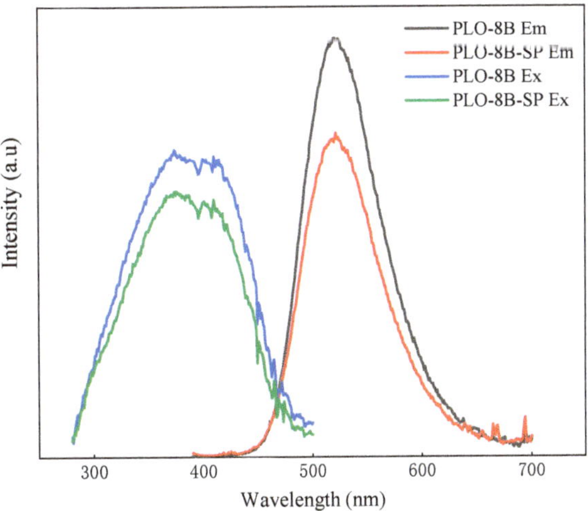

Figure 15. Em and Ex spectra of PLO-8B and PLO-8B-IO.

3.7. X-Ray Diffraction (XRD) Analysis

Figure 16 plots XRD patterns of PLO-8B luminescent powder (PLO-8B), PLO-8B luminescent powder coated with polymer shell (PLO-8B-P), and PLO-8B luminescent powder coated with silica–polymer hybrid shell (PLO-8B-SP). All the diffraction peaks were indexed and are good in agreement with JCPDS card no. 034-0379 with lattice constants a = 8.4424Å, b = 8.822Å, c = 5.1607 Å with interfacial angles α = 90.00°, β = 93.42°, γ = 90.00°, respectively. Hence persistent phosphors of were not changed, which is consistent with the results of Fluorescence excitation and emission test.

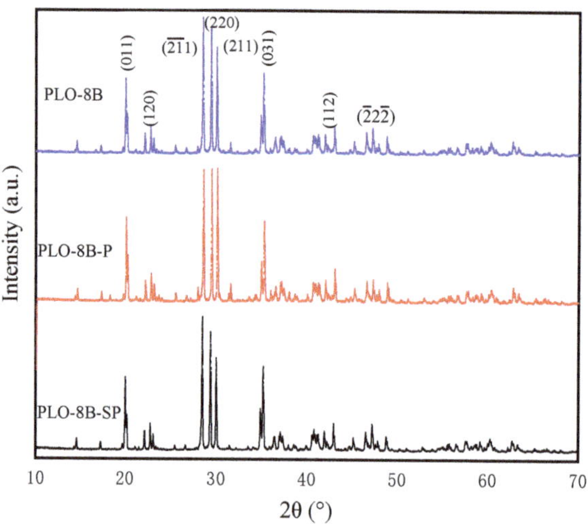

Figure 16. XRD patterns of PLO-8B luminescent powder (PLO-8B), PLO-8B luminescent powder coated with polymer shell (PLO-8B-P), and PLO-8B luminescent powder coated with silica–polymer hybrid shell (PLO-8B-SP) [32].

3.8. Scanning Electron Microscopy (SEM) Analysis

The SEM images of samples are shown in Figure 17. As can be seen under 400× magnification, the surface irregularity of PLO-8B is weakened after coated with shells, which means the surface covered by a complete protective layer. Due to the addition of silica shell, the surface of PLO-8B-SP is smoother than PLO-8B-P. As can be seen under 10,000× magnification, excess silane coupling agent reacted with AA/MMA to form a small amount of protrusion on the surface of PLO-8B-P. A dense and intact protective layer appeared on the surface of PLO-8B-IO, indicating the reduction of silane coupling agent and AA/MMA dosage is suitable. As seen under 20,000× magnification, the coated shells form a honeycomb structure covering the regular and distinct crystal structure of PLO-8B, which protects the luminescent structure from damage.

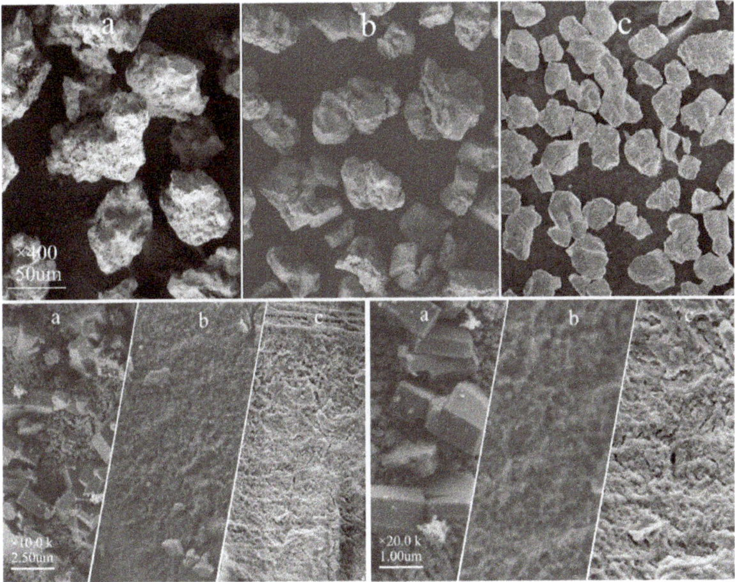

Figure 17. SEM images: (**a**) PLO-8B; (**b**) PLO-8B-P; (**c**) PLO-8B-SP.

4. Conclusions

$SrAl_2O_4$: Eu^{2+}, Dy^{3+} persistent phosphors coated different type of shells were prepared in this paper to improve moisture resistance and organic compatibility. The optimal parameters (coating reaction time, sodium silicate dosage, processing sequence, silane coupling agent types, silane coupling agent dosages, and sodium dodecyl benzene sulfonate dosages) of chemical precipitation and sol-gel method for the preparation of silica–polymer hybrid shell were determined by the moisture resistance test. The silica–polymer hybrid shell coating balances the organic compatibility and thermal stability as compared to the silica or polymer shell coating in the oil absorption test and thermogravimetric analysis. The phase structure of $SrAl_2O_4$: Eu^{2+}, Dy^{3+} persistent phosphors did not change in the coating process based on luminous performance, XRD, and SEM analysis. The optimal silica–polymer hybrid shell coating for $SrAl_2O_4$: Eu^{2+}, Dy^{3+} persistent phosphors improved the moisture resistance and organic compatibility by 71.72% and 33.33% compared to the uncoated $SrAl_2O_4$: Eu^{2+}, Dy^{3+} persistent phosphors. The outstanding durable properties of $SrAl_2O_4$: Eu^{2+}, Dy^{3+} persistent phosphors coated with silica–polymer hybrid shell as shown in this research is crucial for its potential application in waterborne luminescent coatings of road surface marking.

Author Contributions: Formal analysis, L.L.; investigation, L.L.; methodology, L.L. and L.Z.; project administration, J.P.; writing—original draft, L.L.; validation, L.Y. and J.P.; writing—review and editing, L.L., Y.C. and L.Y.; supervision, R.L. All authors have read and agreed to the published version of the manuscript.

Funding: This research was sponsored by the National Key R&D Program of China (Grant No. 2018YFE0103800).

References

1. Naidoo, S.; Steyn, W.J. Performance of thermoplastic road-marking material. *J. S. Afr. Inst. Civ. Eng.* **2018**, *60*, 9–22. [CrossRef]
2. Pan, P.; Wu, S.; Xiao, Y.; Liu, G. A review on hydronic asphalt pavement for energy harvesting and snow melting. *Renew. Sustain. Energy Rev.* **2015**, *48*, 624–634. [CrossRef]

3. Liu, T.; Li, R.; Pei, J.; Xing, X.; Guo, Q. Effect of different fibers reinforcement on the properties of PZT/PVA composites. *Mater. Chem. Phys.* **2020**, *239*, 122063. [CrossRef]

4. Ge, L.; Sun, L.; Zhang, X. New Generation of Spray Polyurea Road Marking Paint. *Paint Coat. Ind.* **2006**, *3*, 36–38.

5. Liu, J.; Wang, B. *The Double-Component Epoxy Resin Road Marking Coating*; TaiYuan University of Technology: TaiYuan, China, 2007.

6. Rojas-Hernandez, R.E.; Rubio-Marcos, F.; Rodriguez, M.Á.; Fernandez, J.F. Long lasting phosphors: SrAl2O4:Eu, Dy as the most studied material. *Renew. Sustain. Energy Rev.* **2018**, *81*, 2759–2770. [CrossRef]

7. Delgado, T.; Afshani, J.; Hagemann, H. Spectroscopic study of a single crystal of $SrAl_2O_4:Eu^{2+}:Dy^{3+}$. *J. Phys. Chem. C* **2019**, *123*, 8607–8613. [CrossRef]

8. Yeilay Kaya, S.; Karacaoglu, E.; Karasu, B. Effect of Al/Sr ratio on the luminescence properties of $SrAl_2O_4:Eu^{2+}$, Dy^{3+} phosphors. *Ceram. Int.* **2012**, *38*, 3701–3706. [CrossRef]

9. Praticò, F.G.; Vaiana, R.; Noto, S. Photoluminescent road coatings for open-graded and dense-graded asphalts: Theoretical and experimental investigation. *J. Mater. Civ. Eng.* **2018**, *30*, 1–9. [CrossRef]

10. Praticò, F.G.; Noto, S.; Moro, A. Optimisation of photoluminescent painting treatments on different surface layers. In Proceedings of the 4th Chinese–European Workshop on Functional Pavement Design CEW 2016, Delft, The Netherlands, 29 June–1 July 2016; pp. 1533–1542.

11. Giuliani, F.; Autelitano, F. Photoluminescent road surface dressing: A first laboratory experimental investigation. *Mater. Tech.* **2014**, *102*, 6–7.

12. Glow in the Dark Roads not Glowing. Available online: https://www.bbc.com/news/technology-27187827 (accessed on 16 January 2020).

13. Bacero, R.; To, D.; Arista, J.P.; Dela Cruz, M.K.; Villaneva, J.P.; Uy, F.A. Evaluation of Strontium Aluminate in Traffic Paint Pavement Markings for Rural and Unilluminated Roads. *J. East. Asia Soc. Transp. Stud.* **2015**, *11*, 1726–1744.

14. Botterman, J.; Smet, P.F. Persistent phosphor $SrAl_2O_4$: Eu, Dy in outdoor conditions: Saved by the trap distribution. *Opt. Express* **2015**, *23*, A868–A881. [CrossRef] [PubMed]

15. Botterman, J.; Smet, P. Glow-in-the-dark traffic markings: Feasible or not? In Proceedings of the 2nd International Workshop on Persistent and Photostimulable Phosphors (IWPPP 2013), Guangzhou, China, 17–21 November 2013.

16. Mohajerani, A.; Bakaric, J.; Jeffrey-Bailey, T. The urban heat island effect, its causes, and mitigation, with reference to the thermal properties of asphalt concrete. *J. Environ. Manag.* **2017**, *197*, 522–538. [CrossRef] [PubMed]

17. Li, H.; Shuen, H.; Yong, P. Research of Silica Solcoated Afterglow Luminescence Materials. *Paint Coat. Ind. China* **2007**, *7*, 18–21.

18. Zhuang, J.; Xia, Z.; Liu, H.; Zhang, Z.; Liao, L. The improvement of moisture resistance and thermal stability of $Ca_3SiO_4Cl_2:Eu^{2+}$ phosphor coated with SiO_2. *Appl. Surf. Sci.* **2011**, *257*, 4350–4353. [CrossRef]

19. Xia, Z.; Li, G.; Chen, D.; Xiao, H. Synthesis and calcination temperature dependent photoluminescence properties of novel bromosilicate phosphors. *Mater. Lett.* **2009**, *63*, 2600–2602. [CrossRef]

20. Guo, B. Study on the Surface Modification of Long Afterglow Phosphors by Amino Silane Coupling Agent. *Rare Earth* **2011**, *32*, 6–10.

21. Xing, W. *Preparation of Silicone-Acrylic Emulsion Strontium Aluminates Luminous Coating*; Shenyang Ligong University: Shenyang, China, 2015.

22. Lü, X.; Zhong, M.; Shu, W.; Yu, Q.; Xiong, X.; Wang, R. Alumina encapsulated $SrAl_2O_4$: Eu^{2+}, Dy^{3+} phosphors. *Powder Technol.* **2007**, *177*, 83–86.

23. Zhang, J.Y.; Zhang, Z.T.; Tang, Z.L.; Wang, T.M. Hydrolysis mechanism and method to improve water resistance of long after-glow phosphor. In *Proceedings of the Materials Science Forum*; Trans Tech Publications Ltd.: Zurich-Uetikon, Switzerland, 2003; Volume 423, pp. 147–150.

24. China Climate Bulletin. 2018. Available online: http://www.cma.gov.cn/root7/auto13139/201903/t20190319_517664.html (accessed on 16 January 2020).

25. Jin, W.X.; Sun, C.H.; Zuo, J.Q.; Li, W.J. Spatiotemporal Characteristics of Summer Precipitation with Different Durations in Central East China. *Clim. Environ. Res.* **2015**, *20*, 465–476.

26. Iso, E. *General Methods of Test for Pigments and Extenders Part 9: Determination of pH Value of an Aqueous Suspension*; German Institute for Standardisation: Berlin, Germany, 1995; ISO787-5-1.

27. Shirokoff, J.; Lye, L. A Review of Asphalt Binders Characterized by X-ray Diffraction. *Innov. Corros. Mater. Sci. Former. Recent Pat. Corros. Sci.* **2019**, *9*, 28–40. [CrossRef]

28. Mazumder, M.; Ahmed, R.; Ali, A.W.; Lee, S.-J. SEM and ESEM techniques used for analysis of asphalt binder and mixture: A state of the art review. *Constr. Build. Mater.* **2018**, *186*, 313–329. [CrossRef]

29. Chang, X.; Zhang, R.; Xiao, Y.; Chen, X.; Zhang, X.; Liu, G. Mapping of publications on asphalt pavement and bitumen materials: A bibliometric review. *Constr. Build. Mater.* **2020**, *234*, 117370. [CrossRef]

30. Xiao, Y.; Wan, M.; Jenkins, K.J.; Wu, S.P.; Cui, P.Q. Using activated carbon to reduce the volatile organic compounds from bituminous materials. *J. Mater. Civ. Eng.* **2017**, *29*, 4017166. [CrossRef]

31. Cui, P.; Wu, S.; Xiao, Y.; Wan, M.; Cui, P. Inhibiting effect of Layered Double Hydroxides on the emissions of volatile organic compounds from bituminous materials. *J. Clean. Prod.* **2015**, *108*, 987–991. [CrossRef]

32. Swati, G.; Chawla, S.; Mishra, S.; Rajesh, B.; Vijayan, N.; Sivaiah, B.; Dhar, A.; Haranath, D. e enhancement and decay characteristics of long afterglow nanophosphors for dark-vision display applications. *Appl. Surf. Sci.* **2015**, *333*, 178–185. [CrossRef]

Article

Thermal Performance of Novel Multilayer Cool Coatings for Asphalt Pavements

Yujing Chen [1], Kui Hu [1,2,*] and Shihao Cao [1,2]

[1]   College of Civil Engineering and Architecture, Henan University of Technology, Zhengzhou 450001, China;
      yujingchen@gmail.com (Y.C.); emailshc@163.com (S.C.)
[2]   National Engineering Laboratory for Wheat & Corn Further Processing, Henan University of Technology,
      Zhengzhou 450001, China
*   Correspondence: emailkuihu@163.com

Received: 25 May 2019; Accepted: 11 June 2019; Published: 13 June 2019

Abstract: Cool coatings are typically used to address high-temperature problems with asphalt pavements, such as rutting. However, research on the effect of the coating structure on the cooling performance remains a major challenge. In this paper, we used a three-layer cool coating (TLCC) to experimentally investigate the effects of the reflective layer, the emissive layer, and the thermal insulation layer on the cooling effect using a self-developed cooling effect evaluation device (CEED). Based on the test results, we further established temperature fields inside uncoated and coated samples, which were used to study how the TLCC affects the inner temperature field. Our results showed that the reflective layer was the main parameter influencing the cooling effect (8.18 °C), while the other two layers were secondary factors that further improved the cooling effect to 13.25 °C. A comparison of the temperature fields showed that the TLCC could effectively change the internal temperature field compared with the uncoated sample, for example, by reducing the maximum temperature inside, whose corresponding position was also deeper. As the depth increased, the cooling effect of the TLCC first increased and then decreased slowly. The results emphasize the importance of considering the effect of the coating structure on the cooling performance. This study provides a reference for effectively alleviating high-temperature distresses on asphalt pavement, which is conducive to the sustainable development of pavements.

Keywords: asphalt pavement; cool coating; coating structure; cooling effect; temperature field

---

1. Introduction

Asphalt pavements have the advantages of high flatness, riding comfort, and low noise and have been widely used in engineering construction. However, asphalt mixtures have a high absorption rate for sunlight, especially in summer, which causes the temperature of the asphalt pavement to rise sharply. On the one hand, high temperature is not good for the asphalt pavement itself. For example, high temperature tends to cause the asphalt mixture to soften easily, and with the continuous rolling of vehicles, it may produce viscous flow and even road distress, such as rutting [1,2]. This greatly increases the maintenance cost and reduces the life of asphalt pavement, which is detrimental to the sustainable development of asphalt pavement. On the other hand, high-temperature asphalt pavements can exacerbate the urban heat island (UHI) effect, thereby accelerating global warming [3,4].

Cool coatings are considered an effective technical means because of passive cooling and resistance to the UHI effect. Initially, cool coatings were popular on roofs as a relatively inexpensive technology used to reduce buildings' energy requirements for cooling during the summer [5]. In recent years, the application of cool coatings on asphalt pavements has attracted increasing attention. At present, its application on roads mainly focuses on the following aspects:

1. Development of high-performance functional fillers [6–8] and pigments, especially nonwhite pigments with high infrared reflectivity [9–11], such as Bi^{3+}-doped and Bi^{3+}/Tb^{3+}-co-doped $LaYO_3$ pigments [12]. Further, Xie et al. noted that coatings doped with chrome are a feasible way to improve the reflectance of sunlight, especially near-infrared reflectivity [13].
2. Durability of the coating. Peng et al. [14] investigated the hydrophobic and anti-icing performance of silicone coating. Li et al. [15] researched the life cycle assessment of reflective coatings for pavements.
3. Impact on the UHI [16–19]. Qin et al. [20] suggested that raising the albedo of pavements could effectively improve the urban canyon albedo only when the canyon has a low aspect ratio (e.g., h/w ≤ 1).

However, research on coating structures is still in its infancy. Although cool coatings for asphalt pavements have achieved good cooling performance, most coatings are single-layer structures. This means that some functional materials do not perform their intended function. For example, some reflective functional materials may be wrapped by other opaque materials, thereby weakening their optical reflection function. Therefore, the single layer structure limits the future development of the cool coating. In other words, it is still a considerable challenge to study the effect of the structure on the cooling performance. Moreover, another important issue is that there are few studies that focus on how cool coatings affect the temperature field inside the asphalt pavement, which is important for the softening of asphalt mixtures [21].

The purpose of this article is to prepare and test the cooling effect of a coating containing three structural layers, that is, the TLCC (three-layer cool coating). First, we developed a cooling effect evaluation device (CEED) and methods to ensure the accuracy of the test results. Second, we tested the effect of each structural layer on the cooling effect of the coating through the above device and determined the content of key materials in each layer. Finally, the influence of the coating on the internal temperature field of the whole sample was analyzed. This paper contributes to alleviating the problems caused by high temperatures, such as rutting.

2. Design Principles of Cool Coatings

Asphalt pavement is in a complex light and heat environment. Figure 1 shows that asphalt pavement can absorb, reflect, and emit radiation and that it can also transmit it downward by conduction and upward by convection. The above factors can be quantitatively described by Equation (1). When the asphalt pavement is in thermal equilibrium, its calculation formula is as shown in Equation (2).

$$q' = \alpha\left(G_S + G_{sky}\right) - G_{sur} - G_h \tag{1}$$

where:

q'—specific rate of heat flow, W/m^2;
α—absorptivity of total solar radiation;
G_S—total solar radiation, W/m^2;
G_{sky}—atmospheric counter radiation, W/m^2;
G_{sur}—thermal radiation, W/m^2;
G_h—convective heat transfer, W/m^2.

When $q' = 0$,

$$\alpha\left(G_S + G_{sky}\right) = G_{sur} + G_h \tag{2}$$

If the heat inside the asphalt pavement is reduced, we can control the following aspects: increase the reflected radiation (α) and increase the emitted thermal radiation (G_{sur}). In addition to the above, it is also necessary to consider reducing the heat transfer downward.

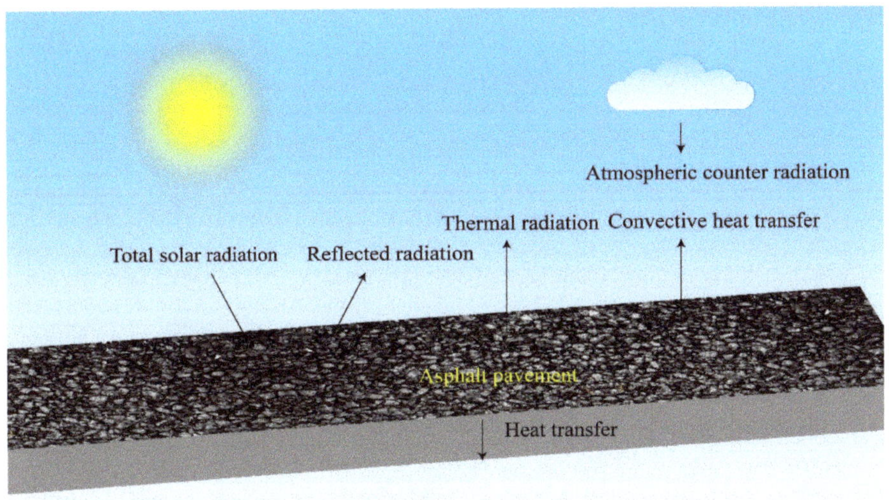

Figure 1. Thermal environment of asphalt pavement.

Therefore, we have proposed a design of the TLCC (three-layer cool coating), the structure of which is shown in Figure 2. The design of the cool coating is as follows: (1) The first layer, the reflective layer, whose main purpose is to reflect visible light and infrared light, ensures that as little light as possible enters the coating and passes down, while the layer's surface glare should be considered. (2) The second layer, also called the emissive layer, is mainly intended to emit radiation. The emissive layer pumps energy into the surrounding space by thermal radiation, which is a passive cooling method that is important for reducing the temperature of the object and mitigating the urban heat island effect. (3) The third layer, the thermal insulation layer, is responsible for preventing heat from being transferred downward into the interior of the asphalt pavement.

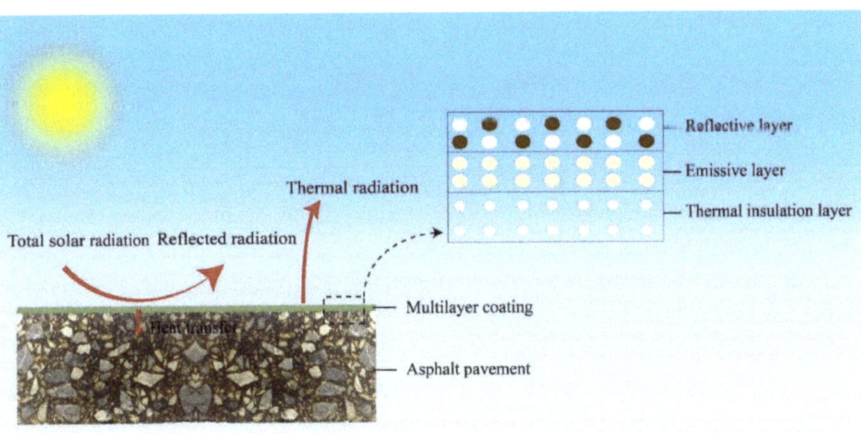

Figure 2. Working principle of the three-layer cool coating (TLCC).

3. Materials and Methods

3.1. Materials

3.1.1. Cool Coatings

According to the design principle of the TLCC with multifunctional layers, key functional materials consist of reflective materials, radiant materials and insulating materials. We chose rutile titanium dioxide [22] as the reflective material, silicon dioxide [23,24] as the radiant material, and hollow glass microbeads as the insulating material to study the influencing factors of the cooling performance of the TLCC, as shown in Table 1, Table 2, and Table 3. Then, epoxy resin was chosen as the adhesive material with high transparency. Furthermore, to form a high-performance coating, some other additives were used, for example, a diluent agent, curing agent, dispersing agent, and antifoaming agent. Finally, carbon black of different proportions was selected to alleviate driving glare.

Table 1. Technical characteristics of titanium oxide.

Appearance	Refractive Index	Average Particle Diameter	Brightness	pH	Density
White	2.8	0.20–0.26 μm	≥ 94.5%	6–9	4.23 g/cc

Table 2. Technical characteristics of silicon dioxide.

Appearance	Refractive Index	Average Particle Diameter	Boiling Point	pH	Density
White	1.553	8 μm	2950 °C	6.5–7.5	2.65 g/cc

Table 3. Technical characteristics of hollow glass microbeads.

Appearance	Density	Particle Size	Thermal Conductivity	pH
White	0.60 g/cc	15–120 μm	0.003–0.01 BTU/in·hr °F	8–9.5

3.1.2. Asphalt and Mixtures

Basalt was chosen as the aggregate, and the gradation of the asphalt mixture was AC-16 type. The optimum content of asphalt binder is 4.6 wt %, which was obtained from the Marshall test. The preparation process of the asphalt mixture conformed to China's specification of JTG F40-2004. The size of the rutting board made of the asphalt mixture was 300 mm × 300 mm × 50 mm.

3.2. Methods

3.2.1. Preparation of Test Samples

To spray coatings evenly on the standard rutting board, an ultrasonic dispersing instrument and stirrer were used for the operation. As shown in Figure 3, the specific preparation process was as follows:

1. Modifying the bonding material. To improve the properties of the adhesive material (epoxy resin), it was necessary to use additives, such as dispersing agents and antifoaming agents.
2. Adding functional materials. Although the functional materials of each layer of the cool coating were different, their operation steps were the same.
3. Spray coating. Coating was applied to the surface of the rutting plate. Each coating layer was 0.3 kg/m^2.
4. Curing. The curing speed is related to the ambient temperature, and the higher the ambient temperature, the faster the curing speed. In this test, the temperature condition of 40 °C was selected.

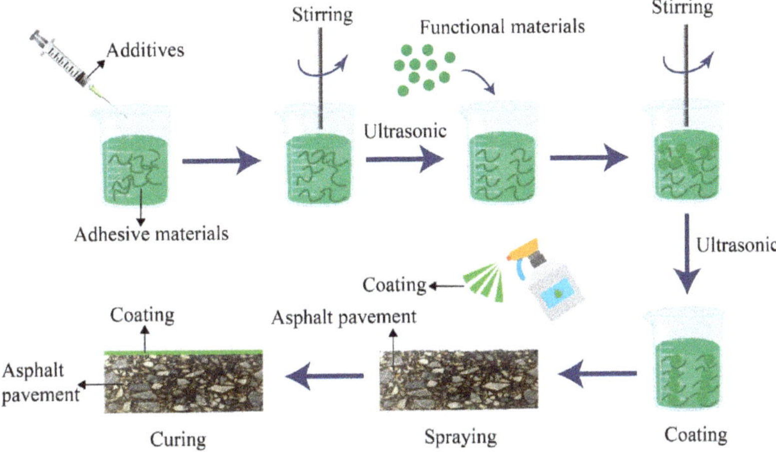

Figure 3. Schematic of the coating preparation process.

Based on the above steps, we investigated the effect of each functional layer on the overall cooling performance of the cool coatings by the control group and different test groups, as shown in Table 4.

Table 4. Coating scheme on test samples.

Scheme	Structures	Materials
Scheme 1	Reflective layer	Rutile titanium dioxide + Carbon black
Scheme 2	Reflective layer Emissive layer	Rutile titanium dioxide + Carbon black Silicon dioxide
Scheme 3	Reflective layer Emissive layer Thermal insulation layer	Rutile titanium dioxide + Carbon black Silicon dioxide Hollow glass microbead

3.2.2. Cooling Effect Test

We designed a test device (CEED) to evaluate the cooling performance of cool coatings, as shown in Figure 4. To ensure the accuracy of the test results, the device mainly had the following design features:

1. First, the device had to be capable of controlling environmental parameters and eliminating the effects of different environmental factors on measurement accuracy, such as air temperature and humidity.
2. Then, to simulate sunlight, we used tungsten halogen lamps, whose spectrum is similar to that of sunlight. We adjusted the angle of incidence to emit parallel light, which was achieved by a suitable lampshade design.
3. Furthermore, to best simulate the actual solar radiation in different weather conditions, we adjusted the intensity of the light striking the surface of the road simulation structure (the sample).
4. Finally, to ensure the accuracy of the test dates, we took measures, such as wrapping the sides and bottom of samples with a 5 cm-thick layer of thermal insulation materials, to reduce the heat loss during the test.

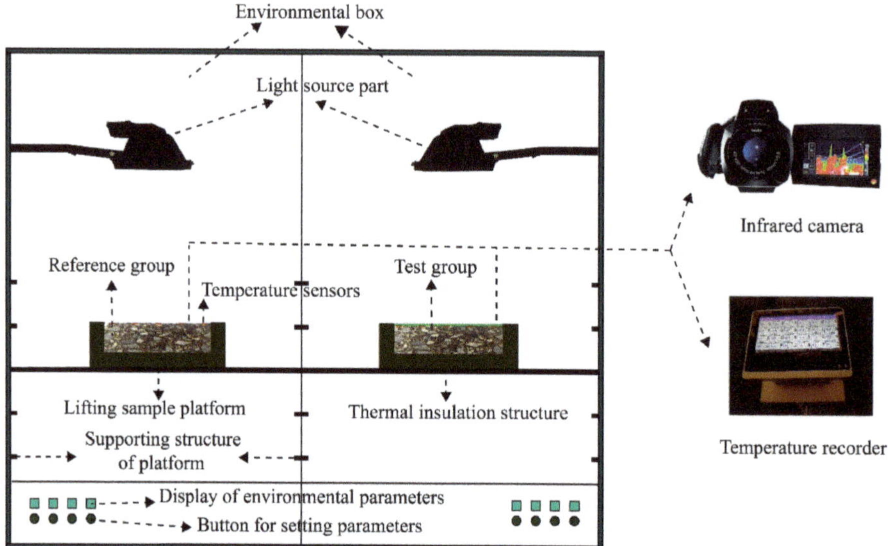

Figure 4. Self-developed cooling effect evaluation device (CEED).

In addition to the indoor CEED, temperature recorders, infrared cameras, weather stations and other devices were applied in this test. These devices worked together to complete the cooling effect test work according to certain process steps. As shown in Figure 5, the specific steps of our test were as follows:

1. The working environment parameters were set, including air temperature, humidity, and radiation intensity, where the temperature was 20 °C, the humidity was 50% RH, and the equivalent radiation intensity was 700 W/m^2.
2. The environmental parameters were checked. At the preparatory stage prior to the test, the weather station helped check and determine environmental parameters, such as radiant intensity.
3. The cooling effect of the coatings was tested. When the light source was turned on, the sample entered the heating stage. The temperature data were collected using a temperature recorder and temperature sensors every 12 minutes. After the samples reached thermal equilibrium, the thermal infrared images were taken on the side of the samples using an infrared camera, followed by temperature field analysis.
4. The cooling effect of the coatings was evaluated. The temperature difference between the coated and uncoated samples was calculated using Equation (3), where ΔT is used to indicate the cooling performance of the coating:

$$\Delta T = Temperature_{uncoated} - Temperature_{coated} \tag{3}$$

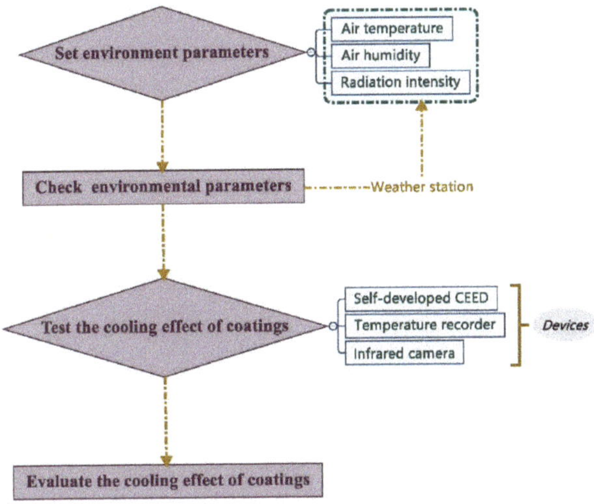

Figure 5. Specific steps of the cooling effect test.

4. Results and Discussion

4.1. Reflective Layer

4.1.1. Content of Titanium Dioxide

To evaluate the titanium dioxide content in the reflective layer, the doses of titanium dioxide were selected to be 4.5%, 9%, 13.5%, and 18%. The time histories of rising temperature are shown in Figure 6a, and the comparative cooling effects are shown in Figure 6b.

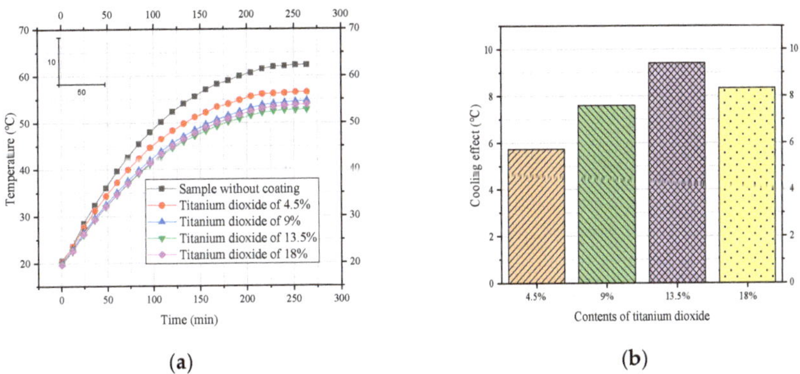

Figure 6. Effect of titanium dioxide on the cooling performance of coatings. (**a**) Process of heating to thermal equilibrium; (**b**) comparison of cooling effects.

As Figure 6 shows, the cooling effect at first increased with an increase in the content of titanium dioxide. When the percentage of titanium dioxide changed from 4.5% to 13.5%, the cooling effect significantly increased, reaching 9.41 °C. Titanium dioxide is a kind of material with a high refractive index and wide band gap that can reflect most visible light and near-infrared light. Therefore, with the increase in titanium dioxide particles, the cooling performance of the coating was improved. However, when the content of the reflective functional material reached 18%, the cooling effect of the coating did not increase but decreased slightly. This may be due to the aggregation of some particles, which resulted in a reduction in the specific surface area of heat dissipation and thus a reduction in scattering efficiency. Therefore, we recommend 13.5% titanium dioxide in the reflective layer.

4.1.2. Carbon Black in the Reflective Layer

To study the effect of carbon black content on the cooling performance of the coating, its proportion was chosen as the only variable at 0.4%, 0.6%, and 0.8%. Other ingredients and their proportions were the same.

Figure 7 shows that the use of carbon black reduced the cooling effect of the coating, and as the amount of carbon black increased, the cooling effect of the coating became lower. The use of carbon black (0.4%) caused a slight decrease in the cooling effect of the coating, from 9.41 °C to 8.91 °C, a decrease of 0.5 °C. When the amount of carbon black was increased from 0.4% to 0.6%, the change in cooling effect was still small, decreasing by 0.73 °C; however, when the dose of carbon black was increased to 0.8%, the cooling effect of the coating was significantly decreased, a drop of 1.49 °C. This trend suggests that carbon black had an adverse effect on the cooling effect of the coating while improving surface glare. Moreover, large doses of carbon black resulted in increased costs. Taking these factors into consideration, we recommend a carbon black content of 0.6%.

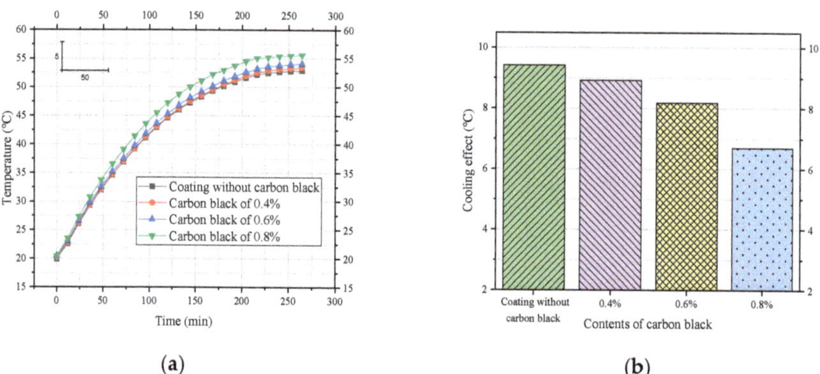

(**a**)   (**b**)

Figure 7. Effect of carbon black on the cooling performance of coatings. (**a**) Process of heating to thermal equilibrium; (**b**) comparison of cooling effects.

4.2. Emissive Layer

A single-layer coating of silicon dioxide is the simplest and most promising coating for enhancing the emissive layer. To determine the effect of the emissive layer on the overall cooling performance of the coating, a coated sample with an emissive layer was used as a test group, while a coated sample without an emissive layer was used as a reference group. Furthermore, to determine the content of silicon dioxide in the emissive layer, its percentage was chosen to be 13%, 18%, and 21%.

Figure 8 shows that the emissive layer had a positive effect on the cooling performance of the coating. In particular, the application of the emissive layer allowed the coating to have a cooling effect of more than 8.18 °C. As the amount of silicon dioxide increased from 13% to 18%, the cooling effect of the coating increased from 9.78 °C to 11.1 °C, an increase of 1.32 °C. When the amount of silicon dioxide was increased to 21%, the increase in the cooling effect of the coating was small, only 0.49 °C. It should be noted that as the filler (silicon dioxide) increased, the viscosity of the coating increased. However, the viscosity must be controlled within a suitable range to maintain its workability. Therefore, it was necessary to control the content of silicon dioxide. Therefore, we recommend a silicon dioxide content of 18%.

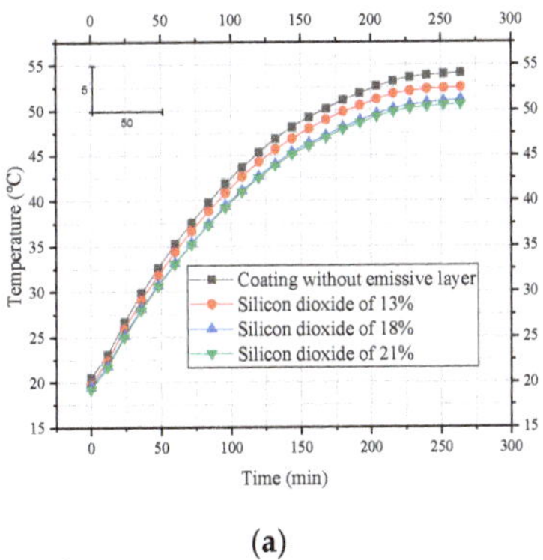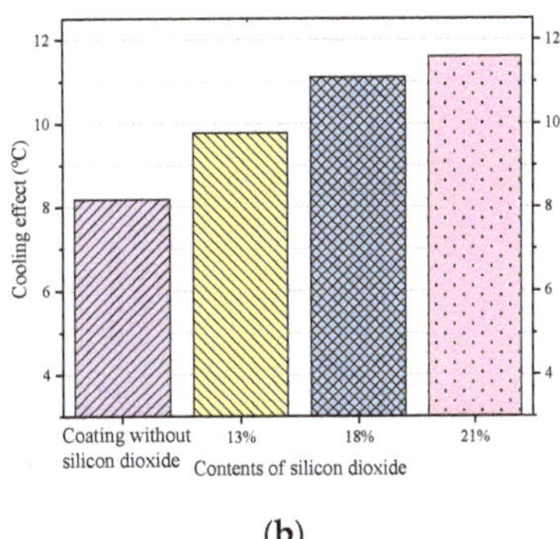

(a)   (b)

Figure 8. Effect of silicon dioxide on the cooling performance of coatings. (**a**) Process of heating to thermal equilibrium; (**b**) comparison of cooling effects.

4.3. Thermal Insulation Layer

The coating layer without the insulating layer was used as a reference group, and 11.5%, 17.5%, and 23.5% of hollow glass microspheres were added to prepare a test group. As shown in Figure 9, the coating with the thermal insulation layer had a better cooling effect. Specifically, the cooling effect of the coating exceeded 11.1 °C due to the thermal insulation layer. A content of 11.5% of the hollow glass beads increased the cooling effect of the coating by 1.18 °C, from 11.1 °C to 12.28 °C. When the content of the hollow glass microspheres continued to increase to 17.5%, the cooling effect was still satisfactory, increasing by 0.97 °C. Since hollow glass beads have a hollow structure and low thermal conductivity, they can improve the thermal insulation performance of the coating. However, when its content was increased to 23.5%, the cooling effect of the coating did not increase much, increasing only by 0.44 °C. Moreover, as the content of hollow glass microspheres increased, the compressive modulus and strength of the composites decreased. Therefore, the use of an appropriate amount of microspheres allows the coating to have a satisfactory overall performance. We recommend using 17.5% hollow glass beads in the thermal insulation layer.

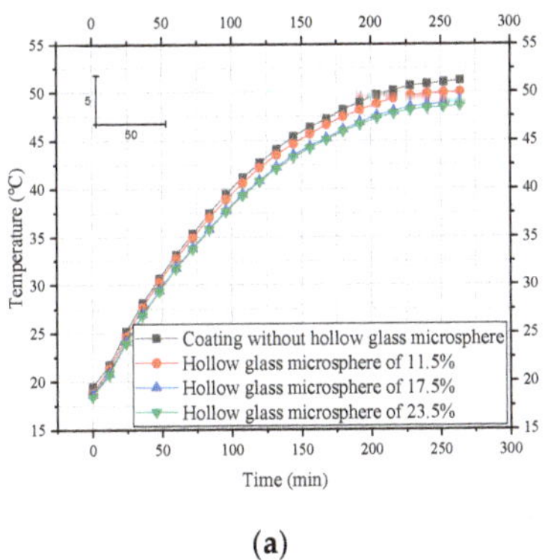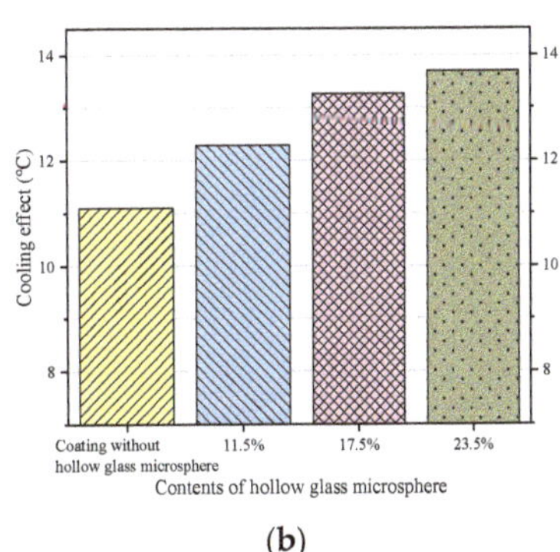

(a)   (b)

Figure 9. Effect of hollow glass microbeads on the cooling performance of coatings. (**a**) Process of heating to thermal equilibrium; (**b**) comparison of cooling effects.

4.4. Temperature Fields of the Sample Inside

To investigate the cooling effect of the coating inside the sample, the coated (TLCC) and the uncoated rutting board were taken as the research samples. An infrared camera was used to collect the data on the side, and the temperature field was analyzed by the relevant data analysis software, such as Infrec Analyzer, as shown in Figure 10.

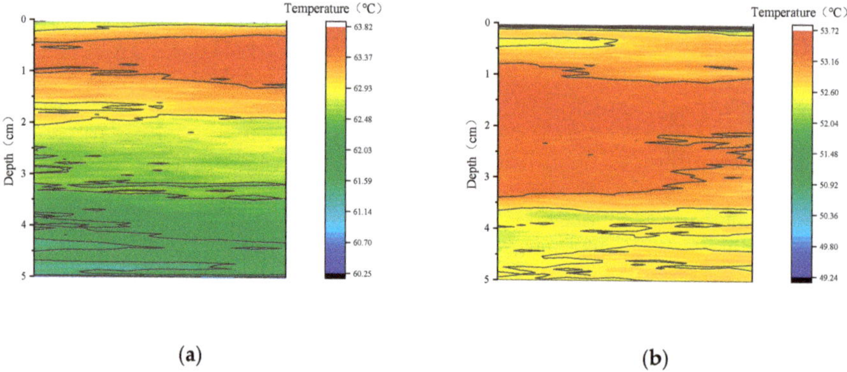

(a)   **(b)**

Figure 10. Temperature fields of samples inside an (**a**) uncoated sample and a (**b**) coated sample.

Figure 11 illustrates that the coating significantly changed the temperature field inside the sample. Specifically, the cool coating had a cooling effect of more than 12 °C within 5 cm below the surface. In addition, we still found that the change of the temperature field had a certain regularity. First, Figure 11a shows that the highest temperature point inside the sample was not at the surface but at a certain distance below the surface, regardless of whether the coating was applied. The cool coating changed the temperature and depth of the hottest point. The cool coating not only reduced the maximum temperature of the asphalt concrete by 13.12 °C but also deepened the corresponding position by 1.24 cm. This means that the cool coating was important for relieving the high-temperature softening problem of asphalt pavements in summer. Second, the cooling effect in Figure 11b shows the temperature difference (Δt) between the coated and uncoated samples at the same position. It is not difficult to see that the cooling effect of the coating increased first with increasing depth, followed by a large peak (13.68 °C). Subsequently, the cooling effect was gradually reduced. However, the cooling effect suddenly appeared with two peaks at approximately 4 cm, which was due to the unevenness caused by the large pores inside the sample. However, this accidental error did not change the tendency of the cooling effect to gradually decrease. In general, the cool coating had a good cooling effect within 5 cm below the surface of the asphalt concrete pavement, and this effect showed a tendency to decrease slightly after the increase.

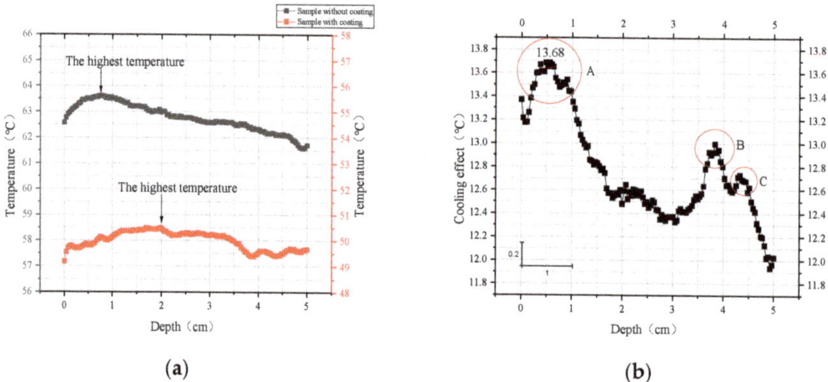

(a)   **(b)**

Figure 11. Analysis of internal temperature data. (**a**) Internal temperature changes of the coated and uncoated samples; (**b**) cooling effect of the TLCC.

5. Conclusions and Outlook

In this paper, a novel cool coating (TLCC) was prepared that comprises three different structural layers: a reflecting layer, an emitting layer, and a thermal insulation layer. Then, we independently designed a CEED (cooling effect evaluation device) and a corresponding method for coatings, which were used to investigate the effect of each structural layer on the cooling effect and determine the content of various materials. Finally, we investigated the effect of this novel coating on the internal temperature field of asphalt concrete samples and studied the cooling effect of the coating inside the sample. According to the analysis of this article, the following results are obtained:

- We developed a CEED that can regulate various environmental parameters, including atmospheric temperature, humidity, and radiation intensity. This device improves the accuracy of the test results. At the same time, we proposed a corresponding design method, which includes the following four steps: Setting the environmental parameters, such as air temperature; checking the environmental parameters using a weather station; testing the cooling effect of the coating using temperature recorders and an infrared camera; and evaluating the cooling effect of the coating. This device and method take into account the environmental factors that influence the cooling effect of the coating and provide a reference for the accurate testing of the coating.

- Using the above device, we investigated the effects of different structural layers in the coating. The reflective layer is the main factor producing the coating cooling effect. On this basis, the addition of the emissive layer and the thermal insulation layer can improve the cooling effect of the coating. The reflective coating is capable of cooling the sample surface by 8.18 °C compared to the uncoated sample. The addition of the emissive layer can increase the cooling effect of the coating to 11.1 °C. The bottom thermal insulation layer can prevent heat from passing downward, which also increases the cooling effect to 13.25 °C. This information provides a reference to further optimize the structure of the coating.

- In the above coating structure, the contents of key materials are determined. The contents of titanium dioxide, carbon black, silicon dioxide, and hollow glass microspheres are 13.5%, 0.6%, 18%, and 17.5%, respectively.

- We analyzed the temperature field inside the samples with and without the cool coating (TLCC). We drew the following three conclusions: First, the cooling effect exceeds 12 °C regardless of any depth of the asphalt mixture sample (thickness of 5 cm); second, TLCC can reduce the maximum temperature inside the asphalt pavement (13.12 °C) and lower the maximum temperature's corresponding position (1.24 cm); third, the cooling effect shows a tendency to increase first and then decrease slowly with an increase in depth. In general, this cool coating of the TLCC significantly changes the temperature field of the asphalt pavement.

This paper provides a reference for the structural optimization of the coating and a better cooling effect. At the same time, this study provides a more accurate and effective technical means for evaluating the cooling effect of the coating compared with other techniques. This study contributes to the reduction of road distresses, such as ruts, and the development of sustainable asphalt pavement. For future research, we propose the following suggestions:

- First, modify materials or find alternative new nanomaterials to develop high-performance coatings.

- Second, carry out finite element simulations and establish the temperature field inside the pavement to study the influence of relevant factors on the cooling effect of the coating, including solar radiation, temperature, and humidity.

- Third, it is necessary to establish a temperature field prediction model for asphalt pavement with a paved coating; the current temperature field prediction model is based on atmospheric temperature and solar radiation and does not consider whether the asphalt pavement surface is coated.

Author Contributions: Conceptualization, Y.C. and K.H.; methodology, Y.C.; formal analysis, K.H.; investigation, Y.C.; data curation, Y.C.; writing—original draft preparation, Y.C.; writing—review and editing, Y.C. and K.H.; software, S.C.; visualization, S.C.; supervision, K.H.; funding acquisition, K.H.

Funding: This research was funded by the National Natural Science Foundation of China (No. 51608045) and the Foundation for Distinguished Young Talents of Henan University of Technology (No. 2017BS035 and No. 2018QNJH09).

Conflicts of Interest: The authors declare no conflict of interest.

References

1. Mejłun, Ł.; Judycki, J.; Dołżycki, B. Comparison of Elastic and Viscoelastic Analysis of Asphalt Pavement at High Temperature. *Procedia. Eng.* **2017**, *172*, 746–753. [CrossRef]
2. Cui, P.; Xiao, Y.; Fang, M.; Chen, Z.; Yi, M.; Li, M. Residual Fatigue Properties of Asphalt Pavement after Long-Term Field Service. *Materials* **2018**, *11*, 892. [CrossRef] [PubMed]
3. Chen, J.Q.; Wang, H.; Zhu, H.Z. Analytical approach for evaluating temperature field of thermal modified asphalt pavement and urban heat island effect. *Appl. Therm. Eng.* **2017**, *113*, 739–748. [CrossRef]
4. Georgakis, C.; Zoras, S.; Santamouris, M. Studying the effect of "cool" coatings in street urban canyons and its potential as a heat island mitigation technique. *Sustain. Cities Soc.* **2014**, *13*, 20–31. [CrossRef]
5. Pisello, A.L. State of the art on the development of cool coatings for buildings and cities. *J. Sol. Energy Eng.* **2017**, *144*, 660–680. [CrossRef]
6. Zheng, M.L.; Han, L.L.; Wang, F.; Mi, H.C.; Li, Y.F.; He, L.T. Comparison and analysis on heat reflective coating for asphalt pavement based on cooling effect and anti-skid performance. *Constr. Build. Mater.* **2015**, *93*, 1197–1205. [CrossRef]
7. Zeyghami, M.; Goswami, D.Y.; Stefanakos, E. A review of clear sky radiative cooling developments and applications in renewable power systems and passive building cooling. *Sol. Energy Mater. Sol. Cells* **2018**, *178*, 115–128. [CrossRef]
8. Hossain, M.M.; Gu, M. Radiative Cooling: Principles, Progress, and Potentials. *Adv. Sci. Technol.* **2016**, *3*, 1500360. [CrossRef]
9. Yuan, L.; Han, A.; Ye, M.; Chen, X.; Ding, C.; Yao, L. Synthesis and characterization of novel nontoxic $BiFe_{1-x}Al_xO_3$/mica-titania pigments with high NIR reflectance. *Ceram. Int.* **2017**, *43*, 16488–16494. [CrossRef]
10. Gao, Q.; Wu, X.; Fan, Y.; Meng, Q. Color performance and near infrared reflectance property of novel yellow pigment based on Fe_2TiO_5 nanorods decorated mica composites. *Dyes Pigm.* **2017**, *146*, 537–542. [CrossRef]
11. Bao, W.; Ma, F.; Zhang, Y.; Hao, X.; Deng, Z.; Zou, X.; Gao, W. Synthesis and characterization of Fe^{3+} doped $Co_{0.5}Mg_{0.5}Al_2O_4$ inorganic pigments with high near-infrared reflectance. *Adv. Powder Technol.* **2016**, *292*, 7–13. [CrossRef]
12. Xiao, Y.; Huang, B.; Chen, J.; Sun, X. Novel Bi^{3+} doped and Bi^{3+}/Tb^{3+} co-doped $LaYO_3$ pigments with high near-infrared reflectances. *J. Alloys Compd.* **2018**, *762*, 873–880. [CrossRef]
13. Xie, N.; Li, H.; Abdelhady, A.; Harvey, J. Laboratorial investigation on optical and thermal properties of cool pavement nano-coatings for urban heat island mitigation. *Build. Environ.* **2019**, *147*, 231–240. [CrossRef]
14. Peng, C.; Zhang, H.; You, Z.; Xu, F.; Jiang, G.; Lv, S.; Zhang, R.; Yang, H. Preparation and anti-icing properties of a superhydrophobic silicone coating on asphalt mixture. *Constr. Build. Mater.* **2018**, *189*, 227–235. [CrossRef]
15. Li, H.; Saboori, A.; Cao, X. Information synthesis and preliminary case study for life cycle assessment of reflective coatings for cool pavements. *Int. J. Transp. Sci. Technol.* **2016**, *5*, 38–46. [CrossRef]
16. Qin, Y.H. A review on the development of cool pavements to mitigate urban heat island effect. *Renew. Sustain. Energy Rev.* **2015**, *52*, 445–459. [CrossRef]
17. Akbari, H.; Cartalis, C.; Kolokotsa, D.; Muscio, A.; Pisello, A.L.; Rossi, F.; Santamouris, M.; Synnefa, A.; Wong, N.H.; Zinzi, M. Local climate change and urban heat island mitigation techniques—The state of the art. *J. Civ. Eng. Manag.* **2016**, *22*, 1–16. [CrossRef]
18. Santamouris, M. Using cool pavements as a mitigation strategy to fight urban heat island—A review of the actual developments. *Renew. Sustain. Energy Rev.* **2013**, *26*, 224–240. [CrossRef]

19. Rossi, F.; Castellani, B.; Presciutti, A.; Morini, E.; Anderini, E.; Filipponi, M.; Nicolini, A. Experimental evaluation of urban heat island mitigation potential of retro-reflective pavement in urban canyons. *Energy Build.* **2016**, *126*, 340–352. [CrossRef]

20. Qin, Y.H. Urban canyon albedo and its implication on the use of reflective cool pavements. *Energy Build.* **2015**, *96*, 86–94. [CrossRef]

21. Li, Y.; Liu, L.; Sun, L. Temperature predictions for asphalt pavement with thick asphalt layer. *Constr. Build. Mater.* **2018**, *160*, 802–809. [CrossRef]

22. Hossain, F.M.; Sheppard, L.; Nowotny, J.; Murch, G.E. Optical properties of anatase and rutile titanium dioxide: Ab initio calculations for pure and anion-doped material. *J. Phys. Chem. Solids* **2008**, *69*, 1820–1828. [CrossRef]

23. Shimazaki, K.; Imaizumi, M.; Kibe, K. SiO_2 and Al_2O_3/SiO_2 coatings for increasing emissivity of Cu(In, Ga)Se$_2$ thin-film solar cells for space applications. *Thin Solid Films* **2008**, *516*, 2218–2224. [CrossRef]

24. Gentle, A.R.; Smith, G.B. Radiative heat pumping from the Earth using surface phonon resonant nanoparticles. *Nano Lett.* **2010**, *10*, 373–379. [CrossRef] [PubMed]

Article

Effect of Inhibitor on Adsorption Behavior and Mechanism of Micro-Zone Corrosion on Carbon Steel

Fengjuan Wang [1], Zhifeng Zhang [1], Shengping Wu [1], Jinyang Jiang [1,*] and Hongyan Chu [2]

[1]   Jiangsu Key Laboratory of Construction Materials, School of Materials Science and Engineering, Southeast University, Nanjing 211189, China; fengjuan19921118@sina.com (F.W.); zzf_cool_0130@126.com (Z.Z.); shpwu1988@163.com (S.W.)
[2]   College of Civil Engineering, Nanjing Forestry University, Nanjing 210037, China; chuhongyan@njfu.edu.cn
*   Correspondence: jiangjinyang16@163.com; Tel.: +86-25-5209-1175

Received: 23 May 2019; Accepted: 10 June 2019; Published: 13 June 2019

Abstract: A new type of inhibitor is studied in this paper. Inhibition efficiency and adsorption behavior of an inhibitor film on the steel surface is tested via the electrochemical method and theoretical calculation to establish the adsorption model. Test results confirm that inhibition efficiency is improved with the addition of an inhibitor, and the inhibitor film is formed firmly by comparing the characteristic peaks of S and N. Moreover, the micro-zone corrosion progress of Fe in 3.5% invasive NaCl-simulated seawater environment is studied. The results further show that corrosion is initiated under the zone without the inhibitor film, while it is prevented under the protection of the film. By the experiments, it is shown that inhibitor can be adsorbed on the surface of steel stably and has excellent protection performance for reinforced rebar, which can be widely used in concrete structure.

Keywords: carbon steel; EIS test; XPS; interfaces; inhibition mechanism

---

1. Introduction

Building materials comprise the most consumable materials in a modern city, especially concrete. However, concrete can be destroyed easily by static or dynamic loads, and concrete corrosion can take place anytime due to an adverse environment such as, fog and humidity, seawater, and alkaline or acidic soils [1]. Concrete structure destruction is mostly induced by the destruction of steel rebar in concrete. Generally, a passive film layer can be produced on the Fe surface in the environment of a high-alkalinity pore solution, and such a passive film can prevent the oxidation–reduction reaction on the steel surface [2]. However, a fragile passive thin film can be easily degraded by carbonization; CO_2 transported to the surface of the steel bar reacts with $Ca(OH)_2$ in the concrete, or chloride ions penetrate the surface of the steel causing corrosion to occur [3,4]. Therefore, to avoid the corrosion, many technological approaches have been developed to protect steel rebar, including coating with organic layers [5,6], polymer coatings [7], the formation of oxide layers [8], cathodic protection [9], and coating with other metals or alloys [10].

Nevertheless, due to the lower bond strength and short service-life durability incurred by using above methods, a new technology that has no impact on the bond strength between the steel and concrete structure and has a high service-life performance is required to ameliorate the current issues. In the past, inhibitors were used to prevent the permeation of detrimental ions from pore solutions into the surface of steel rebar, thus making the passive film immune to erosion by carbonation and chloride ions. Zarrouk et al. [11] studied the relevant inhibition properties, especially adsorption behavior, and theoretically calculated the effect of inhibitors in a hydrochloric acid solution on carbon steel. Studies show that the inhibition efficiency has a positive correlation with the inhibitor concentration, but a negative correlation with the temperature. Zhang et al. [12] researched oxo-triazole derivatives used as a corrosion inhibitor for mild steel in an acidic solution. Liu et al. [13] researched ginger extract as a green inhibitor for carbon steel in simulated concrete pore solutions. Verbruggen et al. [14] researched inhibitor evaluation in different simulated concrete pore solution for the protection of steel rebars. Neither the results of weight-loss measurements, or of electrochemical tests using such materials have shown satisfactory protection of mild steel against corrosion.

Beyond the above-mentioned studies, we used the scanning vibrating electrode technique (SVET) to study the anticorrosion mechanism under an invasive environment with and without the protection of a new inhibitor, bismuth-thiol, and visually display the corrosion process.

2. Experimental Methods

2.1. Materials and Sample Preparation

The molecule of the inhibitor under study, namely 2-(5-mercapto-1,3,thiadizaole-2-yl)-(4-methylbenzene), is shown in Figure 1. In the concentration studies, the concentration of the inhibitor in a simulated concrete solution (SCP) with 3.5% NaCl ranged from 0.1 to 5.0 mmol/L with a natural pH at an ambient temperature. The simulated concrete solution is a mixture of KOH (28.0 g/L), NaOH (8.0 g/L), and $Ca(OH)_2$ (2.2 g/L), and the pH of the simulated concrete solution is 13.6. This solution in the absence of the inhibitor is taken as the blank for comparison.

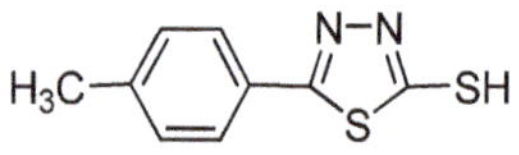

Figure 1. Structure of inhibitor molecule.

Commercial Q235 steel rebar ($10 \times 10 \times 10$ mm^3, which contains 0.14–0.22 C, 0.30–0.65 Mn, <0.60 Si, <0.05 S, <0.045 P (wt%) and Fe balance) is used in this study. Before the experiment, the rebar specimens are polished, degreased ultrasonically, and then dried using the methods described in References [15,16]. All samples tested are immersed in a prepared solution (SCP with 3.5% NaCl) at ambient temperature and exposed to air.

2.2. EIS Measurements

Electrochemical impedance spectrum (EIS) experiments are conducted on a potentiostat/galvanostat/zero-resistance ammeter (Gamry Instruments 1000, Gamry Instruments Corp., USA) using the usual three-electrode test setup. The ac frequency ranged from 1×10^5 Hz to 1×10^{-2} Hz.

2.3. Surface Chemical Composition Measurements

Before conducting X-ray photoelectron spectroscopy (XPS) measurements, as-prepared cubic steel ($4 \times 4 \times 2$ mm^3) is soaked for 3 days in SCP with NaCl and inhibitor with a concentration of 3.5 wt% and 5.0 mmol/L to form the firm inhibitor film. Under the same conditions, carbon steel soaked in a solution without the inhibitor is prepared for comparison.

The XPS measurements of the surface chemical composition are carried out on a scanning microprobe (PHI Quantum 2000, Physical Electronics, Inc., Chanhassen, MN) with an Al $K\alpha$ radiation source. Parameter settings and test procedures are based on Reference [17].

2.4. SVET Measurements

The SVET measurements of the carbon steel ($4 \times 4 \times 2$ mm^3) are conducted using the same process as the XPS measurements. Before the SVET tests, however, scratches are made on the surface of the carbon steel. The tests are then conducted in the invasive environment of a pure 3.5% NaCl solution.

Scanning electrochemical workstation (Applicable Electronics, Inc., Chanhassen, MN) is used to conduct the SVET measurements. Test setup is based on Reference [18].

3. Results and Discussion

3.1. EIS Measurement Results

A set of Nyquist and Bode plots of carbon steel in as-prepared solutions with varied concentrations of the inhibitor is shown in Figure 2a, which visually displays the capacitive and resistive behavior at the interface between the solution and the adsorbed inhibitor layer [15].

As is known, charge transfer resistance R_{ct} represents the developing tendency of capacitive loop, and double layer capacitance of C_{dl} represents the coverage percentage of the inhibitor on the carbon steel surface. A lower value of C_{dl} indicates a high coverage of inhibitor on the surface, which can provide a better anticorrosion/corrosion-protection effect [17]. In Figure 2a, compared to the bare sample, the larger diameter of the capacitive loop and the increase of the capacitive loop with increasing concentration both show that the inhibitor provides stronger corrosion protection for Q235 carbon steel.

Bode plots of the inhibitor at varied concentrations are shown in Figure 2b–d. From the Bode plots of the blank and inhibitor concentrations of 0.1 and 0.5 mmol/L, there is one time constant, which is probably associated with the charge-transfer process [19]. On the contrary, there are obviously two time constants for the sample with concentrations of 2.0 and 5.0 mmol/L.

In order to analyze the impedance characteristics, an equivalent circuit is proposed to fit the impedance spectra, as shown in Figure 2e,f. In the figure, plots with one time constant are fitted by the circuit shown in Figure 2e, while the balance is fitted by the circuit shown in Figure 2f.

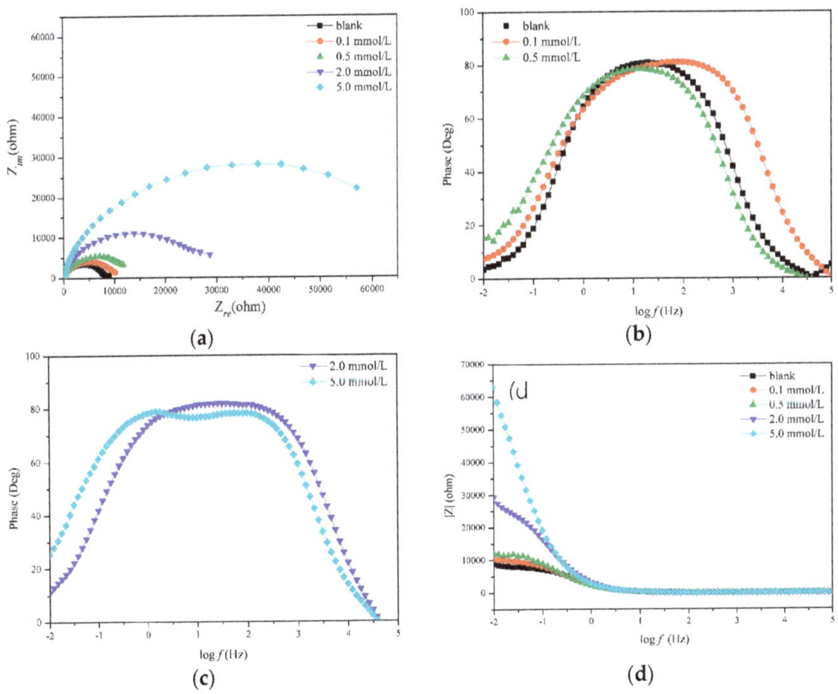

Figure 2. *Cont.*

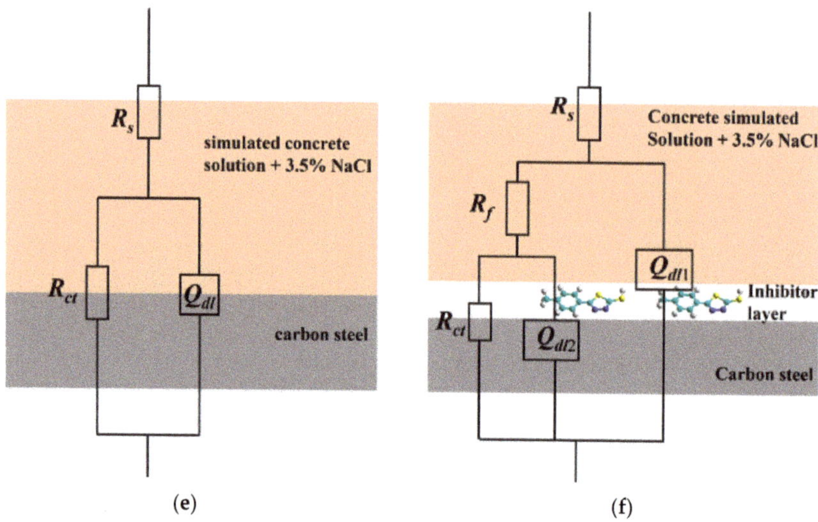

Figure 2. EIS spectra of Q235 carbon steel in simulated concrete solution with 3.5% NaCl in the absence of and with varied inhibitor concentrations: (**a**) Nyquist plots; Bode plots of (**b**) blank sample, 0.1 and 0.5 mmol/L, (**c**) of 2.0 and 5.0 mmol/L, and (**d**) of |Z|; equivalent circuit with (**e**) one-time constant and (**f**) two-time constants.

The impedance values of the equivalent circuit and electrochemical parameters are given in Table 1. When the R_{ct} values are obtained, *IE%* values of the inhibitor are calculated by [15]:

$$IE\% = (R_{ct} - R_{ct0})/R_{ct} \times 100 \tag{1}$$

where, R_{ct} and R_{ct0} are the respective charge-transfer resistance values of the previously described as-prepared solutions containing the inhibitor and inhibitor-free. By comparing the R_{ct} values, it is clear that the more inhibitor molecules there are, the better the inhibition efficiency.

Table 1. EIS parameters for corrosion of Q235 carbon steel in simulated concrete solution (3.5% NaCl) without and with different inhibitor concentrations.

Sample (mmol/L)	R_s (Ω cm^2)	Q_{dl1} (F cm^{-2})	$n1$	R_f (Ω cm^2)	Q_{dl2} (F cm^{-2})	$n2$	R_{ct} (Ω cm^2)	η (%)
Blank	4.382	6.89×10^{-5}	0.9245	–	–	–	7783	–
0.1	1.191	7.316×10^{-5}	0.9004	–	–	–	9691	19.7
0.5	6.545	8.83×10^{-5}	0.8892	–	–	–	12,320	36.8
2.0	1.543	4.831×10^{-5}	0.9374	274	1.890×10^{-5}	0.4388	32,060	75.7
5.0	2.679	4.539×10^{-5}	0.9375	678.4	2.665×10^{-5}	0.7467	70,650	89.0

Owing to the fact that corrosion could not occur at the covered sites, the molecules of inhibitors adsorbed on the Fe surface can supply a barrier that prevents the corrosion process. Therefore, the more barriers that are on the Fe surface covered by the inhibitor molecule film, the higher the inhibition efficiency regarding the corrosion process [17].

3.2. Adsorption Isotherm Analysis

Adsorption isotherm is usually used to build the interaction mode between the inhibitors and the Fe surface [20]. Langmuir adsorption isotherm is cited based on the EIS test results to study the adsorption behavior of the inhibitor on Fe surface. The following formula can be used to express the Langmuir adsorption isotherm [21]:

$$c/\theta = 1/K_{ads} + c \tag{2}$$

in which, c, θ, and K_{ads} represent the concentration of inhibitor (mmol/L), coverage equivalent (defined as $IE\%$), and equilibrium constant of inhibitor adsorption (L/mol), respectively. c/θ versus c is plotted in Figure 3 which is a straight line and the slope is close to 1, showing that Langmuir adsorption isotherm is the fittest to describe the behavior of inhibitor on the Fe surface [22].

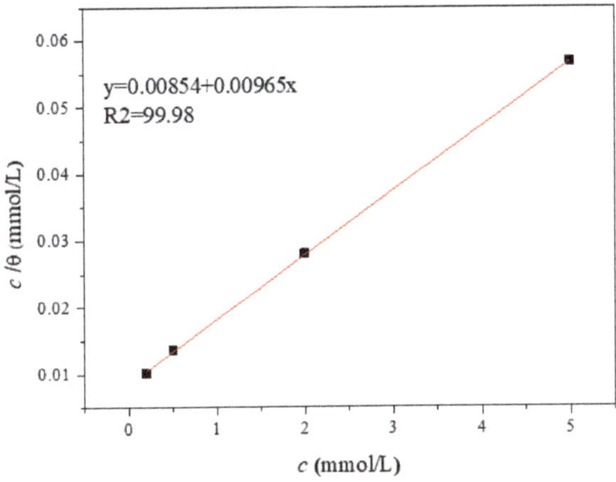

Figure 3. Adsorption isotherm of inhibitor on Q235 carbon steel surface in simulated concrete solution with 3.5% NaCl and different inhibitor concentrations.

By the calculation of K_{ads} and the following equation [23], the standard adsorption free energy (ΔG^0_{ads}) can also be obtained:

$$\Delta G^0_{ads} = -RT \ln(55.5 K_{ads}) \tag{3}$$

where, R, T, and the value 55.5 are the universal gas constant (J·mol/K), temperature (K), and molar concentration (mol/L) of water in solution, respectively.

In this paper, due to the calculated value of ΔG^0_{ads} is −21.25 kJ/mol, we can know that it is spontaneous for the adsorption behavior of inhibitor molecule, which contains physisorption and chemisorption together, caused by electrostatic interactions and covalent bonds, respectively [24].

3.3. XPS Analysis

XPS measurement is a useful tool to identify the combination mode between elements and the substrate. Therefore, the inhibitor adsorbed on the carbon steel surface is measured using XPS to prove that the inhibitor is adsorbed on the surface and to explain the combination process. Figure 4 shows total spectra (Figure 4a), as well as high-resolution C 1s (Figure 4b), O 1s (Figure 4c), N 1s (Figure 4d), and S 2p (Figure 4e) XPS spectra, for samples with inhibitor and inhibitor-free, no other impurity ions are detected. In the C 1s, O 1s, N 1s, and S 2p regions, the deconvolution of multiple peaks is performed to determine the respective binding energies [25].

As shown in Figure 4b,c, it can be seen that there is no obvious difference between the samples with an inhibitor and the blank sample for the high-resolution of O 1s XPS spectra, especially the peak at 288.3 eV is attributed to the sp^2-hybridized carbon [26,27], which we can speculate comes from the inhibitor molecule.

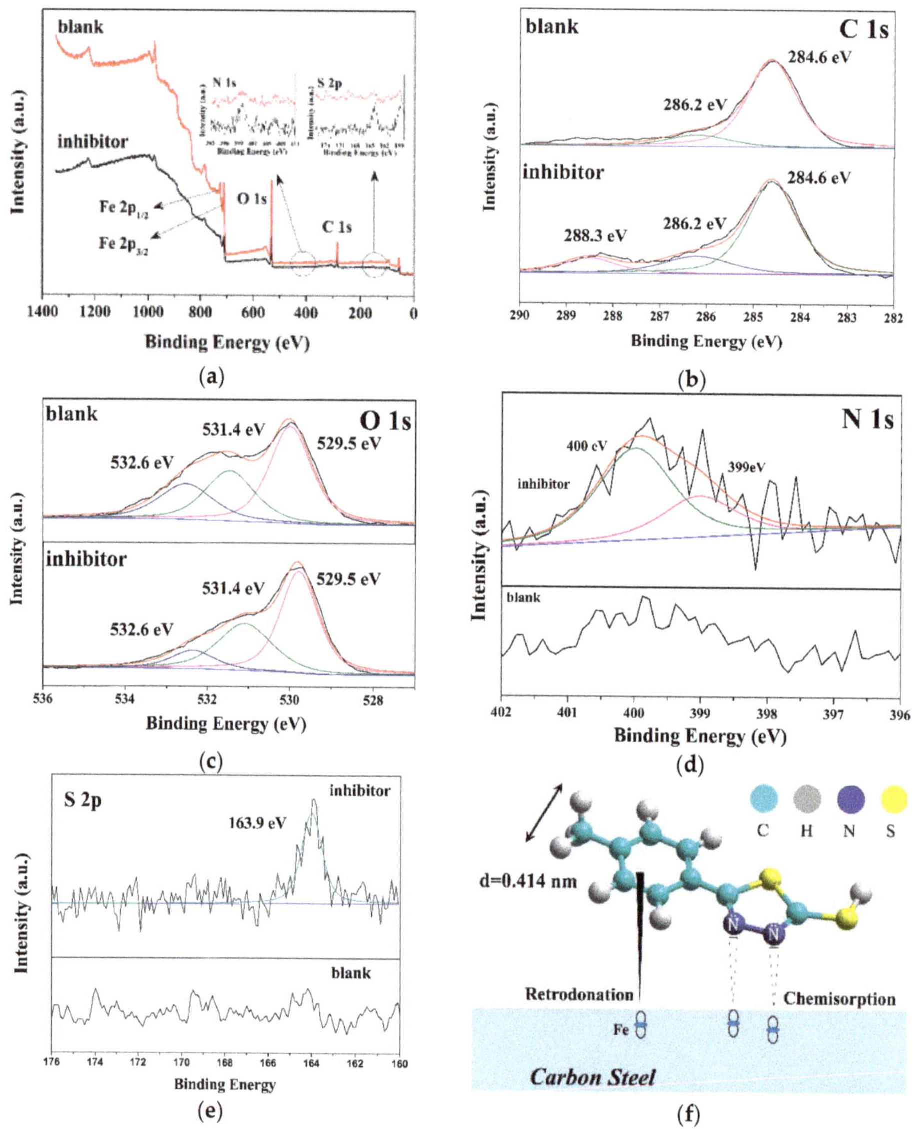

Figure 4. XPS spectra of (**a**) total spectra; high-resolution spectra of (**b**) C 1s, (**c**) O 1s, (**d**) N 1s, and (**e**) S 2p peaks; (**f**) schematic of inhibitor adsorbed on carbon steel surface.

As shown in Figure 4b, two characteristic peaks of 284.6 eV and 286.2 eV are observed in the inhibitor and blank samples, while the binding energy of 288.3 eV is only detected in the sample with inhibitor. The peaks at 284.6 eV and 286.2 eV are ascribed to a C–C bond and C=O bond, respectively, which is attributed to the adventitious hydrocarbon from the XPS instrument itself [28]. The peak at 288.3 eV is attributed to the sp^2-hybridized carbon [26,27], which we can speculate comes from the inhibitor molecule. Figure 4c shows the high-resolution of O 1s XPS spectra, the first component at a binding energy of 529.5 eV is assigned to the Fe oxide, such as FeO and Fe_2O_3 [29,30]. The second peak, with a binding energy of 531.9 eV, can be considered to result from the hydroxide bonds chemisorbed on the surface [31,32]. The last peak, at 533.6 eV, can be attributed to O of the adsorbed water [33]. From Figure 4c, it can be seen that there is no obvious difference between the samples with an inhibitor and the blank sample for the high-resolution of O 1s XPS spectra.

In Figure 4d, a broad N 1s peak in region of 396–402 eV with a maximum located at a binding energy of 399.7 eV is clearly identified for the inhibitor sample. Two Gaussian–Lorentzian peaks of the high-resolution N 1s spectrum exist: One at a binding energy of 399.0 eV that could be attributed to the type of C–N bond [34], and the other at 400 eV that could be assigned to the bond of N adsorbed on the carbon steel surface [35]. In comparison, no obvious N 1s peak is detected in the blank sample. Based on the results of the analysis of the high-resolution N 1s XPS spectrum, we can conclude that it is the formation of chemical bonds between elemental N and Fe that facilitate the adsorption of the inhibitor on the carbon steel surface for this type of inhibitor, which is consistent with the findings of Reference [36]. In Figure 4e, it can be seen that there is a peak located at 163.9 eV, which corresponds to the C–S bond [37]. In contrast, no S 2p peak can be seen in the spectrum of the blank sample.

Therefore, by integrating the results of the adsorption isotherm and XPS analysis, we can further surmise that physisorption and chemisorption coexist, as a result of electrostatic interactions for the former and of N–Fe chemical bonds for the latter. Based on the above analysis, the adsorption mechanism of the inhibitor with coexisting physisorption and chemisorption on carbon steel is depicted in Figure 4f.

3.4. SVET Analysis

SVET is used to measure the potential differences on the Fe surface in a solution with an extremely high resolution ratio due to the fluxes of ionic current, which is caused by the electrochemical reactions occurring at the Fe surface [38]. Distribution of the local current density over a surface can be depicted in SVET as two-(2D) or three-dimensional (3D) current–density maps [39]. Figure 5 depicts the SVET 2D maps obtained from carbon steel samples, each immersed in 3.5% NaCl for 15 min. The first and last photographs in the figure are of the samples before and after the immersion test, in which the valid test area measures $3240 \times 3240\ \mu m^2$, on the surface of which three scratches are made to distinguish the blank steel sample from the steel sample that has adsorbed the inhibitor. The three scratches are designated as Scratch 1, Scratch 2, and Scratch 3.

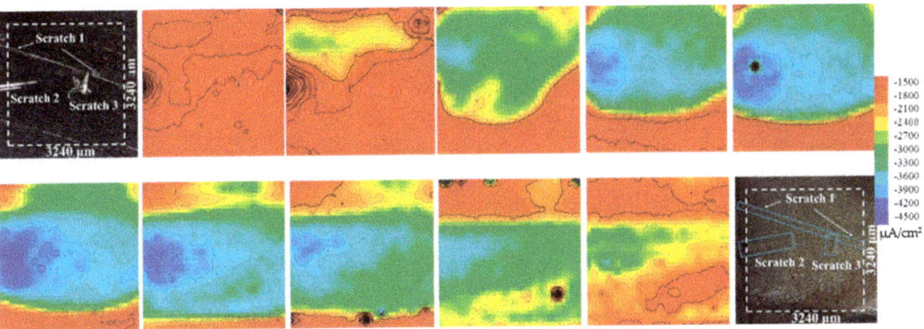

Figure 5. Corrosion progress of sample in NaCl solution measured by scanning vibrating electrode technique (SVET) test.

In the first 15 min, the sample presents a uniform corrosion feature and the current density is nearly $-1500\ \mu A/cm^2$. Notably, the current density at the Scratch 2 location is slightly higher than in any other location. As the corrosion process continues, the current density continuously increases, and corrosion occurs at the locations of all three scratches. After the SVET test, the scratch locations cannot be distinguished clearly. The maximum current density is $-4500\ \mu A/cm^2$, which is three times higher than that recorded in the first 15 min. After immersion in 3.5% NaCl for 2 h, the size of the corrosion-zone area begins to decrease due to the coverage of the corrosion product, which blocks the probe from detecting the current variation. The current density remains invariant in the entire immersion period at the localized zone of the inhibitor film, indicating that the inhibitor film provides the steady protection from corrosion for the carbon steel.

Two matters must be discussed. First, while the scratches are small, the corrosion zone is large. This may have been caused by the fact that, when the scratches are carved on the surface of the carbon steel, the inhibitor film is wiped off along with the removal or loosening of the adjacent film, since the inhibitor film is connected to the adjacent film in the close-packed structure. Second, invasive Cl^- ions cannot break the inhibitor film directly. The process of Cl^- ions invading the carbon steel in the zone of the inhibitor film is, (1) it destroys the carbon steel in the transition zone between the inhibitor film and the blank zone, and (2) then strips the inhibitor film bit by bit until the inhibitor film is completely destroyed. Using 2D maps of the SVET test results, the corrosion process under the protection of an inhibitor is vividly illustrated.

4. Conclusions

The results obtained in this work show that it is effective for an inhibitor of bismuth-thiol to prevent the corrosion process of the carbon steel in the simulated concrete pore solution with erosive NaCl, and that the inhibition efficiency increases with increasing inhibitor concentration. Subsequently, the existence of adsorption layers formed by the inhibitor on the carbon steel is confirmed by the XPS analysis through detecting the characteristic atoms of S and N. The corrosion mechanism and process are also verified by SVET measurements, which show that under the protection of inhibitor the current density of the inhibitor zone is much lower than that of the blank zone. Considering the above results, we determine that it is the inhibitor film that plays the vital role in resisting the corrosion process and, in particular, that the progress of corrosion can be stopped under the protection of an intact protective inhibitor film.

As the rust inhibitor is a kind of small organic molecule that can be present in the pore solution of concrete, it has no effect on the long-term service performance of the concrete. This thesis shows that this rust inhibitor has perfect rust resistance performance for steel bars. In the subsequent researches, it is necessary to develop or select a migration rust inhibitor with similar molecular structure in order to study the repair effect that the rust inhibitor migrates to the surface of the steel through the concrete to repair the damaged passivation film on the steel bars when the concrete structure begins to break.

Author Contributions: Conceptualization, F.W. and J.J.; data curation, F.W. and Z.Z.; formal analysis, F.W. and Z.Z.; investigation, F.W. and H.C.; methodology, F.W. and S.W.; resources, S.W.; writing—original draft, F.W., Z.Z., J.J., and H.C.; writing—review and editing, F.W., Z.Z., J.J., and H.C.

Funding: This work was supported by the National Program on Key Basic Research Project (Grant No. 2015CB655105), and the National Natural Science Foundation of China (Grant No. 51508091) and with no external funding.

Conflicts of Interest: The authors declare no conflict of interest.

References

1. Pei, X.; Noel, M.; Green, M.; Fam, A.; Shier, G. Cementitious coatings for improved corrosion resistance of steel reinforcement. *Surf. Coat. Technol.* **2017**, *315*, 188–195. [CrossRef]
2. Jiang, J.; Wang, D.; Chu, H.; Ma, H.; Liu, Y.; Gao, Y.; Shi, J.; Sun, W. The Passive Film Growth Mechanism of New Corrosion-Resistant Steel Rebar in Simulated Concrete Pore Solution: Nanometer Structure and Electrochemical Study. *Materials* **2017**, *10*, 412. [CrossRef] [PubMed]
3. Revert, A.-B.; De Weerdt, K.; Hornbostel, K.; Geiker, M.-R. Carbonation-induced corrosion: Investigation of the corrosion onset. *Constr. Build. Mater.* **2018**, *162*, 847–856. [CrossRef]
4. Luo, H.; Su, H.; Dong, C.; Li, X. Passivation and electrochemical behavior of 316L stainless steel in chlorinated simulated concrete pore solution. *Appl. Surf. Sci.* **2017**, *400*, 38–48. [CrossRef]

5. Soriano, C.; Alfantazi, A. Corrosion behavior of galvanized steel due to typical soil organics. *Constr. Build. Mater.* **2016**, *102*, 904–912. [CrossRef]

6. Hsu, C.; Cho, Y.; Liu, T.; Chang, H.; Lien, S. Optimization of residual stress of SiO_2/organic silicon stacked layer prepared using inductively coupled plasma deposition. *Surf. Coat. Technol.* **2017**, *320*, 293–297. [CrossRef]

7. Yabuki, A.; Tanabe, S.; Fathona, I.-W. Self-healing polymer coating with the microfibers of superabsorbent polymers provides corrosion inhibition in carbon steel. *Surf. Coat. Technol.* **2018**, *341*, 71–77. [CrossRef]

8. Cole, I.-S. Recent Progress and Required Developments in Atmospheric Corrosion of Galvanised Steel and Zinc. *Materials* **2017**, *10*, 1288. [CrossRef]

9. Park, J.-H.; Park, J.-M. Photo-generated cathodic protection performance of electrophoretically Co-deposited layers of TiO_2 nanoparticles and graphene nanoplatelets on steel substrate. *Surf. Coat. Technol.* **2014**, *258*, 62–71. [CrossRef]

10. Chen, C.L.; Sutrisna. Study of W-Co ODS coating on stainless steels by mechanical alloying. *Surf. Coat. Technol.* **2018**, *350*, 954–961. [CrossRef]

11. Zarrouk, A.; Zarrok, H.; Ramli, Y.; Bouachrine, M.; Hammouti, B.; Sahibed-dine, A.; Bentiss, F. Inhibitive properties, adsorption and theoretical study of 3,7-dimethyl-1-(prop-2-yn-1-yl)quinoxalin-2(1H)-one as efficient corrosion inhibitor for carbon steel in hydrochloric acid solution. *J. Mol. Liq.* **2016**, *222*, 239–252. [CrossRef]

12. Tao, Z.; Zhang, S.; Li, W.; Hou, B. Corrosion inhibition of mild steel in acidic solution by some oxo-triazole derivatives. *Corros. Sci.* **2009**, *51*, 2588–2595. [CrossRef]

13. Liu, Y.; Song, Z.; Wang, W.; Jiang, L.; Zhang, Y.; Guo, M.; Song, F.; Xu, N. Effect of ginger extract as green inhibitor on chloride-induced corrosion of carbon steel in simulated concrete pore solutions. *J. Clean. Prod.* **2019**, *214*, 298–307. [CrossRef]

14. Verbruggen, H.; Terryn, H.; De Graeve, I. Inhibitor evaluation in different simulated concrete pore solution for the protection of steel rebars. *Constr. Build. Mater.* **2016**, *124*, 887–896. [CrossRef]

15. Tian, H.; Cheng, Y.-F.; Li, W.; Hou, B. Triazolyl-acylhydrazone derivatives as novel inhibitors for copper corrosion in chloride solutions. *Corros. Sci.* **2015**, *100*, 341–352. [CrossRef]

16. Rivera-Corral, J.-O.; Fajardo, G.; Arliguie, G.; Orozco-Cruz, R.; Deby, F.; Valdez, P. Corrosion behavior of steel reinforcement bars embedded in concrete exposed to chlorides: Effect of surface finish. *Constr. Build. Mater.* **2017**, *147*, 815–826. [CrossRef]

17. Gao, X.; Zhao, C.; Lu, H.; Gao, F.; Ma, H. Influence of phytic acid on the corrosion behavior of iron under acidic and neutral conditions. *Electrochim. Acta* **2014**, *150*, 188–196. [CrossRef]

18. Coelho, L.-B.; Cossement, D.; Olivier, M.-G. Benzotriazole and cerium chloride as corrosion inhibitors for AA2024-T3: An EIS investigation supported by SVET and ToF-SIMS analysis. *Corros. Sci.* **2018**, *130*, 177–189. [CrossRef]

19. Zhu, Y.; Zhang, H.; Zhang, Z.; Yao, Y. Electrochemical impedance spectroscopy (EIS) of hydration process and drying shrinkage for cement paste with W/C of 0.25 affected by high range water reducer. *Constr. Build. Mater.* **2017**, *131*, 536–541. [CrossRef]

20. Feng, L.; Zhang, S.; Qiang, Y.; Xu, Y.; Guo, L.; Madkour, L.-H.; Chen, S. Experimental and Theoretical Investigation of Thiazolyl Blue as a Corrosion Inhibitor for Copper in Neutral Sodium Chloride Solution. *Materials* **2018**, *11*, 1042. [CrossRef]

21. Unnisa, C.-B.-N.; Devi, G.-N.; Hemapriya, V.; Chitra, S.; Chung, I.; Kim, S.; Prabakaran, M. Linear polyesters as effective corrosion inhibitors for steel rebars in chloride induced alkaline medium—An electrochemical approach. *Constr. Build. Mater.* **2018**, *165*, 866–876. [CrossRef]

22. He, X.; Jiang, Y.; Li, C.; Wang, W.; Hou, B.; Wu, L. Inhibition properties and adsorption behavior of imidazole and 2-phenyl-2-imidazoline on AA5052 in 1.0 M HCl solution. *Corros. Sci.* **2014**, *83*, 124–136. [CrossRef]

23. Zhang, H.; Gao, K.; Yan, L.; Pang, X. Inhibition of the corrosion of X70 and Q235 steel in CO_2-saturated brine by imidazoline-based inhibitor. *J. Electroanal. Chem.* **2017**, *791*, 83–94. [CrossRef]

24. Yurt, A.; Ulutas, S.; Dal, H. Electrochemical and theoretical investigation on the corrosion of aluminium in acidic solution containing some Schiff bases. *Appl. Surf. Sci.* **2006**, *253*, 919–925. [CrossRef]

25. Arman, S.-Y.; Omidvar, H.; Tabaian, S.-H.; Sajjadnejad, M.; Fouladvand, S.; Afshar, S. Evaluation of nanostructured S-doped TiO_2 thin films and their photoelectrochemical application as photoanode for corrosion protection of 304 stainless steel. *Surf. Coat. Technol.* **2014**, *251*, 162–169. [CrossRef]

26. Wang, X.; Yang, W.; Li, F.; Xue, Y.; Liu, R.; Hao, Y. In Situ Microwave-Assisted Synthesis of Porous N-TiO$_2$/g-C3N4 Heterojunctions with Enhanced Visible-Light Photocatalytic Properties. *Ind. Eng. Chem. Res.* **2013**, *52*, 17140–17150. [CrossRef]

27. Thomas, A.; Fischer, A.; Goettmann, F.; Antonietti, M.; Mueller, J.; Schloegl, R.; Carlsson, J.-M. Graphitic carbon nitride materials: Variation of structure and morphology and their use as metal-free catalysts. *J. Mater. Chem.* **2008**, *18*, 4893–4908. [CrossRef]

28. Li, S.; Fu, J. Improvement in corrosion protection properties of TiO$_2$ coatings by chromium doping. *Corros. Sci.* **2013**, *68*, 101–110. [CrossRef]

29. Grosvenor, A.-P.; Kobe, B.-A.; Biesinger, M.-C.; McIntyre, N.-S. Investigation of multiplet splitting of Fe 2p XPS spectra and bonding in iron compounds. *Surf. Interface Anal.* **2004**, *36*, 1564–1574. [CrossRef]

30. Zhang, Y.; Cheng, P.; Yu, K.; Zhao, X.; Ding, G. ITO film prepared by ion beam sputtering and its application in high-temperature thermocouple. *Vacuum* **2017**, *146*, 31–34. [CrossRef]

31. Li, C.; Fan, W.; Lu, H.; Ge, Y.; Bai, H.; Shi, W. Fabrication of Au@CdS/RGO/TiO$_2$ heterostructure for photoelectrochemical hydrogen production. *New J. Chem.* **2016**, *40*, 2287–2295. [CrossRef]

32. Hassan, H.-H.; Amin, M.-A.; Gubbala, S.; Sunkara, M.-K. Participation of the dissolved O-2 in the passive layer formation on Zn surface in neutral media. *Electrochim. Acta* **2007**, *52*, 6929–6937. [CrossRef]

33. Tourabi, M.; Nohair, K.; Traisnel, M.; Jama, C.; Bentiss, F. Electrochemical and XPS studies of the corrosion inhibition of carbon steel in hydrochloric acid pickling solutions by 3, 5-bis(2-thienylmethyl)-4-amino-1, 2, 4-triazole. *Corros. Sci.* **2013**, *75*, 123–133. [CrossRef]

34. Kozak, A.-O.; Porada, O.-K.; Ivashchenko, L.-A.; Scrynskyy, P.-L.; Tomila, T.-V.; Manzhara, V.-S. Comparative investigation of Si-C-N Films prepared by plasma enhanced chemical vapour deposition and magnetron sputtering. *Appl. Surf. Sci.* **2017**, *425*, 646–653. [CrossRef]

35. Prabhawalkar, P.-D.; Raole, P.-M.; Kothari, D.-C.; Nair, M.-R. XPS studies at various depths of low-energy n-$^{2+}$ ions implanted on 304 stainless-steel. *Vacuum* **1986**, *36*, 817–820. [CrossRef]

36. Zhang, W.; Ma, R.; Liu, H.; Liu, Y.; Li, S.; Niu, L. Electrochemical and surface analysis studies of 2-(quinolin-2-yl) quinazoiin-4 (3H) -one as corrosion inhibitor for Q235 steel in hydrochloric acid. *J. Mol. Liq.* **2016**, *222*, 671–679. [CrossRef]

37. Rendon-Nava, D.; Mendoza-Espinosa, D.; Negron-Silva, G.-E.; Tellez-Arreola, J.-L.; Martinez-Torres, A.; Valdez-Calderon, A.; Gonzalez-Montiel, S. Chrysin functionalized NHC-Au (I) complexes: Synthesis, characterization and effects on the nematode Caenorhabditis elegans. *New J. Chem.* **2017**, *41*, 2013–2019. [CrossRef]

38. Lamaka, S.-V.; Taryba, M.; Montemor, M.-F.; Isaacs, H.-S.; Ferreira, M.-G.-S. Quasi-simultaneous measurements of ionic currents by vibrating probe and pH distribution by ion-selective microelectrode. *Electrochem. Commun.* **2011**, *13*, 20–23. [CrossRef]

39. Wang, M.; Liu, M.; Fu, J. An intelligent anticorrosion coating based on pH-responsive smart nanocontainers fabricated via a facile method for protection of carbon steel. *J. Mater. Chem. A* **2015**, *3*, 6423–6431. [CrossRef]

Article

Performances of Cement Mortar Incorporating Superabsorbent Polymer (SAP) Using Different Dosing Methods

Yawen Tan [1], Huaxin Chen [1], Zhendi Wang [2], Cheng Xue [3] and Rui He [1,*]

[1] School of Materials Science and Engineering, Chang'an University, Xi'an 710061, Shanxi, China; tywen1210@126.com (Y.T.); hxchen@chd.edu.cn (H.C.)
[2] State Key Laboratory of Green Building Materials, China Building Materials Academy, Beijing 100024, China; wzhendi@163.com
[3] CCCC Second Highway Engineering CO., LTD, Xi'an 710065, Shaanxi, China; xuecheng3a@163.com
* Correspondence: heruia@163.com; Tel.: +86-135-7244-8396

Received: 10 April 2019; Accepted: 9 May 2019; Published: 17 May 2019

Abstract: Modified cement mortar was prepared by incorporating a superabsorbent polymer (SAP) with two kinds of dosing state, dry powdery SAP and swelled SAP (where the SAP has been pre-wetted in tap water), respectively. The mechanical properties, drying shrinkage and freeze–thaw resistance of the mortars were compared and analyzed with the variation of SAP content and entrained water-to-cement ratios. Additionally, the effect of SAP on the microstructure of mortar was characterized by scanning electron microscopy (SEM). The results indicate that agglomerative accumulation is formed in the voids of mortar after water desorption from SAP and there are abundant hydration products, most of which are C-S-H gels, around the SAP voids. The incorporation of the powdery SAP increases the 28 d compressive strength of the mortars by about 10% to 50%, while for the incorporation of swelled SAP, the 28 d compressive strength of the mortar can be increased by about −26% to 6%. At a dosage of 0.1% SAP and an entrained water–cement ratio of 0.06, the powdery SAP and the swelled SAP can reduce the mortar shrinkage rate by about 32.2% and 14.5%, respectively. Both the incorporation of powdery and swelled SAP has a positive effect on the freeze–thaw resistance of cement mortar. In particular, for powdery SAP with an entrained water-to-cement ratio of 0.06, the mass loss rate after 300 cycles is still lower than 5%.

Keywords: superabsorbent polymer (SAP); compressive strength; drying shrinkage; freeze–thaw resistance

1. Introduction

Superabsorbent polymers (SAPs) are excellent high-performance liquid absorbent and retention polymers, which can absorb and retain several hundred times higher amounts of water or aqueous liquids relative to their own mass in a short time due to their three-dimensional hydrophilic network structures [1–3]. Taking advantage of such specific characteristics, SAP has been widely applied in many fields [4–7], such as concrete, agriculture, personal hygiene, waste treatment, and medical industries [8–10]. In concrete in particular, the typical application of SAP is placed on developing internal curing technology, which refers to storing an amount of water in the interior of cement during the mixing process of the cementitious materials by adding internal curing materials [11,12].

SAP is often introduced to cement-based materials as a promising new internal curing material for concrete [13,14]. When SAP is exposed to cement-based mixtures, a transition in the osmotic pressure gradient and chemical potential occurs between the inside and outside of the three-dimensional network leading to water absorption and storage of the SAP [15–17]. The stored water can be released to compensate for the internal moisture loss of cement-based materials, which can promote further hydration of the cement particles and reduce self-desiccation shrinkage during the hydration of cement or evaporation of the water [18,19]. Therefore, the addition of SAP can significantly mitigate shrinkage in low water-to-cement ratio mixtures and greatly reduce the occurrence of cracking [20]. Additionally, SAP can also increase the degree of hydration, enhance the freeze–thaw resistance capacity, change the rheological behaviors and optimize the pore size distribution [21–23], among other actions.

However, the majority of recent studies about SAP have focused on the effect and theory on the shrinkage, hydration and mechanical performances of cement-based materials due to the variation of SAP content and entrained water-to-cement ratios [24–26]. There are relatively few studies devoted to the influence of dosing methods of SAP and the majority of the studies choose to add dry SAP powder to other components of the cement-based materials for mixing [27–29]. By now, the popular dosing methods of SAP can be classified as dry powdery SAP and swelled SAP (where SAP has been pre-wetted in tap water), which correspond to the dry powder and swelled gel form, respectively, when adding them to the cement-based materials.

When SAP powder is added to cement-based mixtures, although it shows a favorable dispersibility in cement-based materials, a common phenomenon of "competitive absorption" is prone to occur between SAP and cement-based materials during mixing and remarkably influences the workability and rheology of mixtures [30–32]. As for the swelled SAP, it normally releases the water too rapidly in cement-based mixtures, which can increase the total water-to-cement ratio, impair the effectiveness of internal curing, and work against the mechanical performance of cement-based materials [33–35].

In addition, the swelled SAP is generally difficult to disperse uniformly, which can result in the partial internal curing of cement-based mixtures. That said, the dosing methods of SAP can have a significant impact on the water kinetics of SAP during cement hydration and the mechanical performance and durability of cement-based materials. Therefore, the issue of adequate preparation and the adding method of SAP into the concrete is crucial for the fulfillment of internal curing. The current research aims to investigate the appropriate dosing method of SAP (dry powdery SAP or swelled SAP) and provide the benefits of the hydration of cement-based materials.

In this research, two different adding methods were designed to prepare the cement mortar for concrete internal curing. The mechanical properties, shrinkage properties and freeze–thaw resistance of cement mortar were compared and analyzed with the variation of SAP content and entrained water-to-cement ratio. The effect of SAP on the microstructure and pore structure of concrete was characterized by scanning electron microscopy (SEM).

2. Materials and Methods

2.1. Materials

The cement used in this study was P.I 42.5 (from China United Cement Corporation, Beijing, China) and the chemical composition can be found in Table 1. The Chinese ISO standard was used. The superplasticizer was a polycarboxylate-based aqueous solution with a solid content of 35%.

Table 1. Chemical and mineral composition of cement.

Chemical Compositions	CaO	SiO$_2$	Al$_2$O$_3$	Fe$_2$O$_3$	MgO	Na$_2$O	SO$_3$
Mass percentage (%)	65.36	22.15	4.51	3.39	2.31	0.49	0.46
Mineral compositions	C$_3$S	C$_2$S	C$_2$A	C$_4$AF			
Mass percentage (%)	56.54	20.87	6.22	10.31			

The SAP, a copolymer of acrylamide and sodium acrylate, was used in cement mortars. The particle diameters of SAP ranged from 80 μm to 200 μm, and the absorption capacities of the de-ionized water and cement filtrate solution with a swelling time of 5 min were 150.0 ± 2.6 g/g and 30.5 ± 0.9 g/g, respectively [23,34]. Tap water was used as mixing and internal curing water. The particle shape, size and surface characteristics of SAP were observed by SEM, as shown in Figure 1a. The swelled SAP is shown in Figure 1b.

(**a**) SEM micrograph of drying SAP   (**b**) SEM micrograph of drying SAP

Figure 1. The superabsorbent polymer (SAP) particle characteristics: scanning electron microscopy (SEM) micrograph (**a**) and swelled image (**b**).

2.2. Mixture Preparation and Testing Methods

2.2.1. Two Adding Methods of Superabsorbent Polymer (SAP)

Two adding methods with powder and swelled SAP, respectively, were designed in cement-based materials. One was where the powder of SAP was added directly, and the other was described as the detailed procedure in Figure 2. Firstly, the particles of SAP were uniformly paved on a watch glass that was slightly larger than the SAP coverage area (SAP thickness layer was less than 0.2 mm), and then a given volume of water (the amount of water was closely related to SAP content and entrained water-to-cement ratios), which was poured in an injector and uniformly sprayed on the particles of SAP to make each particle of SAP fully swelled. Finally, the swelled SAP was added into other dry components of a concrete mix.

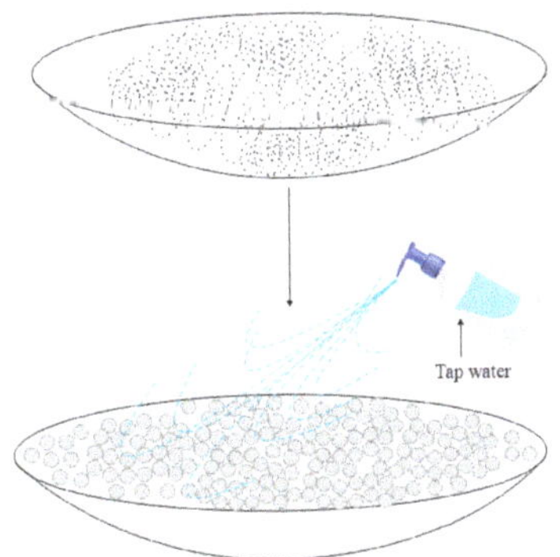

Figure 2. Preparation process of swelled SAP.

2.2.2. Preparation and Testing of Internal Curing Mortar

The mix composition of mortar is shown in Table 2. Two reference cement mortars with a water-to-cement ratio (W/C) of 0.42 and 0.48, respectively, were designed. The dosage of sand was 2 times that of cement. For the consistence of fresh mixture, superplasticizer was added in an amount of 0.5 wt% and 0.2 wt% (mass-% of cement weight), respectively, to ensure practical workability for the 0.42 and 0.48 mixtures, which were denoted as Ref-1 and Ref-2, respectively. The amount of superplasticizer was adjusted to ensure practical fluidity (210 ± 10 mm).

Table 2. Design of SAP-modified cement mortars.

Code	Cement/g	Sand/g	Effective w/c	Entrained w/c	Total w/c	SAP
R-1	495	990	0.42	0	0.42	0
R-2	495	990	0.48	0	0.48	0
$P_{0.1\%}$-0	495	990	0.42	0	0.42	0.10%
$P_{0.2\%}$-0	495	990	0.42	0	0.42	0.20%
$P_{0.3\%}$-0	495	990	0.42	0	0.42	0.30%
$P_{0.4\%}$-0	495	990	0.42	0	0.42	0.40%
$P_{0.1\%}$-0.06	495	990	0.42	0.06	0.48	0.10%
$P_{0.2\%}$-0.06	495	990	0.42	0.06	0.48	0.20%
$P_{0.3\%}$-0.06	495	990	0.42	0.06	0.48	0.30%
$P_{0.4\%}$-0.06	495	990	0.42	0.06	0.48	0.40%
$S_{0.1\%}$-0.06	495	990	0.42	0.06	0.48	0.10%
$S_{0.2\%}$-0.06	495	990	0.42	0.06	0.48	0.20%
$S_{0.3\%}$-0.06	495	990	0.42	0.06	0.48	0.30%
$S_{0.4\%}$-0.06	495	990	0.42	0.06	0.48	0.40%
$P_{0.3\%}$-0.02	495	990	0.42	0.02	0.44	0.30%
$P_{0.3\%}$-0.04	495	990	0.42	0.04	0.46	0.30%
$P_{0.3\%}$-0.08	495	990	0.42	0.08	0.50	0.30%
$S_{0.3\%}$-0.02	495	990	0.42	0.02	0.44	0.30%
$S_{0.3\%}$-0.04	495	990	0.42	0.04	0.46	0.30%
$S_{0.3\%}$-0.08	495	990	0.42	0.08	0.50	0.30%

Cement mortar with a water-to-cement ratio of 0.42 is prone to self-shrinkage. Based on the model of Powers and Brownyard [13], the amount of entrained water needed in the SAP can be calculated and amounts to an entrained water-to-cement ratio $(w/c)_e$ of 0.06 [16]. Therefore, the effective water-to-cement ratio of cement mortar is 0.42 $(w/c)_{ref}$ and the total water-to-cement ratio is 0.48 $(w/c)_{tot}$. Furthermore, an amount of about 0.2% SAP can also be calculated based on the absorption capacity of the cement filtrate solution in Section 2.1.

The cement mortars containing 0.1%, 0.2%, 0.3%, and 0.4% (mass-% of cement weight) SAP in the form of powders and which did not have an additional water-to-cement ratio, were denoted as $P_{0.1\%}$-0, $P_{0.2\%}$-0, $P_{0.3\%}$-0, and $P_{0.4\%}$-0, respectively. The cement mortars $P_{0.1\%}$-0.06, $P_{0.2\%}$-0.06, $P_{0.3\%}$-0.06, and $P_{0.4\%}$-0.06 respectively represented the cement mortars with the powders of 0.1%, 0.2%, 0.3%, and 0.4% (mass-% of cement weight) SAP and an additional water-to-cement ratio of 0.06 $(w/c)_e$. Likewise, the mortars $S_{0.1\%}$-0.06, $S_{0.2\%}$-0.06, $S_{0.3\%}$-0.06, and $S_{0.4\%}$-0.06, respectively, represented the cement mortars with the 0.1%, 0.2%, 0.3%, and 0.4% (mass-% of cement weight) swelled SAPs and an additional water-to-cement ratio of 0.06 $(w/c)_e$. In addition, when the cement mortars contained 0.3% SAP, the additional water-to-cement ratio was considered from 0.02 to 0.08 to investigate the effects of the additional water-to-cement ratio with different adding methods, and which were denoted as $P_{0.3\%}$-0.02, $P_{0.3\%}$-0.04, $P_{0.3\%}$-0.08, $S_{0.3\%}$-0.02, $S_{0.3\%}$-0.04, and $S_{0.3\%}$-0.08, respectively. The amount of superplasticizer in SAP cement mortars was the same as for Ref-1, in order to decrease the influence on the setting time [16]. The mixing program of raw materials is shown in Figure 3.

Specimens with the shape of $160 \times 40 \times 40$ mm^3 were prepared to study the compressive strength. Three specimens were prepared for each mortar for each of the performed tests. The specimens molded were kept in the environment of 20 °C and 98% humidity for 24 h, and were then demolded and cured under the conditions for 27 days. The strength compressive testing was carried out according to the Chinese standard (GB 17671), and the specimens were tested with a controlled load rate of 2.4 ± 0.2 kN/s.

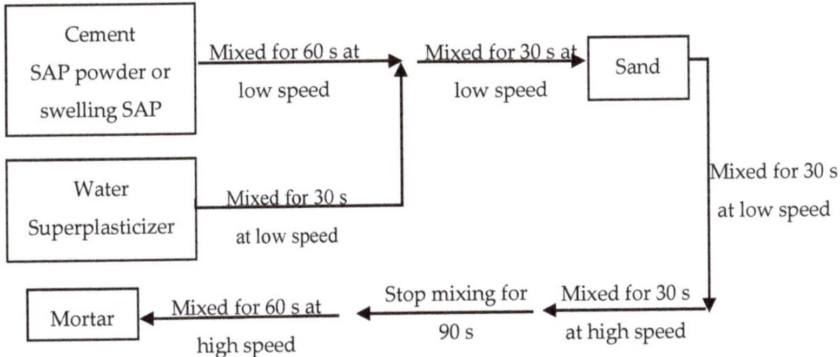

Figure 3. Mixing procedure of cement mortar with SAP.

Specimens with a size of $25 \times 25 \times 280$ mm^3 were used to evaluate the drying shrinkage, and the testing process of shrinkage specimens was performed in accordance with the Chinese standard (JC/T 603). The self-desiccation shrinkage of the specimens was considered to be mainly generated during the first three days [32]. Therefore, the deformation after three days was recorded as the drying shrinkage of the specimens in the paper. The drying shrinkage specimens molded were also kept in the environment of 20 °C and 98% humidity for 24 h, and were then demolded and cured under water for 2 d. Following this, they were removed from the water, the surface water was wiped off with a damp cloth and a comparator with an accuracy of 0.01 mm was used to measure the initial value (l_0). Finally, the specimens were moved to a room of 20 °C and 60% ± 5% humidity until the desired ages (i.e., 7 d, 14 d, 38 d and 35 d) to measure the data (l_i) for that age. The drying shrinkage ratio can be calculated according to Equation (1).

$$s = (l_i - l_0)/250 \times 100 \tag{1}$$

Specimens with a size of $160 \times 40 \times 40$ mm^3 were used to explore the freeze–thaw resistance performance. Furthermore, six specimens were prepared for each mortar for each of the performed tests. All specimens molded were kept in the environment of 20 °C and 98% humidity for 24 h, and were then demolded and cured under the conditions for 23 days. Following this, specimens were cured in water for another four days. Finally, specimens aged 28 days were used for the freezing and thawing test. The specific procedure of the freeze–thaw test was performed in accordance with the Chinese standard (GB/T 50082-2009). The mass and dynamic elasticity modulus were tested once after an interval of 25 times cycles, and the mass loss and relative dynamic elastic modulus were calculated. Every freeze–thaw cycle continued for 3.5 h, with the highest temperature of 18 ± 2 °C and lowest temperature of 20 ± 2 °C. Furthermore, the standard requirements are that the freeze-thaw test can be stopped when it reaches any of the following three conditions:

① The number of freeze–thaw cycles reaches 300;
② The relative dynamic elastic modulus of the specimen is less than 60% of the initial value;
③ The rate of mass loss of the specimen is more than 5%.

3. Results and Discussion

3.1. The Microstructure around SAP Voids

No matter how the SAP was added into the mixture, given SAP voids, were reserved in the hardened mortar after losing the stored water of powdery or swelled SAP. The SAP voids are closely correlated with the pore structure, strength, drying shrinkage, and permeability of cement mortars [20,32]. Therefore, it is necessary to observe the void characteristics reserved from SAP.

The microscopic structure of $P_{0.3\%}$-0.06 pores is shown in Figure 4. It can be observed from Figure 4a that the shape of the SAP void is irregular and the outer diameter is approximately 30–40 μm. It can also be observed that the SAP pore diameter is smaller than the SAP particle size mentioned in Section 2.1, which may be due to the filling or covering of the partial SAP voids by the hydration product. The formation of the large SAP void has some adverse effects on the compressive strength of the specimen [25]. However, SAP particles can be curled easily and agglomerated to form an agglomerative accumulation after the desorption of water in the mortar [32]. Therefore, SAP voids in mortar are completely different from normal air voids (Figure 4c). Figure 4b shows the observed interface between the SAP and hardened matrix. It can be seen from Figure 4b that there are abundant hydration products, especially C-S-H gels, around the SAP void, which is different from the loose hydration products around the normal air void (Figure 4c).

Figure 4. SEM images of cement paste with the addition of powdery SAP at 28 days: (**a**) microscopic structure of the SAP void; (**b**) microscopic interface between the SAP and hardened mortar; (**c**) microscopic structure of the air void.

3.2. Mechanical Properties

It is known that many efforts have been devoted to studying the mechanical properties of cement-based materials with SAP, while the influence of SAP addition on the compressive strength of cement-based materials is still controversial. The addition of SAP introduces a certain amount of SAP voids into the cement-based materials, which may decrease the mechanical properties of cement-based materials reported in the majority of existing literature [25,36]. Conversely, the other studies thought that SAP did not negatively affect the mechanical properties of cement-based materials due to the increased hydration of cement [37]. The 28 d compressive strength of cement mortars which had been designed by the corresponding adding methods of SAP and entrained w/c were evaluated, and the results are shown in Figures 5 and 6. In addition, Figure 5 also shows the effect of SAP on the macroscopic pores visible to the human eye, which was measured using a Vernier caliper with an accuracy of 0.02 mm on the fracture surface of mortar after the end of the compressive strength test.

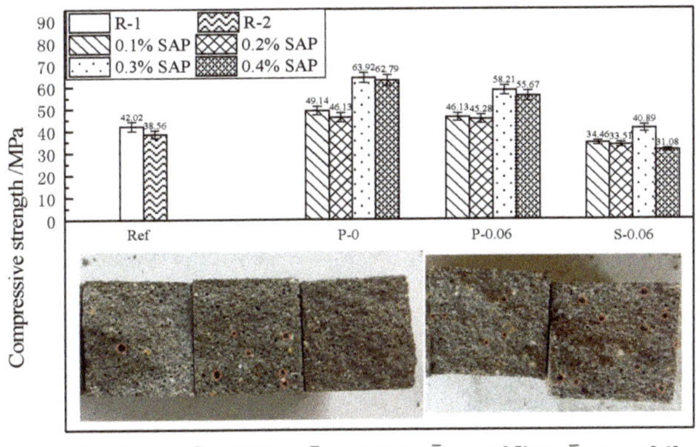

$\bar{D}_{Ref1}=1.4\,mm \qquad \bar{D}_{Ref2}=3.1\,mm \qquad \bar{D}_{P0.3\%-0}<1.00\,mm \qquad \bar{D}_{P0.3\%-0.06}=2.70\,mm \qquad \bar{D}_{S0.3\%-0.06}=5.62\,mm$

Figure 5. Compressive strength of cement mortar with different adding methods of SAP.

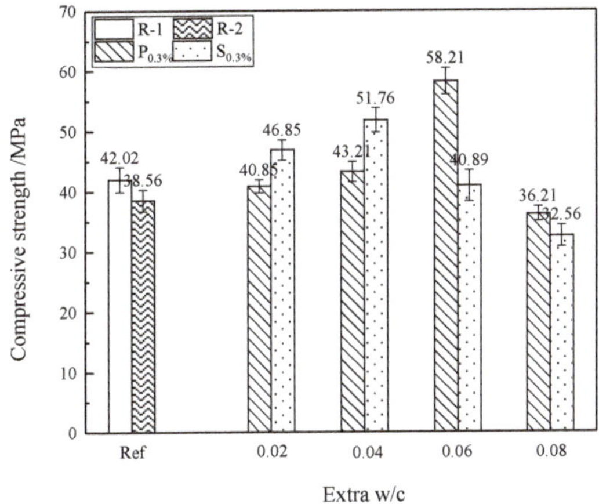

Extra w/c

Figure 6. Compressive strength of cement mortar with different entrained water-to-cement ratio (w/c) values.

As expected, the reference mortar with w/c = 0.48 shows a lower compressive strength than that with w/c = 0.42 at the age of 28 d, shown in Figure 5. Furthermore, compared with the reference mortars (R-1 and R-2), the incorporation of the powdery SAP raises the 28 d compressive strength of the mortars by about 10% to 50%, while for the incorporation of swelled SAP, the 28 d compressive strength of the mortar can be increased by about −26% to 6%. It indicates that the 28 d compressive strength of the mortars is closely related to the dosing method of SAP, and the incorporation of powdery SAP is more contributed to the development of the 28 d compressive strength of mortar than that of swelled SAP. This can be attributed to the "competitive water absorption" between the SAP and cemented materials when SAP is added to the mortar in powder form [30,32].

Compared with other dosages of SAP, the mortar with 0.3% SAP has the highest compressive strength for the two adding methods (Figure 5). Thus, the macroscopic pore characteristic in the broken section of the specimens, for which 0.3% SAP and 0.06 (w/c)$_e$ were added, was studied specifically. The average size of the visible pores in the broken section of the specimen (P$_{0.3\%}$-0 and P$_{0.3\%}$-0.06) containing powdery SAP is smaller than that containing swelled SAP, and the average pore size of the mortar without additional w/c (P$_{0.3\%}$-0) is always less than 1mm. However, the average size of mortar with swelled SAP (S$_{0.3\%}$-0.06) is 5.62 mm, which may result in a significant reduction in the compressive strength of the mortar.

In addition, it can also be observed from Figure 6 that the compressive strength of $P_{0.3\%}$ at an additional water–cement ratio of 0.06 for 28 d is 58.21 MPa, whereas the compressive strengths of R-2 and $S_{0.3\%}$ with the same total water-to-cement ratio (w/c = 0.48) are 38.56 MPa and 40.89 MPa, which are 0.66 and 0.7 times the of compressive strength of $P_{0.3\%}$ respectively. Furthermore, the mortar containing 0.3% powdery SAP and an entrained water-to-cement ratio of 0.06 exhibits a higher compressive strength than the Ref_1, Ref_2 and other mortars containing SAP. It may be explained that an appropriate entrained w/c can positively promote the potential of internal curing for the further hydration of cement-based materials, as well as refine the voids and improve the mechanical properties of the mortar.

3.3. Drying Shrinkage

Dry shrinkage is a type of reduction in the volume that is due to the loss of moisture in hardened cement mortar. Therefore, it is well established that moisture in hardened cement mortar is of great significance to dry shrinkage [38]. The results of drying shrinkage of cement mortars with SAP are presented in Figures 7 and 8.

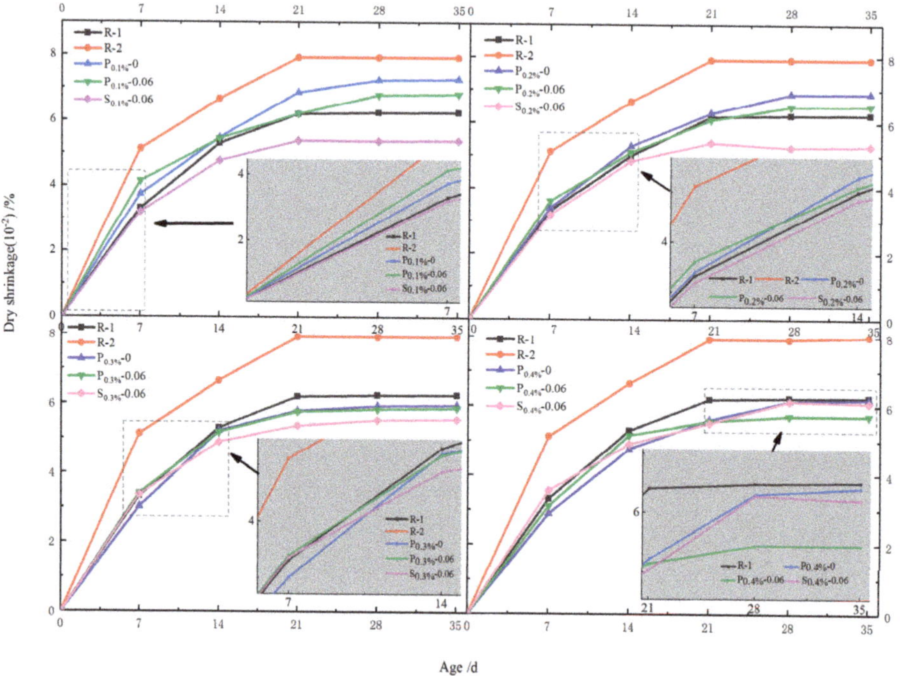

Figure 7. Drying shrinkage of cement mortar with different adding methods of SAP.

Figure 7 shows that for the specified four different dosages of SAP, the rate of shrinkage greatly varies before 21 days, and then slightly increases. As existing studies have reported [39,40], the drying shrinkage is closely related to the total water–cement ratio and generally raises with the increase of the total water-to-cement ratio. Therefore, the reference mortar with w/c = 0.48 shows obviously higher drying shrinkage than that with w/c = 0.42 during the testing process. Noticeably, the rate of drying shrinkage is significantly affected by the methods of adding SAP. The drying shrinkage of R-2 at a dosage of 0.1% for 28 d is 7.93×10^{-4}, whereas the drying shrinkage of $P_{0.1\%}$-0.06 and $S_{0.1\%}$-0.06 with the same total water–cement ratio (w/c = 0.48) are only 5.38×10^{-4} and 6.78×10^{-4}, which reduced the drying shrinkage rate by about 32.2% and 14.5%, respectively. Furthermore, the rate of drying shrinkage with swelled SAP was even lower than that of R-1 (w/c = 0.48) throughout the test. It can be seen that, when the moisture in the pores of mortar decreased due to evaporation and hydration, the water reserved in SAP could release and re-fill the pores to alleviate the surface tension [41]. In addition, the water stored by swelled SAP could release water more easily and earlier to re-fill the pores than that of the powdery SAP, and thus the reduction of drying shrinkage caused by swelled SAP was much more obvious.

It can be also seen from Figure 7 that, despite the adding methods, compared to 0.1% and 0.2% SAP content, both 0.3% and 0.4% contents show a more evident reduction in drying shrinkage of cement mortars. Additionally, the drying shrinkage of cement mortars that added 0.3% and 0.4% SAP was lower than those without SAP (R-1 and R-2) after 14 days. Furthermore, compared with 0.4% SAP, the dosage of 0.3% SAP can better reduce the drying shrinkage of the mortar, which indicates that the drying shrinkage of mortar is also closely related to the amount of swelled SAP.

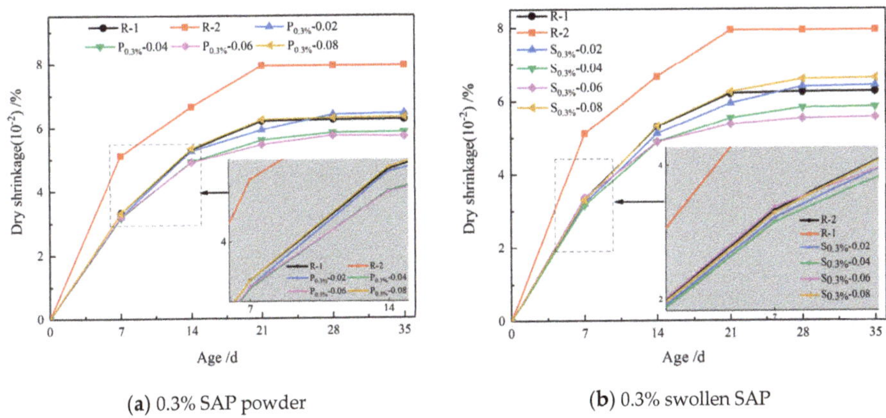

| (a) 0.3% SAP powder | (b) 0.3% swollen SAP |

Figure 8. Drying shrinkage of cement mortar with different entrained w/c values.

Figure 8a,b depict the effect of the entrained w/c on the drying shrinkage of mortar for a given 0.3% powder state and swelled SAP, respectively. It can be found that the entrained w/c is closely related to the drying shrinkage of mortar [39]. It can be seen from Figure 8a that the drying shrinkage of R-1 at a dosage of 0.3% for 28 d is 6.24×10^{-4}, whereas the drying shrinkage of $P_{0.3\%}$-0.02, $P_{0.3\%}$-0.04, $P_{0.3\%}$-0.06, and $P_{0.3\%}$-0.08 with the same effective water-cement ratio (w/c – 0.42) are 6.4×10^{-4}, 5.83×10^{-4}, 5.74×10^{-4}, and 6.29×10^{-4}, respectively. Therefore, when the entrained w/c is within the range of 0.04 to 0.06, the drying shrinkage can be decreased with the addition of powdery SAP; furthermore, a similar result can be found in Figure 8b. However, when the entrained w/c is 0.08, the effect of powdery and swelled SAP on reducing dry shrinkage was obviously decreased during the whole test process compared with the entrained water-to-cement ratios of 0.02, 0.04, and 0.06. It can be explained by the fact that, too much entrained water increases the water–cement ratio of mortar, which can facilitate the drying shrinkage of mortar.

3.4. Freeze–Thaw Resistance

It can be observed that if the dosage of powdery and swelled SAP and the entrained water-to-cement ratio in the mortar are 0.3% and 0.06, respectively, the cement mortar exhibits good mechanical properties (Section 3.2) and drying shrinkage resistance (Section 3.3). Consequently, for a given entrained water-to-cement ratio of 0.06 and 0.3% SAP, the freeze-thaw resistance of the mortar was investigated by measuring the mass loss rate and relative dynamic elastic modulus of the mortar under different freeze-thaw cycles in Figure 9.

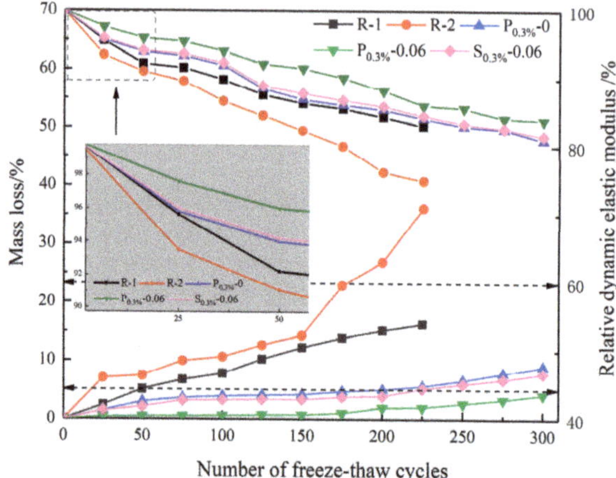

Figure 9. Mass loss and relative dynamic elastic modulus of cement mortar with different adding methods of SAP in freeze-thaw cycles.

It can be found from Figure 9 that the mass losses of R-1 and R-2 are 16.3% and 36.2% after 225 freeze–thaw cycles, whereas the mass losses of $P_{0.3\%}$-0, $P_{0.3\%}$-0.06 and $S_{0.3\%}$-0.06 are 5.7%, 2.1%, and 5.2%, respectively. Therefore, the introduction of SAP can improve the freeze–thaw resistance of the mortar, especially the powdery SAP. Similar findings can also be observed in the relative dynamic elastic modulus curves. It can be explained by the fact that, when the dosage of SAP and the entrained water-to-cement ratio are selected properly, the introduction of SAP may introduce a number of SAP voids in the cement mortar, which has a positive impact on the frost resistance of cement mortar. Furthermore, the addition of SAP can increase the air content of the fresh mortar [42]. However, the scaling takes place on the surface rather than on the inside of the specimen during the freeze–thaw cycles. Compared with powdery SAP, swelled SAP is prone to release water prematurely during mixing, and some of the released water could perspire on the surface of the specimen. Therefore, the mass loss of the mortar containing the swelled SAP increases during the freeze-thaw cycles and the strength drops obviously.

4. Conclusions

The performances of cement mortar incorporating SAP by different adding methods were studied. Based on the results presented, the following conclusions can be drawn:

(1) SAP voids in mortar are completely different from normal air voids and SAP particles can be easily curled and agglomerated to form an agglomerative accumulation after the desorption of water in the mortar. There are abundant hydration products around SAP voids, especially C-S-H gels;

(2) Compared with the reference mortars (R-1 and R-2), the incorporation of the powdery SAP increases the 28 d compressive strength of the mortars by about 10% to 50%, while for the incorporation of swelled SAP, the 28 d compressive strength of the mortar can be increased by about −26% to 6%;

(3) At a dosage of 0.1% SAP and an entrained water–cement ratio of 0.06, the powdery SAP and the swelled SAP can reduce the mortar shrinkage rate by about 32.2% and 14.5%, respectively. When the entrained w/c is within the range of 0.02 to 0.06, the drying shrinkage decreased with the increase of entrained w/c for the two adding methods of SAP;

(4)   The addition of SAP can improve the freeze–thaw resistance of the mortar, especially for the mortar with the powdery SAP and an entrained water-to-cement ratio of 0.06, and the mass loss rate after 300 cycles is still lower than 5%.

In the concrete industry, the primary merit of SAP is to reduce the shrinkage of cement-based materials. Thus, a further research effort is needed to characterize the influence of dry powdery SAP and swelled SAP (where the SAP has been pre-wetted in tap water) on the self-shrinkage and chemical shrinkage of cement-based materials.

Author Contributions: Y.T. performed data curation, formal analysis and contributed in writing original draft; R.H. contributed in methodology, writing review, editing, supervision and validation. H.C. contributed in project administration and supervision; Z.W. and C.X. performed conceptualization, funding acquisition, and resources.

Funding: The research were supported by the National Key R. & and D. Program of China (No. 2017YFB0309903), the National Natural Science Foundation of China (No. 51508030), the Basic Research Project in Qinghai Province (No. 2017-ZJ-715), Key Research and Development Program of Shaanxi Province (Nos. 2018SF-403, 2019JQ-559) and the Fundamental Research Funds for the Central Universities of China (Nos. 300102318401, 300102318402, 300102318501, 300102218502).

Conflicts of Interest: The authors declare no conflict of interest.

References

1.   Wehbe, Y.; Ghahremaninezhad, A. Combined effect of shrinkage reducing admixtures (SRA) and superabsorbent polymers (SAP) on the autogenous shrinkage, hydration and properties of cementitious materials. *Constr. Build. Mater.* **2017**, *138*, 151–162. [CrossRef]

2.   Ramazani-Harandi, M.J.; Zohuriaan-Mehr, M.J.; Yousefi, A.A. Rheological determination of the swollen gel strength of superabsorbent polymer hydrogels. *Polym. Test.* **2006**, *25*, 470–474. [CrossRef]

3.   Wang, F.; Yang, J.; Cheng, H.; Wu, J.; Liang, X. Study on mechanism of desorption behavior of saturated superabsorbent polymers in concrete. *ACI Mater.* **2015**, *112*, 463–470. [CrossRef]

4.   Yu, F.Q.; Sun, N.; Deng, C. The Effect of Super-Absorbent Polymer as a Self-Curing Admixture on the Performance of Mortars. *Adv. Mater. Res.* **2011**, *6*, 287–290. [CrossRef]

5.   Bentz, D.P.; Snyder, K.A. Protected paste volume in concrete: Extension to internal curing using saturated lightweight fine aggregate. *Cem. Concr. Res.* **1999**, *29*, 1863–1867. [CrossRef]

6.   Hajibabaee, A.; Ley, M.T. The impact of wet curing on curling in concrete caused by drying shrinkage. *Mater. Struct.* **2016**, *49*, 1629–1639. [CrossRef]

7.   Shen, D.; Jiang, J.; Jiao, Y. Early-age tensile creep and cracking potential of concrete internally cured with pre-wetted lightweight aggregate. *Constr. Build. Mater.* **2017**, *135*, 420–429. [CrossRef]

8.   Islam, M.R.; Xue, X.; Li, S. Effectiveness of Water-Saving Superabsorbent Polymer in Soil Water Conservation for Oat Based on Ecophysiological Parameters. *Commun. Soil Sci. Plan Anal.* **2011**, *42*, 2322–2333. [CrossRef]

9.   Liu, T.; Qian, L.; Li, B. Homogeneous synthesis of chitin-based acrylate superabsorbents in NaOH/urea solution. *Carbohydr. Polym.* **2013**, *94*, 261–271. [CrossRef]

10.   Hussien, R.A.; Donia, A.M.; Atia, A.A. Studying some hydro-physical properties of two soils amended with kaolinite-modified cross-linked poly-acrylamides. *Catena* **2012**, *92*, 172–178. [CrossRef]

11.   Jensen, O.M.; Hansen, P.F. Water-entrained cement-based materials: I. Principles and theoretical background. *Cem. Concr. Res.* **2001**, *31*, 647–654. [CrossRef]

12.   Bentz, D.P.; Jensen, O.M. Mitigation strategies for autogenous shrinkage cracking. *Cem. Concr. Compos.* **2004**, *26*, 677–685. [CrossRef]

13.   Viejo, I.; Esteves, L.P.; Laspalas, M. Numerical modelling of porous cement-based materials by superabsorbent polymers. *Cem. Concr. Res.* **2016**, *90*, 184–193. [CrossRef]

14.   Lee, H.X.D.; Wong, H.S.; Buenfeld, N.R. The potential of superabsorbent polymer for self-sealing cracks in concrete. *Adv. App. Ceram.* **2010**, *109*, 296–302. [CrossRef]

15.   Lee, H.X.D.; Wong, H.S.; Buenfeld, N.R. Self-sealing of cracks in concrete using superabsorbent polymers. *Cem. Concr. Res.* **2016**, *79*, 194–208. [CrossRef]

16.   Snoeck, D.; Pel, L.; De, N.B. The water kinetics of superabsorbent polymers during cement hydration and internal curing visualized and studied by NMR. *Sci. Rep.* **2017**, *7*, 9514. [CrossRef] [PubMed]

17. Craeye, B.; Geirnaert, M.; Schutter, G.D. Super absorbing polymers as an internal curing agent for mitigation of early-age cracking of high-performance concrete bridge decks. *Constr. Build. Mater.* **2011**, *25*, 1–13. [CrossRef]

18. Senff, L.; Modolo, R.C.E.; Ascensão, G. Development of mortars containing superabsorbent polymer. *Constr. Build. Mater.* **2015**, *95*, 575–584. [CrossRef]

19. Liu, H.; Bu, Y.; Sanjayan, J.G. Suitability of polyacrylamide superabsorbent polymers as the internal curing agent of well cement. *Constr. Build. Mater.* **2016**, *112*, 253–260. [CrossRef]

20. Jensen, O.M.; Hansen, P.F. Water-entrained cement-based materials. *Cem. Concr. Res.* **2002**, *32*, 973–978. [CrossRef]

21. Wang, W.B.; Liu, J.P.; Tian, Q. Effects of Water Absorption and Size of Superabsorbent Polymers on Internal Curing of Cement Paste. *Mater. Sci. Forum* **2013**, *5*, 743–744. [CrossRef]

22. Mechtcherine, V. Effect of Internal Curing by Using Superabsorbent Polymers (SAP) on Autogenous Shrinkage and Other Properties of a High-performance Fine-grained Concrete: Results of a RILEM Round-robin Test, TC 225-SAP. *Mater. Struct.* **2014**, *47*, 541–562. [CrossRef]

23. Snoeck, D.; Jensen, O.M.; Belie, N.D. The influence of superabsorbent polymers on the autogenous shrinkage properties of cement pastes with supplementary cementitious materials. *Cem. Concr. Res.* **2015**, *74*, 59–67. [CrossRef]

24. Craeye, B.; Cockaerts, G.; De Maeijer, P.K. Improving freeze-thaw resistance of concrete road infrastructure by means of superabsorbent polymers. *Infrastructures* **2018**, *3*, 4. [CrossRef]

25. Laustsen, S.; Hasholt, M.T.; Jensen, O.M. Void structure of concrete with superabsorbent polymers and its relation to frost resistance of concrete. *Mater. Struct.* **2015**, *48*, 357–368. [CrossRef]

26. Farzanian, K.; Pimenta Teixeira, K. The mechanical strength, degree of hydration, and electrical resistivity of cement pastes modified with superabsorbent polymers. *Constr. Build. Mater.* **2016**, *109*, 156–165. [CrossRef]

27. Woyciechowski, P.P.; Kalinowski, M. The Influence of Dosing Method and Material Characteristics of Superabsorbent Polymers (SAP) on the Effectiveness of the Concrete Internal Curing. *Materials* **2018**, *11*, 1600. [CrossRef]

28. Schröefl, C.; Mechtcherine, V.; Vontobel, P. Sorption kinetics of superabsorbent polymers (SAPs) in fresh Portland cement-based pastes visualized and quantified by neutron radiography and correlated to the progress of cement hydration. *Cem. Concr. Res.* **2015**, *75*, 1–13. [CrossRef]

29. Klemm, A.J.; Sikora, K.S. The effect of Superabsorbent Polymers (SAP) on microstructure and mechanical properties of fly ash cementitious mortars. *Constr. Build. Mater.* **2013**, *49*, 134–143. [CrossRef]

30. Lee, H.X.D.; Wong, H.S.; Buenfeld, N.R. Effect of alkalinity and calcium concentration of pore solution on the swelling and ionic exchange of superabsorbent polymers in cement paste. *Cem. Concr. Compos.* **2018**, *88*, 150–164. [CrossRef]

31. Goncalves, H.; Goncalves, B.; Silva, L. The influence of porogene additives on the properties of mortars used to control the ambient moisture. *Energy Build.* **2014**, *74*, 61–68. [CrossRef]

32. Ma, X.; Liu, J.; Wu, Z. Effects of SAP on the properties and pore structure of high performance cement-based materials. *Constr. Build. Mater.* **2017**, *131*, 476–484. [CrossRef]

33. Azarijafari, H.; Kazemian, A.; Rahimi, M. Effects of pre-soaked super absorbent polymers on fresh and hardened properties of self-consolidating lightweight concrete. *Constr. Build. Mater.* **2016**, *113*, 215–220. [CrossRef]

34. Snoeck, D. The effects of superabsorbent polymers on the microstructure of cementitious materials studied by means of sorption experiments. *Cem. Concr. Res.* **2015**, *77*, 26–35. [CrossRef]

35. Snoeck, D.; Schaubroeck, D.; Dubruel, P. Effect of high amounts of superabsorbent polymers and additional water on the workability, microstructure and strength of mortars with a water-to-cement ratio of 0.50. *Constr. Build. Mater.* **2014**, *72*, 148–157. [CrossRef]

36. Soliman, A.M.; Nehdi, M.L. Effect of partially hydrated cementitious materials and superabsorbent polymer on early-age shrinkage of UHPC. *Constr. Build. Mater.* **2013**, *41*, 270–275. [CrossRef]

37. Beushausen, H.; Gillmer, M.; Alexander, M. The influence of superabsorbent polymers on strength and durability properties of blended cement mortars. *Cem. Concr. Compos.* **2014**, *52*, 73–80. [CrossRef]

38. Valcuende, M.; Marco, E.; Parra, C. Influence of limestone filler and viscosity-modifying admixture on the shrinkage of self-compacting concrete. *Cem. Concr. Res.* **2012**, *42*, 583–592. [CrossRef]

39. Schröfl, C.; Mechtcherine, V.; Gorges, M. Relation between the molecular structure and the efficiency of superabsorbent polymers (SAP) as concrete admixture to mitigate autogenous shrinkage. *Cem. Concr. Res.* **2012**, *42*, 865–873. [CrossRef]
40. Kong, X.M.; Zhang, Z.L.; Lu, Z.C. Effect of pre-soaked superabsorbent polymer on shrinkage of high-strength concrete. *Mater. Struct.* **2015**, *48*, 2741–2758. [CrossRef]
41. Mo, J.C.; Ou, Z.W. Influence of superabsorbent polymer on shrinkage properties of reactive powder concrete blended with granulated blast furnace slag. *Constr. Build. Mater.* **2017**, *146*, 283–296. [CrossRef]
42. Mechtcherine, V.; Secrieru, E. Effect of superabsorbent polymers (SAPs) on rheological properties of fresh cement-based mortars—Development of yield stress and plastic viscosity over time. *Cem. Concr. Res.* **2015**, *67*, 52–65. [CrossRef]

Article

Preparation and Evaluation of Exhaust-Purifying Cement Concrete Employing Titanium Dioxide

Rui He [1,2,*], Xin Huang [1], Jiansong Zhang [2,*], Yao Geng [1] and Haidong Guo [3]

[1]   School of Materials Science and Engineering, Chang'an University, Xi'an 710061, China
[2]   School of Construction Management Technology, Purdue University, West Lafayette, IN 47907, USA
[3]   Qinghai Academy of Transportation Sciences, Xining 810008, China
*   Correspondence: heruia@163.com (R.H.); zhan3062@purdue.edu (J.Z.)

Received: 10 June 2019; Accepted: 4 July 2019; Published: 7 July 2019

Abstract: To address the increasing air pollution caused by vehicle exhaust, environment-friendly pavement materials that possesses exhaust-purifying properties were prepared using common cement concrete and porous cement concrete as the base of photocatalyst nano-titanium dioxide (TiO_2), respectively. Firstly, Fe^{3+}-doped TiO_2 powder was prepared by applying planetary high-energy ball milling in order to improve the efficiency of the semiconductor photocatalyst for degrading vehicle exhausts. Two nano-TiO_2, namely the original and modified nanomaterials, were adopted to produce the photocatalytic cement concretes subsequently. The physicochemical properties of the modified powder, as well as the mechanical and photocatalytic properties of TiO_2-modified concrete, were characterized using a suite of complementary techniques, including X-ray diffraction (XRD), scanning electron microscopy (SEM), energy dispersive spectroscopy (EDS), compressive strength and degradation efficiency tests. The results show that the ball milling method not only successfully doped Fe^{3+} into catalysts but also caused significant changes in: (1) decreased particle sizes, (2) more amorphous morphology, (3) decreased percentage of the most thermodynamically stable crystal facet, and (4) increased percentage of other high gas sensing crystal facets. Both the original and modified nano-TiO_2 can improve the concrete strength while the strengthening effect of modified nanomaterials is superior. It is pronounced that the photocatalytic property of the modified nano-TiO_2 is much better than that of the original nano particles, and the degradation rate of porous concrete is also better than common concrete when exposed to the same photocatalyst content. In a comprehensive consideration of both mechanical performance and degradation efficiency, the recommended optimum dosage of TiO_2 is 3% to 4% for exhaust-purifying concrete.

Keywords: photocatalytic pavement; vehicle exhaust; photocatalytic concrete; porous concrete; degradation efficiency

---

1. Introduction

With the acceleration of the urbanization process and car ownership, vehicle exhaust has become one of the main causes of air pollutions all over the world, especially in the developing and densely populated countries, such as China, for which the vehicle exhaust that comes from fossil fuels has accounted for about 25% of the total industrial exhaust emissions [1,2]. Air pollutants, mainly represented by nitrogen oxides (NO_x) and sulfur oxides (SO_x) in the emissions, can cause photochemical smog, haze, and acid rains. This poses great threat to ecosystems as well as the wellbeing of human and other species if it reaches a certain level of concentration in the atmosphere [3–5]. During the past decade, various countermeasures have been adopted to reduce vehicle emissions, such as using cleaner fuel, installing a gas-cleaning device in vehicles, and even developing electromobiles [2]. Nevertheless, vehicle exhaust is still a major problem, especially due to the rapid growth of the transportation sector.

As the nearest contact point after exhaust emissions, the vicinity of the road is where NO_x is prone to accumulate due to the gas flows through the exhaust devices and its higher density than oxygen. In street canyons or urban areas, the NO_x concentrations can be considerably higher [6,7]. Faced with this dramatic issue, innovative exhaust-purifying pavements, also known as photocatalytic or functional pavements, have been proposed to degrade NO_x and improve the air quality of road environment, by leveraging heterogeneous photocatalytic oxidations. As of now, photocatalytic pavements have become the most promising strategy to solve the road traffic pollutions, as confirmed by the large number of laboratory-scale tests and an increased amount of real scale studies over the past ten years [8,9].

Recently, photocatalytic oxidation was ranked among the most popular green chemical methods for sustainable development [10,11]. Amongst all the photocatalytic oxidations for NO_x degradation in photocatalytic pavements, nano-titanium dioxide (TiO_2), most often anatase TiO_2, is the most frequently used catalyst due to its low price, robustness, nontoxicity, and mild reaction conditions [9]. When activated by ultraviolet (UV) light and water, nano-TiO_2 can convert the NO_x into harmless nitrates. It can degrade almost all organic matters in the vehicle emissions [3,8].

However, the anatase-type TiO_2, which is proved to have strong photocatalytic degradation ability, is only active under high energy UV irradiation due to the large band gap of 3.2 eV, which composes less than 5% of the solar spectrum [8,12,13]. It means that the efficiency of this photocatalytic process for outdoor photocatalytic pavement is very slow. To address this issue, different methods have been attempted to improve the degradation efficiency, such as changing the TiO_2 concentration or type, increasing the contact area between TiO_2 and sunlight, and so on [6,8,14]. In spite of such exploration, the low efficiency still makes photocatalytic pavements difficult to be promoted.

In the research field of semiconductor materials, numerous studies have been undertaken to induce visible light activity into normally purely UV-active TiO_2 by hydrothermal, sol-gel, the precipitation method, and other chemical and physical processes [13,15]. Doping with transition metal ions such as V, Cr, Fe, Co, and Ni by high voltage acceleration in the range of 50–200 keV could enable a large shift in the absorption band of these photocatalysts toward the visible light range, with differing levels of effectiveness [16]. This principle is expected to have a more promising outlook in improving the photocatalytic performances of TiO_2. Thus, a large number of processing methods have been proposed for the modification of TiO_2 according to the literature [17,18]. Among state-of-the-art techniques, the ball milling method makes mass production of catalysts more viable over other conventional chemical methods. This method has the advantages of low cost, environmental friendliness, and high efficiency and controllability in the laboratory, even in industry [16,19].

Porous concrete has a higher porosity than common concrete and is more susceptible to sunlight, thereby accelerating the photocatalytic reaction. In view of this situation, common cement concrete and porous cement concrete were both used to be the base of nano-TiO_2 to prepare the green pavement material with photocatalytic function. In this paper, the commercial nano-TiO_2 was first modified using the planetary ball milling technique to produce an Fe^{3+}-doped TiO_2 photocatalyst. Following that, two nano-TiO_2, namely the original and modified nanomaterials, were adopted to produce the TiO_2-modified cement concretes. The mechanical and photocatalytic properties of TiO_2-modified concrete were studied. Promising results were obtained showing enhanced photocatalytic efficiency and compressive strength in the use of modified TiO_2 comparing to original TiO_2.

2. Experiments

2.1. Materials

In this study, an anatase-type commercial nano-sized TiO_2, labelled as original TiO_2, was provided by Jiangsu Hehai Nanometer Science & Technology Co. Ltd (Taizhou, China). The detailed physical characteristics of this TiO_2 are shown in Table 1. The modifier used for the Fe^{3+}-doped TiO_2 was analytically pure ferric nitrate ($Fe(NO_3)_3 \cdot 9H_2O$).

The Chinese 42.5 ordinary Portland cement from Jidong (Tangshan, China) was used to prepare concrete samples, along with coarse aggregate (maximum size of 20.0 mm) and fine aggregate (clean, natural sand, maximum size of 4.75 mm, fineness modulus of 2.3). The chemical agent superplasticizer from Subote (Nanjing, China), with a 25% solid concentration, was used at a dosage of 0.8% by mass of cement.

Table 1. Technical index of nano-TiO_2.

Appearance	Average Particle Size/nm	Specific Surface Area/m^2·g^{-1}	Crystal Type	TiO_2 Content/%	Photocatalytic Efficiency/%	pH Value
White powder	10	70.2	Mixed crystal type (anatase, rutile)	>92	≥68	Partial acidity

The grading of porous concrete coarse aggregate is the key factor affecting its permeability and mechanical properties. After determining the target porosity and water–cement ratio of concrete, the mixing ratio was calculated according to the volume method. The coarse aggregate gradation used in this study, combined with those in previous researches, is shown in Table 2 [20–22].

Table 2. Coarse aggregate gradation of porous concrete.

Sieve/mm	2.36	4.75	9.5	13
Passing Rate/%	0	6.5	85.7	100

2.2. Methods

2.2.1. Preparation of Modified TiO_2

In contrast to the commonly used chemical modification method for nano-TiO_2 that requires complex raw materials and operation processes, the planetary high energy ball milling method was used in this study to prepare the iron-doped nano-TiO_2 photocatalyst, of which the process is simpler and the production is continuous. During the milling process, TiO_2 powder undergoes severe plastic deformation due to strong collision, resulting in serious distortion of the lattice of TiO_2, a large increase of internal defects, and improved catalyst activity. At the same time the ball milling process will produce a lot of heat, coupled with mechanical force to induce the separation of Fe^{3+} from ferric nitrate and its diffuse into the distorted TiO_2 lattice, resulting in the formation of a new iron-doped photocatalyst which can reduce the probability of electron recombination and finally achieve the improvement of the photocatalytic performance [19,23].

Based on the above principle, the preparation process involves firstly weighing nano-TiO_2 and ferric nitrate in the amount of 100 g and 15 g, respectively. Then the starting materials were manually premixed and poured into the ball mill jar. The rotational speed and the ball-to-powder weight ratio were 300 rpm and 30:1, respectively. After 3 h of ball milling process, the mixtures were put into a muffle furnace, and calcined at 400 °C for 3 h. Subsequently, the obtained samples were washed by deionized water three times to remove any residual ferric nitrate. At last, the samples were dried at 90 °C for 6 h and then ground into powders. The high energy ball milling setup is shown in Figure 1. The original and modified nano-TiO_2 are shown in Figure 2.

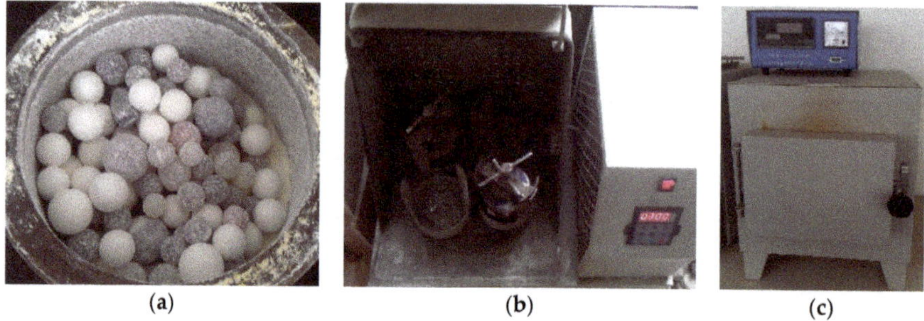

(a)   (b)   (c)

Figure 1. Planetary high energy ball milling setup. (**a**) Ball mill jar, (**b**) Planetary high energy ball mill, and (**c**) Muffle furnace.

(a)   (b)

Figure 2. The original nano-TiO_2 and modified nano-TiO_2. (**a**) Original TiO_2 and (**b**) Modified TiO_2.

2.2.2. Preparation of Concrete Samples

Two types of photocatalytic cement concretes were prepared, namely, the common concrete and the porous concrete, and both of the two TiO_2 products (original and modified) were used for each type of concrete. That is to say, a total of four sets of photocatalytic concretes were prepared. Nanomaterials were used to replace the quantity of cement in the amount of 0%, 1%, 2%, 3%, 4%, and 5%, respectively. The common concrete reference ratio is cement:water:sand:gravel:water-reducing agent, 395:178:621:1205:4. The porous concrete reference ratio is cement:water:gravel:water-reducing agent, 252:100:1542:2. The experiment program is summarized in Table 3, and the contents for the two TiO_2 products were the same.

The preparation process of concrete in this study is described as follows: Firstly, mixing the powder and coarse aggregate for 1 min in dry situation, then adding water slowly and continually mixing for 2 min, and finally forming a 100 mm × 100 mm × 100 mm cube test block. After 24 h of curing, cube test blocks were removed from molds and put into a standard curing room to keep curing at 20 ± 2 °C and relative humidity ≥ 95%. When the curing age arrived at 28 days, the blocks were taken out to carry on experiments of mechanical and photocatalytic properties. The compressive strength of each group of concrete samples was tested in accordance with T 0553 in JTG E30-2005 "Test Methods of Cement and Concrete for Highway Engineering" [24].

Table 3. Experiment program.

Concrete Type	Common Concrete						Porous Concrete					
Nano-TiO_2 content/%	0	1	2	3	4	5	0	1	2	3	4	5
Code	C-0	C-1	C-2	C-3	C-4	C-5	P-0	P-1	P-2	P-3	P-4	P-5

2.2.3. Test of Photocatalytic Degradation Efficiency

The photocatalytic degradation efficiency was tested in a custom-designed testing system that contains a continuous gas flow reactor according to the Italian Standard UNI 11247 [7,25], as shown in Figure 3 [26]. The test samples were the pre-prepared concrete blocks. The device mainly consisted of the following four parts: A sample reaction chamber, a simulation photo source, a standard gas source, and a gas analyzer. Among these four parts, the reaction chamber was made of borosilicate glass in order to prevent the chamber from absorbing pollutant gas and UV light. The surface of the samples placed in the reaction chamber was kept parallel to the lamp to ensure that the surfaces of samples were uniformly illuminated. The simulation photo source comprised of a mercury vapor lamp and an incandescent lamp on the upper side of the reaction chamber, with the wavelength range of 300 to 700 nm, which covers most of the visible range of sunlight [27]. The radiant energy was 48 W/m^2, which is close to the sun radiation intensity at mid-latitudes [3]. On both side of the reaction chamber, there was a gas inlet and a gas outlet, respectively, which were used to supply gas for tests and extract gas for the analyzer. The gas analyzer was located at the same side with gas outlet. A small fan was placed at the corner of the reaction chamber to ensure test gas reacts completely on the specimen surface.

According to the main components of automobile exhaust gas, NO was used as the main pollutant model of photocatalytic degradation experiment. The concentration of NO used during the test was 200.0 ppb according to former studies [4,9]. The gas analyzer NHA-508 (Anhua, Foshan, China) was used as the detection device for NO concentration.

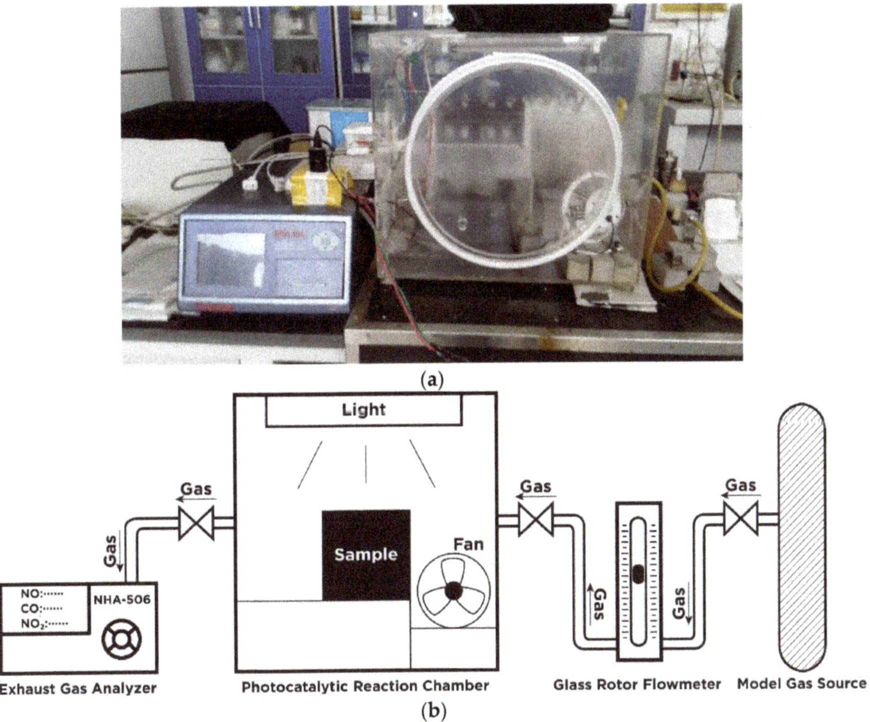

Figure 3. Photocatalytic degradation test equipment [26]. (**a**) Device appearance and (**b**) Device principle.

2.2.4. Microstructure Analysis

The morphologies of the powdered TiO_2 and the inner structure of concrete samples were observed with scanning electron microscopy (SEM, Hitachi S-4800, Tokyo, Japan). The surface elemental analysis of the samples was performed using energy dispersive X-ray spectroscopy (EDS) [13,28]. After the mechanical test, the samples subject to SEM analysis were taken from the fragments caused by the mechanical test, and conductive coating with gold was applied after drying treatment to prevent charging effects. In addition, the powder X-ray diffraction method (XRD, Bruker D8 Advance, Berlin, Germany) was employed to investigate the influence of modification on the crystal phase of the powdered TiO_2 with Ni-filtered Cu Kα radiation from 15° to 80°.

3. Results and Discussion

3.1. Characterization of TiO_2 Samples

The effect of the planetary high energy ball milling treatment on the catalyst morphology and crystal structure was investigated by SEM and XRD tests, prior to preparing concrete samples. The SEM results are shown in Figure 4 and its energy spectrum is shown in Figure 5. It can be seen from Figure 4 that the amorphous nanoparticles were distributed unevenly and there were obvious agglomerations in both samples. However, the TiO_2 particles were more prone to agglomeration after ball milling, which can be partially attributed to the reduction of particle sizes, as well as the high surface energy that was caused by ball milling. Therefore, the modified TiO_2 appears to be more amorphous and finer, providing more surface area for the activation of photocatalytic activity. The EDS spectrum verifies that the ball milling procedure successfully introduces iron into TiO_2 while the spectrum of original TiO_2 has no iron.

According to the XRD patterns of the original and modified TiO_2 shown in Figure 6, the diffraction peaks of both samples showed fingerprint features of an anatase TiO_2 structure with no other characteristic peaks observed, indicating the phase purity of the samples. The percentage of diffraction intensity at different crystal facets is depicted in Table 4. It can be seen that the percentage of the (101) peak, which is the most thermodynamically stable and has relatively low chemical activity [19], decreases sharply from 51.84% to 48.47% after applying the ball milling procedure. The percentages of the other peaks for Fe^{3+}-doped TiO_2, such as (004), (105), (211), and (116), which have high gas sensing activity [29,30], are all increased to some degree. Correspondingly, the percentage of (101) crystal facet is decreased, which indicates that the preferred orientation of (101) has been compromised.

On the basis of these data, the average crystal sizes were calculated according to Scherrer's equation [31]. For the original and modified TiO_2, the results were 21.16 nm and 20.78 nm, respectively. It demonstrates that ball milling favors the formation of small crystallites, leading to more intense diffraction peaks, with the decrease of (101) crystal facet. In summary, it can be concluded that the planetary high energy ball mill can dope Fe^{3+} ions into nano-TiO_2, making it successfully modified.

(a)   (b)

Figure 4. Microtopography of nano-TiO_2. (**a**) Original TiO_2 and (**b**) Modified TiO_2.

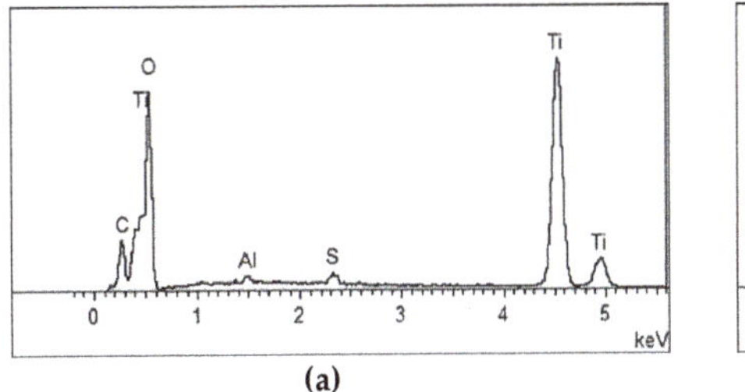

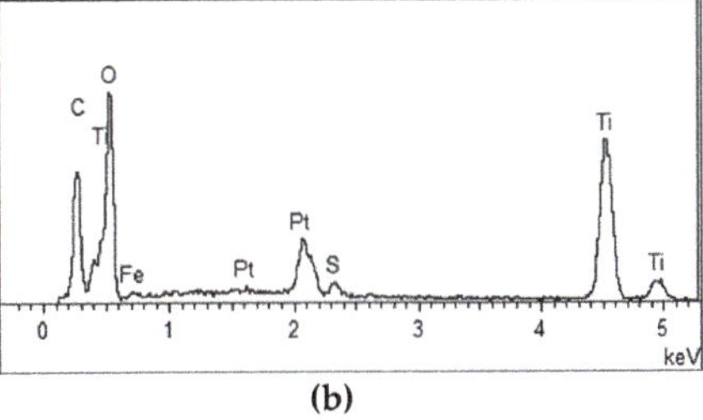

Figure 5. Energy Spectrum Analysis of nano-TiO_2. (**a**) Original TiO_2 and (**b**) Modified TiO_2.

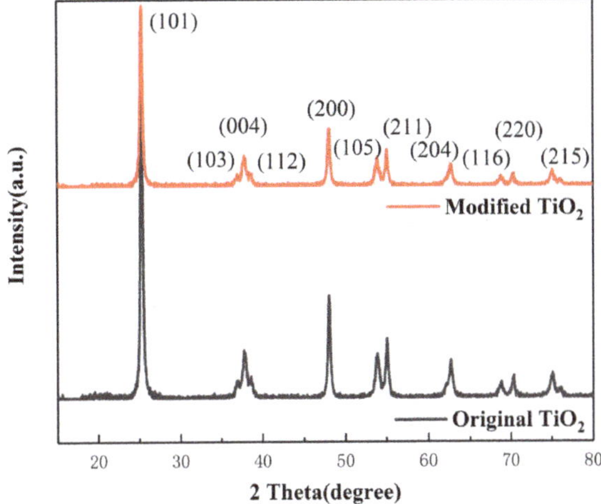

Figure 6. X-ray diffraction pattern of nano-TiO_2.

Table 4. The percentage of diffraction intensity at different crystal facets.

Crystal Facets	Percentage (%)								
	(101)	(103)	(004)	(112)	(200)	(105)	(211)	(204)	(116)
Original TiO_2	51.84	1.35	5.75	1.45	16.23	6.48	9.02	5.75	2.13
Modified TiO_2	48.47	2.57	7.51	2.52	15.03	7.03	9.21	5.48	2.18

3.2. Mechanical Property Analysis

The compressive strength test was conducted to ensure that the incorporation of nano-TiO_2 did not compromise the mechanical performance of concrete, as shown in Figure 7. It can be seen from the figure that the compressive strength of two types of concrete both increased firstly and then reduced with the continuous increment of nano-TiO_2 content. It indicates that the incorporation of nano-TiO_2 is beneficial to the improvement of concrete strength, which is mainly due to the micro-filling effect of nano-TiO_2, making the internal structure of concrete more compact and leading to the improvement of strength. The maximum increase in strength is about 3% and 5%, respectively. As the particle size of modified TiO_2 is smaller than that of the original TiO_2, the strength of modified TiO_2 concrete is higher than that of the original TiO_2 concrete with the same content of TiO_2. In addition, the incorporation of Fe^{3+} can motivate the formation of iron-containing hydration products [32,33], which makes the internal structure more compact, and therefore improves the strength.

Nonetheless, when the nano-material content exceeds a certain range, the strength of each group begins to decline. At first, it can be ascribed to the agglomeration tendency as illustrated in Section 3.1, which could easily cause uneven distribution within the structure. On the other hand, the replacement of cement by TiO_2 makes the corresponding cement content decrease when the nano-materials content increases excessively. In terms of the material strength, the best nano-TiO_2 content of cement were 3% and 4% for the two types of concrete, respectively.

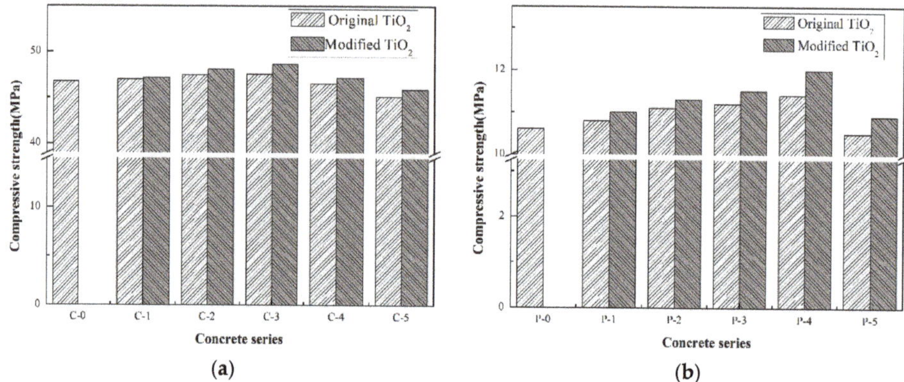

Figure 7. Results of 28 days compressive strength for each mix, (**a**) common concrete and (**b**) porous concrete.

3.3. Photocatalytic Property Analysis

The photocatalytic degradation test of concrete was carried out using the custom-designed test device proposed in this study. It should be noted that the device was subjected to a blank control verification prior to the start of the formal test, and the change of NO concentration was recorded. During the equipment verification, NO concentration was almost unchanged, proving that the sealing of the reaction chamber was good. The parameters used in the testing process are shown in Table 5. Keeping the above test parameters remaining unchanged, photocatalytic degradation tests of all samples were conducted.

Table 5. Photocatalytic degradation test parameters.

Total Amount of Pollutant Gas	Flow Rate	Relative Humidity	Detection Time	Concrete Age
12 L	1.2 L/min	50%	10 min/once total 1 h	28 days

The residual rate of NO concentration after different periods of photocatalysis was calculated and displayed in Figure 8. It should be pointed out that the NO concentration results of concrete samples without TiO_2 almost stayed constant. Thus, these results were not depicted in Figure 8. As it shows, the degradation rate of all the samples increases with the extension of irradiation time. Overall, the degradation rate of samples with modified nano-TiO_2 was much higher than that of the original nano-TiO_2 samples. The reason can be partially attributed to the smaller particle size that causes more contact area with pollutants. Also, the doped Fe^{3+} in the crystal of TiO_2, by the mechanical chemistry theory of ball milling as mentioned above, can act as an acceptor which can trap photogenerated charges and reduce the metal ion species into a lower oxidation state. Then, due to the close energy level of the Ti^{4+}/Ti^{3+} couple, the transformation of Ti^{4+} into Ti^{3+} is accelerated, which has been reported to improve the photodegradation by acting as a hole trap [15,34]. Another point is that Fe^{3+} dopant, which has been reported to be effective to the batho-chromic shift, promotes a possible extension of the activity of TiO_2 towards visible light range [35,36]. Therefore, the photocatalytic activities of TiO_2 are enhanced significantly.

For further comparison, the photocatalytic degradation rates of all the samples at 60 min are shown in Figure 9. At the same content of nano materials, the degradation rate of concretes with modified nano-TiO$_2$ is about 13% to 20% higher than that with original nano-TiO$_2$. At the same time, the photocatalytic degradation rate of porous cement concrete is better than that of common cement concrete, which is because of the greater contact area with light caused by the connected pores of porous concrete. In the case of photocatalytic property, the optimized content of TiO$_2$ is suggested to be 4%.

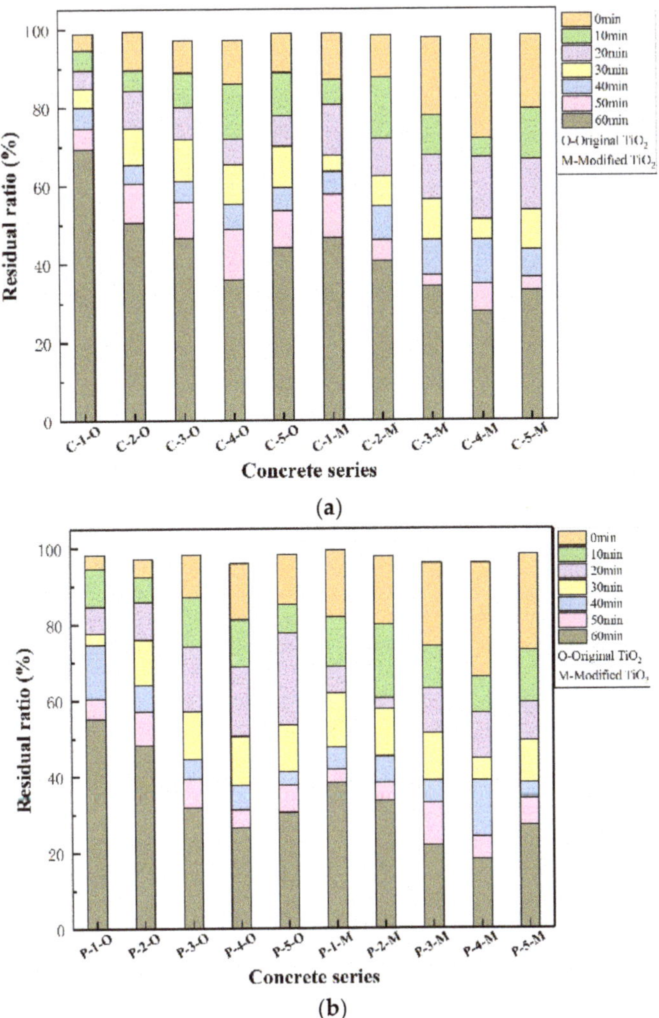

Figure 8. Photocatalytic degradation results of concrete in each group. (**a**) Common concrete and (**b**) porous concrete.

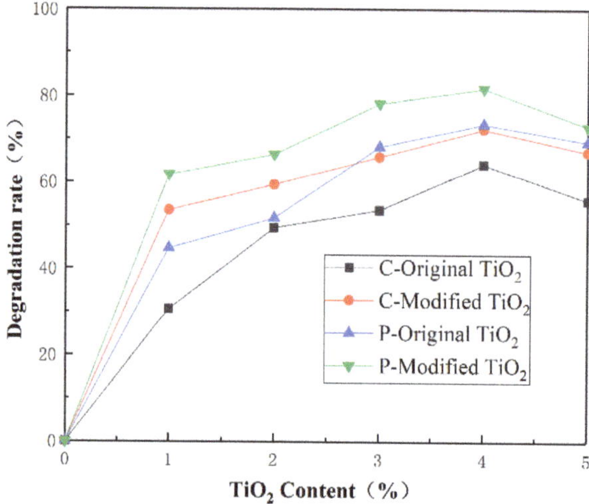

Figure 9. Comparison of photocatalytic degradation rate test results.

3.4. Microstructure Analysis of Modified Concrete

The microstructural test of C-0 and C-4 are shown in Figure 10. It can be seen from Figure 10 that the internal structure of C-0 concrete is looser, and the cracks or pores can be clearly seen between the hydration products. C-4 concrete is relatively dense inside, and agglomerated particulate matters are filled between the hydration products. In order to further confirm the composition of the material, the two samples were analyzed by EDS. The results are shown in Figure 11. It can be seen from Figure 11 that the main elements in C-0 are calcium, silicon, and oxygen, whereas for C-4, calcium, silicon, and oxygen elements, as well as a lot of titanium. It indicates that the inner structure of C-4 has titanium-containing substances except common hydration products, which is TiO$_2$. Therefore, based on the above analysis, it can be seen that the introduction of modified nano-TiO$_2$ can be filled in the cracks or pores between the hydration, so that the structural density increased, which in turn improves the strength of C-4.

In this paper, we can see that the introduction of nano-TiO$_2$, especially the modified nano-TiO$_2$, has a positive impact on the mechanics and degradation properties of concrete. But there is an optimal amount. Considering the effects of mechanical properties and degradation efficiency, the recommended optimum dosage is 3% to 4%.

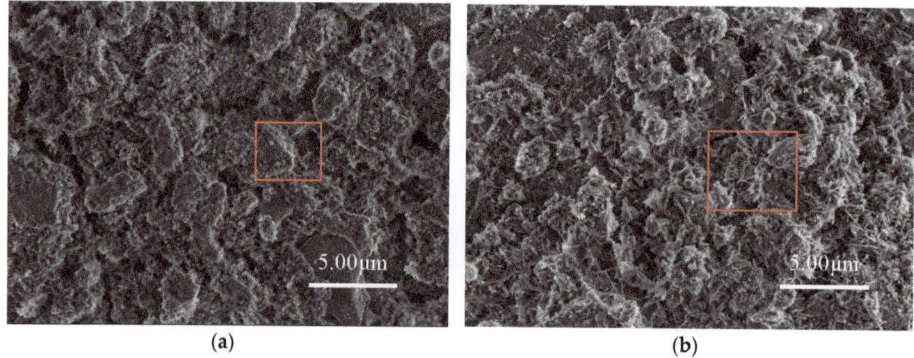

Figure 10. Microstructure analysis of concrete specimens. (**a**) C-0 and (**b**) C-4 with modified TiO$_2$.

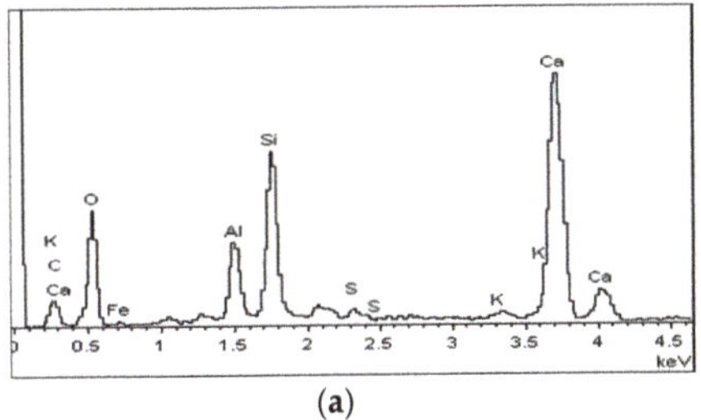 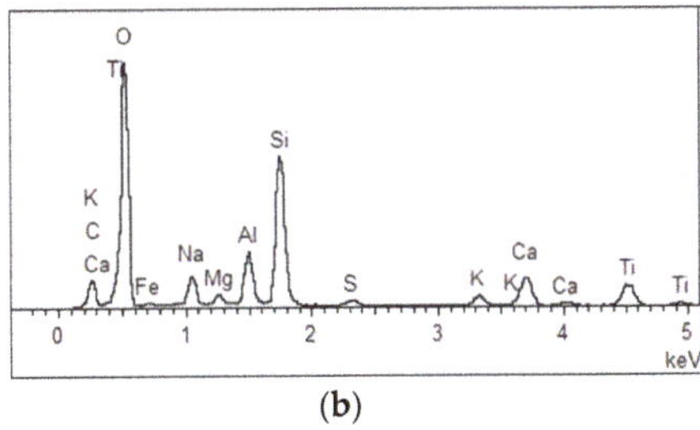

(a) (b)

Figure 11. Energy spectrum analysis of concrete specimen. (a) C-0 and (b) C-4 with modified TiO_2.

4. Conclusions

This paper highlights the feasibility of using the planetary high energy ball milling method in modifying TiO_2, as well as subsequent preparation of exhaust-purifying cement concrete. The photocatalytic efficiency and mechanical performance of exhaust-purifying cement concrete were both evaluated. The main conclusions can be summarized as follows:

(1) The modified nano-TiO_2 doped with Fe^{3+} was successfully prepared by planetary high energy ball milling. The ball milling process caused significant changes to the feature parameters of TiO_2, including decreased particle size, the more amorphous morphology, the compromised percentage of the most thermodynamically stable crystal facet, and the increased percentage of other high gas sensing crystal facets.

(2) The incorporation of nano-TiO_2 is beneficial to the improvement of concrete strength because of the micro-filling effect of nano particles. The incorporation of Fe^{3+} can motivate the formation of iron-containing hydration products, leading to the further increase of concrete strength. In terms of the material strength, the best nano-TiO_2 content of cement was 3% and 4% for the two types of concrete, respectively.

(3) The employment of nano-TiO_2 endow the concrete the functional properties of purifying exhaust gas. Due to the doping of Fe^{3+} in the crystal, the photocatalytic degradation effect of modified nano-TiO_2 is better than that of the original nano-TiO_2. The degradation rate of porous concrete is much higher than that of common concrete with the same content of photocatalyst. For both types of concrete, the photocatalytic degradation rate comes to the peak when the nano-TiO_2 content is 4%.

(4) The internal structure of the concrete was modified to be more compact with the adoption of nano TiO_2, so the strength is improved. In a comprehensive consideration of both mechanical performance and degradation efficiency, the recommended optimum dosage of TiO_2 is 3% to 4% for concrete. It is the most promising strategy to develop functional pavements using photocatalytic porous concrete to purify the urban traffic pollutions.

At last, it is strongly suggested that the modified nano-TiO_2 doped with Fe^{3+} may be further fused with other granular metamaterial components in concrete, to explore potential enhancement in both mechanical and chemical properties in further studies on the basis of the dispersive behavior of granular materials [37,38].

Author Contributions: R.H. contributed in methodology, writing review, editing and supervision; X.H. performed data curation, formal analysis and contributed in writing original draft; J.Z. contributed in revising and improving the manuscript; Y.G. contributed in literature search, figures, and validation; H.G. contributed in project administration and supervision.

Funding: This research was funded by the National Natural Science Foundation of China (No. 51508030), the National Key R. and D. Program of China (No. 2017YFB0309903), the Basic Research Project of Natural Science in Qinghai Province (No. 2017-ZJ-715), Key Research and Development Program of Shaanxi Province (No. 2018SF-403, 2018SF-380), the Basic Research Project of Natural Science in Shaanxi Province (Nos. 2019JQ-559, 2019JQ-380, 2018JQ5146) and the Fundamental Research Funds for the Central Universities of China (No. 300102318401).

Conflicts of Interest: The authors declare no conflict of interest.

References

1. Yang, W.; Li, L. Efficiency evaluation of industrial waste gas control in China: A study based on data envelopment analysis (DEA) model. *J. Clean. Prod.* **2018**, *179*, 1–11. [CrossRef]
2. Ze-dong, Z.; Hao, L.; Yong-dong, L.; Zheng, X. Electrical highway application. *China J. Highw. Transp.* **2019**, *32*, 132–141.
3. Wang, D.; Leng, Z.; Hüben, M.; Oeser, M.; Steinauer, B. Photocatalytic pavements with epoxy-bonded TiO_2-containing spreading material. *Constr. Build. Mater.* **2016**, *107*, 44–51. [CrossRef]
4. Wang, D.; Leng, Z.; Yu, H.; Hüben, M.; Kollmann, J.; Oeser, M. Durability of epoxy-bonded TiO_2-modified aggregate as a photocatalytic coating layer for asphalt pavement under vehicle tire polishing. *Wear* **2017**, *382–383*, 1–7. [CrossRef]
5. Cui, P.; Wu, S.; Xiao, Y.; Wan, M.; Cui, P. Inhibiting effect of Layered Double Hydroxides on the emissions of volatile organic compounds from bituminous materials. *J. Clean. Prod.* **2015**, *10*, 987–991. [CrossRef]
6. Sikkema, J.K.; Ong, S.K.; Alleman, J.E. Photocatalytic concrete pavements: Laboratory investigation of NO oxidation rate under varied environmental conditions. *Constr. Build. Mater.* **2015**, *100*, 305–314. [CrossRef]
7. Folli, A.; Pade, C.; Hansen, T.B.; de Marco, T.; Macphee, D.E. TiO_2 photocatalysis in cementitious systems: Insights into self-cleaning and depollution chemistry. *Cem. Concr. Res.* **2012**, *42*, 539–548. [CrossRef]
8. Liu, W.; Wang, S.; Zhang, J.; Fan, J. Photocatalytic degradation of vehicle exhausts on asphalt pavement by TiO_2/rubber composite structure. *Constr. Build. Mater.* **2015**, *81*, 224–232. [CrossRef]
9. Ballari, M.M.; Brouwers, H.J.H. Full scale demonstration of air-purifying pavement. *J. Hazard. Mater.* **2013**, *254–255*, 406–414. [CrossRef]
10. Schneider, J.; Matsuoka, M.; Takeuchi, M.; Zhang, J.; Horiuchi, Y.; Anpo, M.; Bahnemann, D.W. Understanding TiO_2 Photocatalysis: Mechanisms and Materials. *Chem. Rev.* **2014**, *114*, 9919–9986. [CrossRef]
11. Liao, L.; Heylen, S.; Sree, S.P.; Vallaey, B.; Keulemans, M.; Lenaerts, S.; Roeffaers, M.B.J.; Martens, J.A. Photocatalysis assisted simultaneous carbon oxidation and NO_x reduction. *Appl. Catal. B* **2017**, *202*, 381–387. [CrossRef]
12. Zhao, Q.; Wang, M.; Yanga, H.; Shia, D.; Wang, Y. Preparation, characterization and the antimicrobial properties of metal iondoped TiO_2 nano-powders. *Ceram. Int.* **2018**, *44*, 5145–5154. [CrossRef]
13. Wu, D.; Li, C.; Zhang, D.; Wang, L.; Zhang, X.; Shi, Z.; Lin, Q. Enhanced photocatalytic activity of Gd^{3+} doped TiO_2 and Gd_2O_3 modified TiO_2 prepared via ball milling method. *J. Rare Earths* **2019**, *37*, 845–852. [CrossRef]
14. Kuang, Y.; Zhang, Z.; Ji, X.; Zhang, X. Study on nano-TiO_2 photocatalytic cement paste with automobile exhaust degradation capacity. *J. Funct. Mater.* **2017**, *48*, 2241–2246.
15. Malengreaux, C.M.; Pirard, S.L.; Pirard, G.L.; Mahy, J.G.; Herlitschke, M.; Klobes, B.; Hermann, R.; Heinrichs, B.; Bartlett, J.R. Study of the photocatalytic activity of Fe^{3+}, Cr^{3+}, La^{3+} and Eu^{3+} single-doped and co-doped TiO_2 catalysts produced by aqueous sol-gel processing. *J. Alloys Compd.* **2017**, *691*, 726–738. [CrossRef]
16. Xiaoyan, P.; Dongmei, J.; Yan, L.; Xueming, M. Structural characterization and ferromagnetic behavior of Fe-doped TiO_2 powder by high-energy ball milling. *J. Magn. Magn. Mater.* **2006**, *305*, 388–391. [CrossRef]
17. Ivanov, E.; Suryanarayana, C. Materials and process design through mechanochemical routes. *J. Mater. Synth. Process.* **2000**, *8*, 235–244. [CrossRef]
18. Suryanarayana, C. Mechanical alloying and milling. *Prog. Mater. Sci.* **2001**, *46*, 1–184. [CrossRef]
19. Miao, J.; Zhangd, R.; Zhang, L. Photocatalytic degradations of three dyes with different chemical structures using ball-milled TiO_2. *Mater. Res. Bull.* **2018**, *97*, 109–114. [CrossRef]
20. Liu, Q.; Cao, D. Research on Material Composition and Performance of Porous Asphalt Pavement. *J. Mater. Civ. Eng.* **2009**, *21*, 135–140. [CrossRef]
21. Xiao, Y.; Wang, F.; Cui, P.; Lei, L.; Lin, J.; Yi, M. Evaluation of Fine Aggregate Morphology by Image Method and Its Effect on Skid-Resistance of Micro-Surfacing. *Materials* **2018**, *11*, 920. [CrossRef] [PubMed]
22. Lim, E.; Tan, K.H.; Fwa, T.F. High-Strength High-Porosity Pervious Concrete Pavement. *Adv. Mater. Res.* **2013**, *723*, 361–367. [CrossRef]

23. Zhong, M.; Wei, Z.H.; Si, Z.P.; Li, H.; Wei, G.Y.; Ge, H.L.; Han, G.R. Preparation of iron-doped titania nanocrystalline grains and its photocatalytic property. *J. Chin. Ceram. Soc.* **2010**, *38*, 69–73.

24. *People's Republic of China Transportation Industry Standard JTG E30–2005 Test Methods of Cement and Concrete for Highway Engineering*; Ministry of Communications of PRC: Beijing, China, 2005.

25. Pierpaoli, M.; Favoni, O.; Fava, G.; Ruello, M.L. A Novel Method for the Combined Photocatalytic Activity Determination and Bandgap Estimation. *Methods Protoc.* **2018**, *1*, 22. [CrossRef] [PubMed]

26. Xia, H.Y.; Zhang, R.; Song, L.F.; Gao, L.N.; Liu, G.Y. Test Device for Photocatalytic Coating Degradation Performance of Automobile Exhaust Gas. Chinese Patent CN201621076982.3, 23 September 2016.

27. Kochany, J.; Maguire, J. Sunlight Photodegradation of Metolachlor in Water. *J. Agric. Food Chem.* **1994**, *42*, 406–412. [CrossRef]

28. Chen, Z.; Wu, S.; Xiao, Y.; Zeng, W.; Yi, M.; Wan, J. Effect of hydration and silicone resin on Basic Oxygen Furnace slag and its asphalt mixture. *J. Clean. Prod.* **2016**, *112*, 392–400. [CrossRef]

29. Liu, C.; Lu, H.; Zhang, J.; Gao, J.; Zhu, G.; Yang, Z.; Yin, F.; Wang, C. Crystal facet-dependent p-type and n-type sensing responses of TiO_2 nanocrystals. *Sens. Actuators B* **2018**, *263*, 557–567. [CrossRef]

30. Han, X.G.; Han, X.; Sun, L.Q.; Gao, S.G.; Li, L.; Kuang, Q.; Xie, Z.X.; Wang, C. Synthesis of trapezohedral indium oxide nanoparticles with high-index {211} facets and high gas sensing activity. *Chem. Commun.* **2015**, *51*, 9612–9615. [CrossRef]

31. Kibasomba, P.M.; Dhlamini, S.; Maaza, M.; Liu, C.; Rashad, M.M.; Rayan, D.A.; Mwakikunga, B.W. Strain and grain size of TiO_2 nanoparticles from TEM, Raman spectroscopy and XRD: The revisiting of the Williamson-Hall plot method. *Results Phys.* **2018**, *9*, 628–635. [CrossRef]

32. Han, F.; Song, S.; Liu, J.; Huang, S. Properties of steam-cured precast concrete containing iron tailing powder. *Powder Technol.* **2019**, *345*, 292–299. [CrossRef]

33. Hertel, T.; Neubauer, J.; Goetz-Neunhoeffer, F. Study of hydration potential and kinetics of the ferrite phase in iron-rich CAC. *Cem. Concr. Res.* **2016**, *83*, 79–85. [CrossRef]

34. Arata, K. Preparation of superacids by metal oxides for reactions of butanes and pentanes. *Appl. Catal. A* **1996**, *146*, 3–32. [CrossRef]

35. Miao, S.; Lin-hai, Y.; Zhe-de, X. Photocatalytic activity of iron doping TiO_2 prepared by several methods. *Acta Phys. Chim. Sin.* **2001**, *17*, 282–285.

36. Aba-Guevara, C.G.; Medina-Ramírez, I.E.; Hernández-Ramírez, A.; Jáuregui-Rincón, J.; Lozano-Álvarez, J.A.; Rodríguez-López, J.L. Comparison of two synthesis methods on the preparation of Fe, N-Co-doped TiO_2 materials for degradation of pharmaceutical compounds under visible light. *Ceram. Int.* **2017**, *43*, 5068–5079. [CrossRef]

37. Nejadsadeghi, N.; Placidi, L.; Romeo, M.; Misra, A. Frequency band gaps in dielectric granular metamaterials modulated by electric field. *Mech. Res. Commun.* **2019**, *95*, 96–103. [CrossRef]

38. Misra, A.; Nejadsadeghi, N. Longitudinal and transverse elastic waves in 1D granular materials modeled as micromorphic continua. *Wave Motion* **2019**, *90*, 175–195. [CrossRef]

 *materials* 

Article

Effects of Polypropylene Fibre and Strain Rate on Dynamic Compressive Behaviour of Concrete

Meng Chen [1,2], Chenhui Ren [1], Yangbo Liu [1], Yubo Yang [1], Erlei Wang [3,*] and Xiaolong Liang [1]

[1] School of Resource and Civil Engineering, Northeastern University, Shenyang 110819, China; chenmeng@mail.neu.edu.cn (M.C.); renchenhui97@163.com (C.R.); 20161924@stu.neu.edu.cn (Y.L.); 20161993@stu.neu.edu.cn (Y.Y.); liangliforever@aliyun.com (X.L.)

[2] State Key Laboratory of Silicate Materials for Architectures, Wuhan University of Technology, Wuhan 430070, China

[3] Design & Research Institute of Wuhan University of Technology, Wuhan 430070, China

* Correspondence: cmwhut@163.com; Tel.: +86-138-4041-8369

Received: 7 May 2019; Accepted: 31 May 2019; Published: 3 June 2019

Abstract: This paper presents an experimental study on the dynamic compressive behaviour of polypropylene (PP) fibre reinforced concrete under various strain rates using split Hopkinson pressure bar (SHPB) equipment. The effects of PP fibre content and strain rate on the dynamic compressive stress-strain relationship and failure patterns were estimated. The results indicated that the addition of PP fibre enhanced the dynamic compressive properties of concrete mixtures although it resulted in a significant reduction in workability and a slight decrease in static compressive strength. Considering the workability, static compressive strength and dynamic compressive behaviour, the optimal PP fibre content was found to be 0.9 kg/m^3 as the mixture exhibited the highest increase in dynamic compressive strength of 5.6%, 40.3% in fracture energy absorption and 11.1% in total energy absorption; further, it showed the least reduction (only 5.8%) in static compressive strength among all mixtures compared to the reference mixture without fibre. For all mixtures, the dynamic compressive properties, energy absorption capacity, strain at peak stress, ultimate strain and dynamic increase factor (DIF) were significantly influenced by strain rate, i.e., strain rate effect. When the strain rate was relatively low, PP fibres were effective in controlling the cracking, and the dynamic compressive properties of PP fibre reinforced mixtures were improved accordingly.

Keywords: Fibre reinforced concrete; polypropylene; impact loading; constitutive model; failure mode

---

1. Introduction

Concrete is the most widely used construction material in the world [1]. However, normal concrete is inherently brittle when subjected to static (including tensile and flexural) and high-velocity dynamic loadings, and is susceptible to cracking induced by unsuitable curing conditions, freeze-thawing and shrinkage [2–5]. As a result, normal concrete exhibits a brittle failure pattern under dynamic impact loadings, posing a serious threat to the integrity and safety of concrete infrastructure in highly seismic (e.g., earthquake) or marine zones (e.g., wave impact and wind load). Moreover, high-strength concrete is usually supplied by the protective structures under dynamic impact loading. In addition, many cracks may appear on the concrete surfaces under impact loadings and thus provide transport pathways for some aggressive ions (e.g., chloride ions), which would facilitate the corrosion of reinforcing steel and thus adversely affect the durability of reinforced concrete structures.

In order to overcome the aforementioned shortcomings, many researchers have attempted to incorporate randomly distributed short fibres (i.e., steel, polypropylene and basalt fibres) into the concrete matrix. The inclusion of these short fibres into the concrete results in a significant increase of ductility, impact resistance and energy absorption capacity [6–9]. The enhancement of these material properties can be ascribed to the uniform distribution of short fibres in the concrete matrix [10,11]. Generally, these materials are classified as fibre reinforced concrete (FRC) [12]. Among all of these fibres, steel fibres were first introduced as the reinforcement in concrete [13] and showed high efficiency in improving the dynamic properties of concrete [9,12,14,15]. Karimi et al. [16] explored the fracture energy of three types asphalt concrete and concluded that the mixtures with a 0.2% and 1.5% SWF (steel wool fibre) showed lower fracture energy than the neat mixture. Wu et al. [9] investigated the dynamic behaviour of hybrid steel fibre reinforced ultra-high-performance concrete (UHPC) and concluded that UHPC with hybrid long (1.5%) and short (0.5%) steel fibre reinforcement exhibited the best dynamic mechanical properties. Hao et al. [12] explored the dynamic compressive behaviour of concrete reinforced with randomly distributed spiral steel fibres and found that the energy absorption capacity of the tested concrete was significantly increased with increasing steel fibre dosage. However, some drawbacks of steel fibres may limit the large-scale engineering application of steel fibre reinforced concrete around coastal zones. For instance, as steel fibres rust easily and its weight may cause problems such as fibre balling during the mixing process [17]. When the coastal concrete structures suffer from impact such as wave impact and wind load, the drawbacks of steel fibres facilitate the corrosion of RC structures. Therefore, it is vital to incorporate other types of fibres such as synthetic fibres into normal concrete to improve its dynamic compressive properties as well as corrosion resistance, thereby improving the durability of RC structures near marine zones made by this material.

In the past decades, polypropylene (PP) fibre has been increasingly used in concrete owing to its advantages of lightweight and high corrosion resistance [18]. It was observed that PP fibre was effective in reducing the plastic shrinkage and improving the impact resistance of the matrix [3,18,19]. In recent years, several studies have been conducted on the dynamic properties of PP fibre reinforced concrete. Zhang et al. [20] investigated the dynamic behaviour of PP fibre reinforced mortar under compressive impact loading using SHPB. It was found that the dynamic behaviour of PP fibre reinforced mortar was significantly influenced by strain rate. Moreover, PP fibres can improve the impact toughness of the matrix. Jahanbakhsh and Karimi et al. [21–23]. considered the fracture energy as a material property that is independent of sample size and geometry. Moreover, they calculated the fracture energy according to RILEM TC50-FMC specifications by dividing the fracture work. Zhang et al. [24] conducted drop weight test to estimate the flexural impact of PP fibre reinforced concrete and suggested that adding PP fibres into the concrete can enhance its impact resistance though a brittle failure pattern was observed. Fu et al. [2] also found that the addition of PP fibres improved the energy absorption capacity of the concrete matrix. However, different volume fractions of PP fibre should be considered in order to determine the optimal PP fibre content in normal concrete matrix under the dynamic loading with high strain rates. Although Zhang et al. [17] concluded that the optimal content of PP fibre was 1.5 kg/m^3, which was drawn mainly based on the experimental results of static mechanical tests. In addition, it is stated that the optimal fibre volume fraction for the macro-fibre is 0.5% [25]. Nevertheless, the optimal content for PP fibre is still not determined. Moreover, a number of studies examining the dynamic properties of PP fibre reinforced concrete utilised SHPB equipment with a high strain rate (i.e., normally in the range between 10^2 s^{-1} and 10^3 s^{-1} [12,26–28]), as the previous study mainly focused on the drop weight test. However, the method of drop weight test is greatly restricted by the specimen size and configuration [9]. In conclusion, the dynamic compressive properties of PP fibre reinforced concrete mixtures with different PP fibre volume fractions under high strain rate have not been extensively studied. Therefore, it is important to investigate the dynamic compressive behaviour of concrete reinforced with different dosages of PP fibres.

The main purpose of this paper is to provide a better understanding of the effects of PP fibre dosage and strain rate on the dynamic compressive behaviour of concrete. Portland cement was used as the main binder material and the compressive strength of all mixture was around 60 MPa. PP fibres with various content (i.e., 0.9, 1.8, 2.7 and 3.6 kg/m^3) were used. The PP fibre content of 0.9 kg/m^3 approximately equals to 0.1% fibre volume fraction (V_f). Thus, fibre content of 1.8, 2.7, 3.6 kg/m^3 roughly corresponds to 0.2, 0.3, 0.4% V_f. Firstly, the slump test was conducted to determine the workability of fresh mixture and the static compressive strength was evaluated. Subsequently, the dynamic compressive properties including stress-strain curve and failure mode were examined using SHPB. The dynamic increase factor (DIF), fracture energy, total energy, the strain at peak stress and ultimate strain were discussed in detail by comparing the difference for all mixtures. The fitted DIF functions for each concrete mixture were derived based on statistical study for providing guidance on incorporating PP fibres to the cement matrix. Finally, PP fibre reinforced mixtures in the form of scanning electron microscopy (SEM) images were presented to further understand the enhancement of dynamic compressive behaviour after incorporating fibres.

2. Experimental Program

2.1. Raw Materials

Portland cement (P.I. 42.5), bought from Liaoning Shan Shui Gong Yuan, Co., Ltd. (Liaoning, China), was used as the main binder material with chemical compositions listed in in Table 1. The specific gravity of cement is 3.09. Natural river sand with a nominal maximum size of 4.75 mm was used as the fine aggregate, the specific gravity of which is 2.56. Coarse aggregates were prepared by mixing two different maximum particle sizes of crushed granite, which were 5 mm and 20 mm, respectively. Fine and coarse aggregates were used in saturated surface dry (SSD) conditions according to ASTM C128-15 [29] and ASTM C127-15 [30]. The particle size distributions of fine and coarse aggregates are presented in Tables 2 and 3, respectively. Multifilament polypropylene (PP) fibre (see Figure 1), bought from Beijing Kenye Trade Co., Ltd. (Beijing, China), was used and its properties are shown in Table 4. Polycarboxylate-based superplasticisers (SPs) were used to improve the workability of the mixture.

Table 1. Chemical compositions of cement.

Oxide	SiO_2	Al_2O_3	CaO	Fe_2O_3	SO_3	MgO	Na_2O	K_2O	LOI
Cement	21.35	5.98	60.56	2.91	2.05	2.22	0.21	0.75	3.97

Table 2. The gradation compositions of the fine aggregate.

Sieve Size (mm)	4.75	2.36	1.18	0.60	0.30	0.15	0.075
Total Percentage Retained (%)	0	10.5	27.8	51.5	76.2	90.8	98.6

Table 3. The gradation compositions of the coarse aggregate.

Sieve Size (mm)	20.0	19.0	16.0	13.2	9.5	4.75
Total Percentage Retained (%)	0	2.6	18.3	55.3	85.4	98.1

Figure 1. Photograph of PP fibre.

Table 4. Properties of polypropylene fibre.

Fibre Type	Length (mm)	Diameter (μm)	Fibre Strength (MPa)	Density (kg/m^3)	Modulus of Elasticity (GPa)
PP	19	26	376	910	3.79

2.2. Mixture Proportions

The water-to-binder ratio (w/b) was adjusted and selected 0.24 for the plain concrete matrix in order to ensure its compressive strength higher than 60 MPa. Based on the previous study [25], the highest fibre content was adopted as 3.6 kg/m^3 (equivalent to 0.4% fibre volume fraction). The mixture proportions of concrete used in this study are given in Table 5 and marked with specific symbols. The label 'PP' represents polypropylene fibre, while '00', '09', '18', '27' and '36' stand for the dosage of the corresponding fibre. For example, the fibre content for the mixture No.1 is 0.0 kg/m^3, while the fibre content for the mixture No. 2 is 0.9 kg/m^3.

Table 5. Mixture proportions.

No.	Symbol	Cement (kg/m^3)	Water (kg/m^3)	FA (kg/m^3)	CA (kg/m^3)	SPs (kg/m^3)	PP Fibre (kg/m^3)	V_f (%)
1	PP00	569	136	645	1100	5.69	0.0	/
2	PP09	569	136	645	1100	5.69	0.9	0.1
3	PP18	569	136	645	1100	5.69	1.8	0.2
4	PP27	569	136	645	1100	5.69	2.7	0.3
5	PP36	569	136	645	1100	5.69	3.6	0.4

Note: FA (Fine Aggregate); CA (Coarse Aggregate); SPs (Superplasticisers); PP (Polypropylene).

2.3. Specimen Preparation

The mixing procedure is conducted using a SJD30 concrete horizontal-axle mixer. Cement, fine aggregate and coarse aggregate were first dry mixed for 2 min. Water and SPs were then added to the mixture after 2 min dry mix. Since multifilament fibres are difficult to be dispersed uniformly in the mixture [31], PP fibres were gradually added to the wet mix in order to ensure uniform dispersion of fibres. All the fresh mixtures were then cast into moulds for static and dynamic compressive tests. All specimens were demoulded after 24 h and then stored in a standard curing room (20 ± 2 °C and 95% RH) for 28 days. Table 6 shows the experimental program including the number of static specimens, dynamic specimens and strain rate.

Table 6. Number of static specimens, dynamic specimens and strain rate.

Symbol	Number of Specimens		Strain Rate Rype (s^{-1})				
	Static Specimens	Dynamic Specimens	50	70	90	110	120
PP00	3	15	3	3	3	3	3
PP09	3	15	3	3	3	3	3
PP18	3	15	3	3	3	3	3
PP27	3	15	3	3	3	3	3
PP36	3	15	3	3	3	3	3

2.4. Test Methods

2.4.1. Workability and Static Compressive Test

The slump test was conducted in accordance with ASTM C143-15a [32] to determine the workability of all fresh concrete mixture, where the test involved in measuring the vertical displacement difference between the top of the testing mould and top surface of the fresh concrete specimen. Three 150 mm side-length cubic specimens of each mixture were used for static compression test using a universal testing machine (Zhejiang Luda Machinery Instrument Co., Ltd., Zhejiang, China) at the testing age of 28 d according to Chinese standard GB/T50081-2002 [33].

2.4.2. Dynamic Compressive Test

In this study, the dynamic compressive properties of all concrete mixtures were evaluated by 100 mm diameter SHPB testing equipment at Northeastern University, Liaoning, China. The SHPB testing system is shown in Figure 2 The striker, incident, transmission and absorbing bars are all made of 100 mm diameter superior alloy steel. The lengths of striker, incident, transmission and absorbing bars are 600, 5000, 3500 and 1200 mm, respectively. Cylindrical specimens 100 mm in diameter by 50 mm in height were used for each mixture in this study.

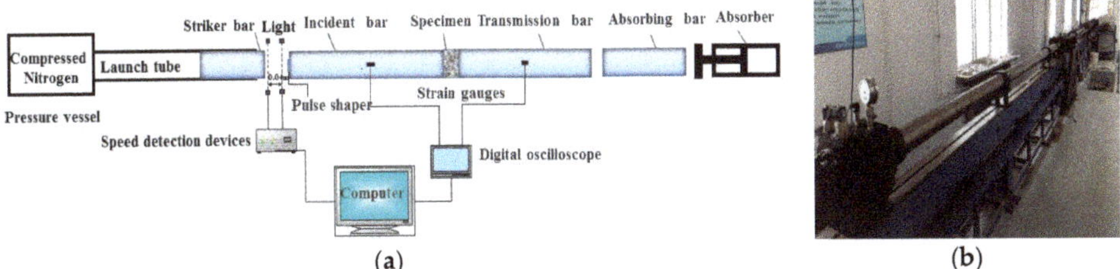

(a) (b)

Figure 2. Schematic diagram and image of SHPB testing system: (**a**) Schematic diagram of SHPB testing system; (**b**) Image of SHPB testing apparatus.

Regarding the testing procedure, the specimen was first placed between the incident and transmission bars, as shown in Figure 2a when the 'start' button in the computer system was pressed, the 100 mm diameter striker bar impacted the incident bar under the set compressing nitrogen to generate the incident pulse ($\varepsilon_i(t)$), known as incident wave. Afterwards, the specimen started to deform when the incident pulse reached it. At that time, some of the incident pulse was reflected as the reflected pulse ($\varepsilon_r(t)$). The remaining pulse passed through the specimen and propagated to the transmission bar, as transmission pulse ($\varepsilon_t(t)$). The whole testing process was recorded in the computer. In addition, the strain gauges shown in Figure 2a were used to measure the change of strain values of the incident and transmission bars, and thus the curves of strain against time were plotted.

The pulse shaping technique was applied to increase the accuracy of the results obtained from SHPB test. The main purpose was to ensure that the tested specimen has sufficient time to reach stress equilibrium by prolonging rising time of incident pulse [7]. A rubber-made slice with a 50 mm in diameter by 2 mm in height was used as a pulse shaper. Figure 3 shows typical waveforms of incident, reflected and transmitted pulses with a pulse shaper.

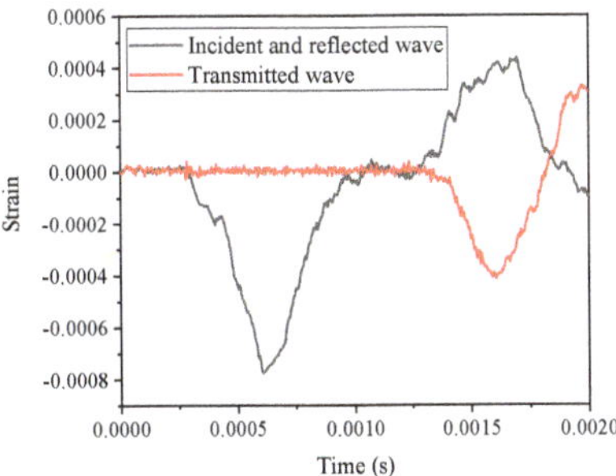

Figure 3. Typical waveforms of incident, reflected and transmitted pulses with a pulse shaper.

The stress-strain curve of each mixture can be obtained by inputting the incident pulse ($\varepsilon_i(t)$), reflected pulse ($\varepsilon_r(t)$) and transmission pulse ($\varepsilon_t(t)$). Tri-wave method was used to calculate the stress ($\sigma_s(t)$), strain ($\varepsilon_s(t)$) and strain rate ($\dot{\varepsilon}_s(t)$) [34–36].

2.4.3. SEM Test

The SEM test (Ultra-plus SEM, Jena, Germany) was performed on the fracture pieces after SHPB test to explore the damage morphology of PP fibre. The selected pieces were coated with a gold film before the test. Regarding the image acquisition parameters, the working distance for the SEM test was set in the range of 11.4 to 31.8 mm, while the acceleration voltage was 15 kV. The magnification was in the wide range of 200×.

3. Results and Discussion

3.1. Workability

The slump value is normally used to represent the workability of concrete. Figure 4 shows the slump value of all mixtures. It is clear that the slump value of all mixtures was significantly influenced by the PP fibre content. The slump value decreased sharply with increasing PP fibre content in the mixture, which agreed well with the previous studies [3,18,37]. The slump value of PP09, PP18, PP27 and PP36 reduced 14.6%, 20.6%, 36.0% and 47.1%, respectively as compared to the reference mixture (PP00) with a slump value of 184.7 mm. The reduction in workability can be ascribed to the following reasons: (1) adding fibres into the concrete matrix resulted in the increase of viscosity especially when the surface area of fibres are larger [18]; (2) PP fibres exhibit hydrophobic characteristic in nature (i.e., repelling water) and therefore the compactness of mixtures reduced with increasing PP fibre content; (3) clogging of PP fibres would occur, which may affect the uniform dispersion of the fibres in the concrete mixtures [38]; and (4) the internal microstructure of mixtures would be changed due to the addition of PP fibres and thus restrict the flow of mixtures [9]. Since the compactness of concrete may affect its mechanical properties, the dosage of SPs should be adjusted accordingly in order to ensure that all PP fibre reinforced concrete mixtures are workable.

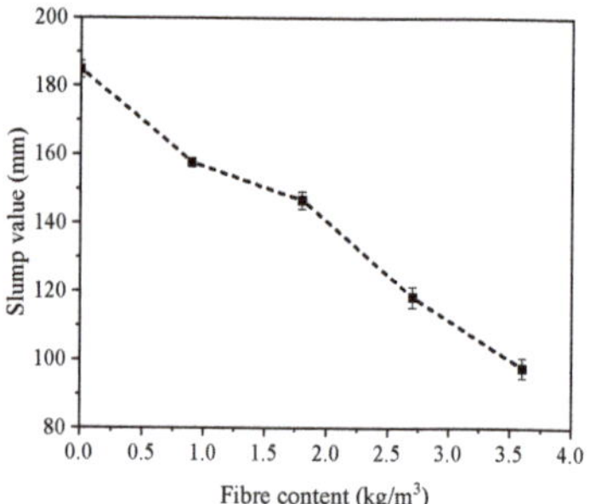

Figure 4. Slump of fibre reinforced concrete with various polypropylene fibre content.

3.2. Static Compressive Strength

Figure 5 displays the static compressive strength of all concrete mixtures at testing age of 28 d. It can be seen that the 28 d compressive strengths of PP fibre reinforced concrete mixtures, i.e., PP09 (63.4 MPa), PP18 (60.8 MPa), PP27 (58.9 MPa) and PP36 (56.8 MPa), were lower than that of PP00 (67.3 MPa). This implies that the PP fibre is not capable of enhancing the compressive strength of concrete matrix, which is consistent with previous studies [18,39] pointing out that PP fibres with low stiffness cannot improve the compressive strength of the plain concrete matrix. The reduction of compressive strength became more obvious with the increase of PP fibre dosage. Compared to that of PP00, the compressive strength of PP36 was reduced by 15.6%, which was the highest among all mixtures. In addition, the compressive strengths of PP09, PP18 and PP27 were reduced by 5.8%, 9.7% and 12.5%, respectively as compared to PP00. Therefore, the optimal PP fibre content is determined to be 0.9 kg/m^3 in terms of static compressive strength. Generally, the reduction in compressive strength after adding PP fibres can be attributed to the hydrophobic characteristic of PP fibre, which affected the dispersion of the fibres in the concrete mixtures and thus the compactness of all PP fibre reinforced concrete was adversely influenced.

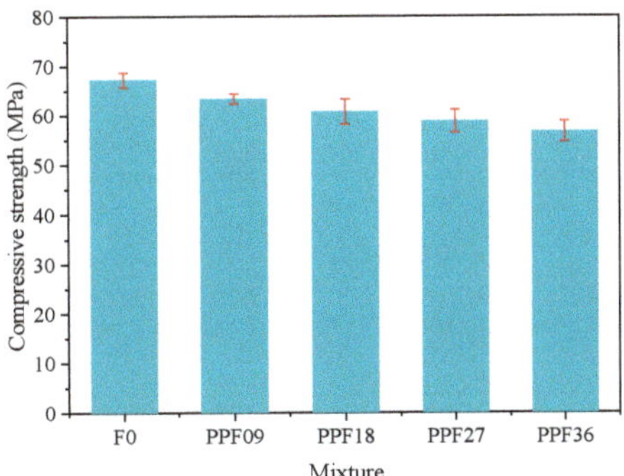

Figure 5. Compressive strength of all mixtures at 28 d.

3.3. Dynamic Compressive Stress-Strain Relationship

3.3.1. Effect of Strain Rate

Figure 6 shows an example of how the final curves of each mixture presented in Figure 7 was derived. Basically, the final curve was obtained by averaging the curves of two or three specimens, and then the average strain rate was correspondingly determined. The effect of strain rates varying from 50–120 s^{-1} on dynamic compressive stress-strain curves of all mixtures at 28 d are presented in Figure 7. All mixtures are sensitive to strain rate in terms of dynamic compressive stress, known as "strain rate effect". The dynamic compressive strength increased with the increase of strain rate, which agreed well with previous studies [2,7,9,12,14,17,40]. For instance, peak stress for PP09 was 68.4 MPa with average strain rate of 51.4 s^{-1} and increased to 115.7 MPa with the strain rate 122.8 s^{-1}. The increase of dynamic compressive strength due to the "strain rate effect" can be attributed to the lateral inertial effect to the contact surface of the specimens under high-velocity impact loading [9,41,42]. Ross et al. [43] mentioned that sensitivity to strain rate is highly dependent on the moisture condition of concrete. Fu et al. [17] suggested that the "strain rate effect" is mainly ascribed to the combining effect caused by the action of the crack propagation, viscous effect of free water and inertial effect. Therefore, "strain rate effect" can be explained as follows: when the strain rate is lower, the concrete mixtures have sufficient accumulated energy and time for the micro-cracks to propagate into macro-cracks before the final failure of the specimens. The energy required for the propagation of the cracks is lower than that of generating cracks [40], implying that the concrete mixtures need more energy under impact loading, i.e., more cracks appeared under high-velocity impact. However, the time needed for the final failure of concrete mixtures is very short under impact loading. Thus, the accumulated energy of the specimens is lower. The compressive strength is increased only for providing the required energy of generating cracks.

Table 7 summarizes the dynamic and static compressive strength, and DIF of all mixtures, where DIF is defined as the ratio of the dynamic compressive strength to the static compressive strength and is widely used to describe the "strain rate effect". As seen in Figure 8, DIF values of all mixtures increased gradually with the increase of strain rate, showing good agreement with previous studies [9,40]. As the average strain rate increased from 50 s^{-1} to 120 s^{-1}, DIF of PP00, PP09, PP18, PP27 and PP36 was increased by 64.1%, 68.5%, 55.3%, 50.4% and 43.7%, respectively. The results suggest that the PP fibre content of 0.9 and 1.8 kg/m^3 was more efficient in improving dynamic compressive strength of concrete under high-velocity impact loading.

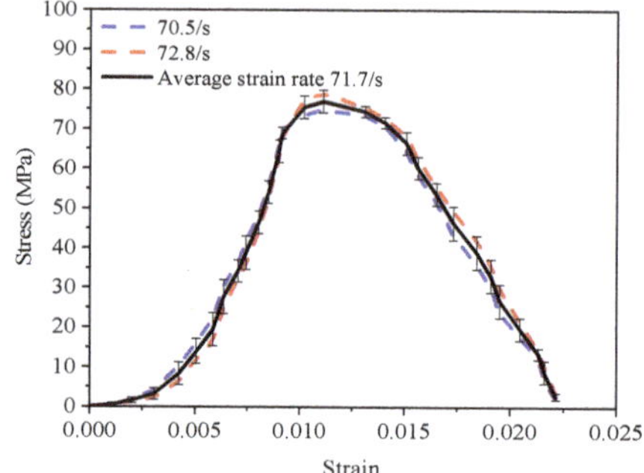

Figure 6. Example of obtaining the dynamic compressive stress-strain curve.

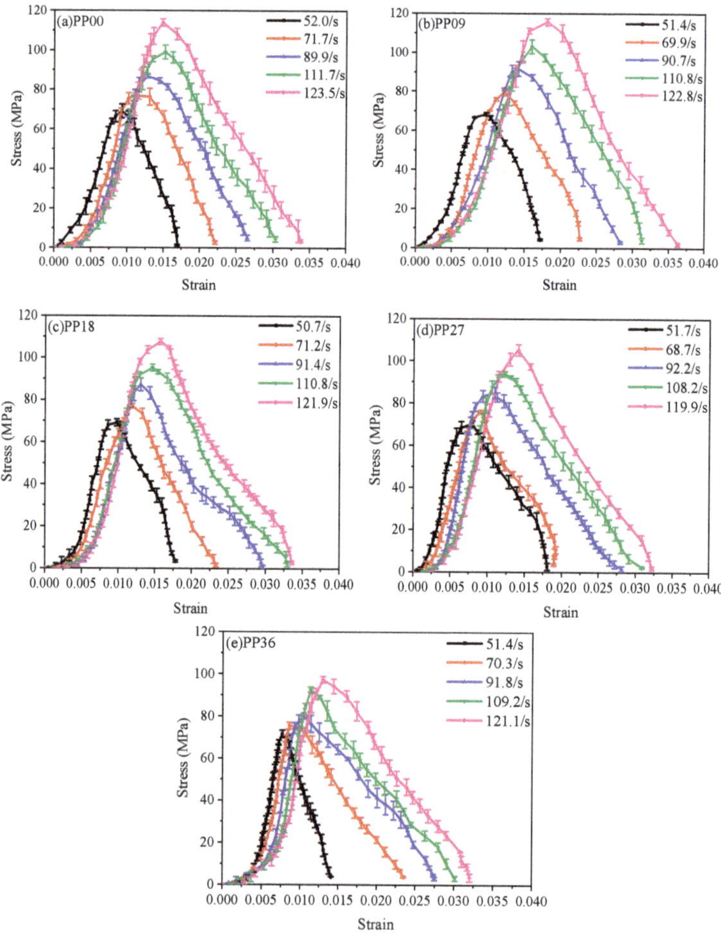

Figure 7. Effect of strain rate on dynamic compressive stress-strain relationship of (**a**) PP00; (**b**) PP09; (**c**) PP18; (**d**) PP27; (**e**) PP36.

Table 7. Summary of dynamic and static properties of PP fibre reinforced concrete.

Symbol	Average Strain Rate (s^{-1})	Maximum Dynamic Stress (MPa)	Strain at Peak Stress (×10^{-3})	Ultimate Strain (×10^{-3})	Static Compressive Strength (MPa)	DIF	Fracture Energy (J)	Total Energy (J)
PP0	52.0	69.1 ± 3.45	9.25	16.95	67.3 ± 1.46	1.03 ± 0.051	108.7 ± 5.92	237.4 ± 3.81
	71.7	76.9 ± 2.86	11.11	22.14	67.3 ± 1.46	1.14 ± 0.042	109.0 ± 6.97	335.6 ± 6.50
	89.9	86.8 ± 1.00	12.82	26.54	67.3 ± 1.46	1.29 ± 0.015	154.8 ± 8.67	436.5 ± 1.97
	111.7	99.1 ± 3.46	15.22	30.35	67.3 ± 1.46	1.47 ± 0.051	233.4 ± 2.05	546.6 ± 26.44
	123.5	113.9 ± 1.80	14.88	33.78	67.3 ± 1.46	1.69 ± 0.027	231.3 ± 13.33	688.5 ± 15.02
PP09	51.4	68.4 ± 0.86	9.78	17.25	63.4 ± 1.06	1.08 ± 0.014	117.1 ± 0.39	242.9 ± 3.97
	69.9	79.8 ± 1.97	12.19	22.68	63.4 ± 1.06	1.26 ± 0.031	143.2 ± 12.53	351.6 ± 21.94
	90.7	91.7 ± 3.74	14.13	28.33	63.4 ± 1.06	1.45 ± 0.059	216.8 ± 0.18	485.1 ± 11.15
	110.8	103.1 ± 3.54	15.95	31.31	63.4 ± 1.06	1.63 ± 0.056	224.4 ± 9.08	595.9 ± 8.41
	122.8	115.7 ± 1.84	18.09	36.29	63.4 ± 1.06	1.82 ± 0.029	324.4 ± 12.66	741.5 ± 10.19
PP18	50.7	69.2 ± 1.69	9.76	17.83	60.8 ± 2.52	1.14 ± 0.028	102.8 ± 4.16	243.4 ± 0.77
	71.2	77.1 ± 1.05	11.89	23.38	60.8 ± 2.52	1.27 ± 0.017	133.0 ± 6.83	310.7 ± 3.68
	91.4	87.3 ± 3.05	13.04	29.69	60.8 ± 2.52	1.44 ± 0.050	138.0 ± 6.66	411.3 ± 23.52
	111.5	95.3 ± 1.60	14.63	32.95	60.8 ± 2.52	1.57 ± 0.026	204.6 ± 8.19	540.0 ± 20.91
	121.9	107.7 ± 1.41	15.71	36.62	60.8 ± 2.52	1.77 ± 0.023	246.9 ± 8.97	625.4 ± 23.76
PP27	51.7	70.1 ± 1.55	7.99	18.06	57.9 ± 2.31	1.21 ± 0.027	102.7 ± 2.68	243.8 ± 10.60
	68.7	76.0 ± 0.64	9.08	23.44	57.9 ± 2.31	1.31 ± 0.011	114.7 ± 7.33	287.4 ± 15.03
	92.2	84.8 ± 2.64	10.98	28.14	57.9 ± 2.31	1.46 ± 0.046	119.9 ± 6.08	422.8 ± 22.37
	108.2	93.7 ± 1.27	12.24	30.88	57.9 ± 2.31	1.61 ± 0.022	164.4 ± 3.44	530.3 ± 16.80
	119.9	105.2 ± 2.66	14.07	32.22	57.9 ± 2.31	1.82 ± 0.046	192.2 ± 1.80	569.8 ± 15.11
PP36	49.8	67.6 ± 1.78	7.67	14.09	56.8 ± 2.06	1.19 ± 0.031	56.6 ± 3.70	157.8 ± 10.10
	70.3	73.6 ± 1.58	8.60	23.50	56.8 ± 2.06	1.30 ± 0.028	66.8 ± 3.93	283.9 ± 15.14
	91.8	80.2 ± 1.40	10.48	27.50	56.8 ± 2.06	1.41 ± 0.025	101.3 ± 12.50	408.3 ± 22.85
	109.2	90.7 ± 1.27	11.35	30.12	56.8 ± 2.06	1.60 ± 0.022	109.5 ± 7.99	423.0 ± 11.74
	121.1	97.3 ± 1.48	12.99	32.04	56.8 ± 2.06	1.71 ± 0.026	151.4 ± 10.98	507.2 ± 37.78

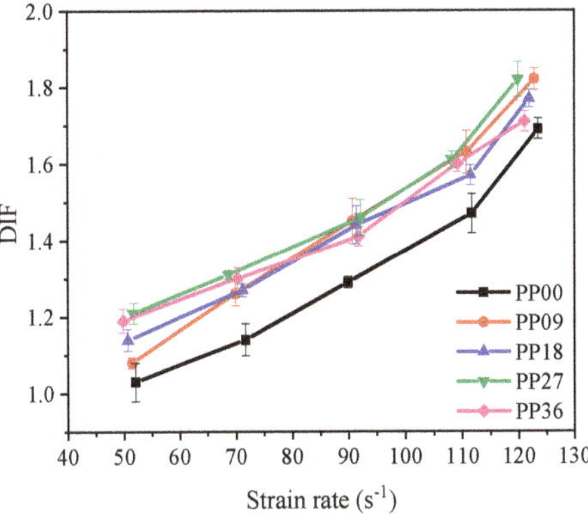

Figure 8. Effect of strain rate on DIF.

The fracture energy is defined as the energy absorbed by the specimen up to the peak stress. It can be observed from Table 7 that fracture energy was enhanced by the increase of strain rate. Figure 9a shows that PP09 has the maximum fracture energy at the similar strain rate during the impact test. The total energy absorption of all mixtures under impact loading is shown in Figure 9b. For each mixture, the total energy was calculated by multiplying the strain energy density (i.e., the area under the stress-strain curve) by the specimen volume [12]. For each mixture, the energy absorption capacity increased gradually with increasing strain rate, which can also be explained by the "strain rate effect", that is, the concrete mixtures absorbed more energy under rapid impact loading because of a larger number of cracks. The fracture energy absorption of PP00, PP09, PP18, PP27 and PP36 was enhanced by about 127.8%, 177.0%, 140.2%, 87.1% and 167.5%, respectively when the strain rate increased from around 50 s^{-1} to 120 s^{-1}. The total energy absorption of PP00, PP09, PP18, PP27 and PP36 was enhanced by about 190.0%, 205.3%, 156.9%, 133.7% and 221.4%, respectively when the strain rate increased from around 50 s^{-1} to 120 s^{-1}.

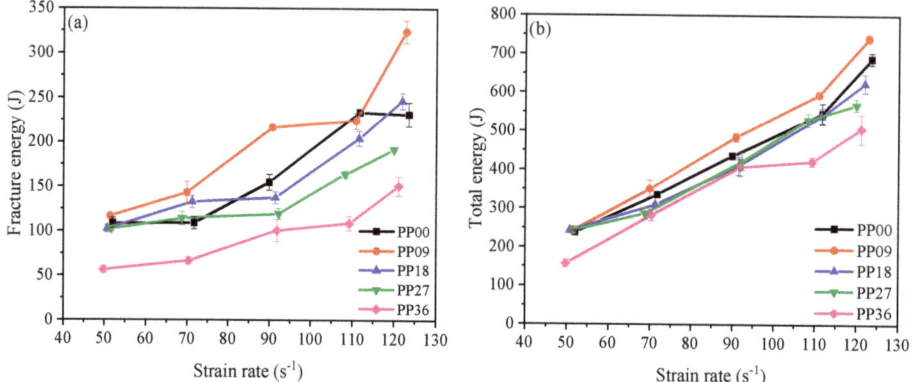

Figure 9. Energy absorption capacity of concrete at different strain rates: (**a**) Fracture energy; (**b**) Tolal energy.

The deformation capacity of concrete materials can be represented by the strain at peak stress and ultimate strain listed in Table 7. The strain at peak stress and ultimate strain were improved with increasing strain rate. The ultimate strain of each mixture was higher than its strain at peak stress ranging from 83.2% to 173.3%. The ultimate strain of PP09 and PP18 in this study was higher than that of PP00. This can be attributed to the bridging action of the fibres in cracks, which prolongs the strain-softening region and subsequently improves the toughness of FRC.

3.3.2. Effect of Polypropylene Fibre Content

The effect of PP fibre content on DIF of all concrete mixtures is presented in Figure 10. At the strain rate of around 50 s^{-1}, the DIF of PP09, PP18, PP27 and PP36 increased 4.9%, 10.7%, 17.5% and 15.5%, respectively as compared to PP00. When the strain rate was around 70 s^{-1}, the DIF of PP09, PP18, PP27 and PP36 enhanced by 10.5%, 11.4%, 14.9% and 14.0% compared to PP00. When the strain rate was around 120 s^{-1}, the DIF of PP09, PP18, PP27 and PP36 enhanced by 7.7%, 4.7%, 7.7% and 1.2% compared to PP00. Under the strain rate of 50 s^{-1} and 70 s^{-1}, the effect of PP fibre contents on DIF of all mixtures was remarkable. PP09 was found more efficient in controlling the cracking under high strain rate impact loading, which can also be attributed to the short impact loading time under high-velocity impact and thus the contribution of fibres was not remarkable.

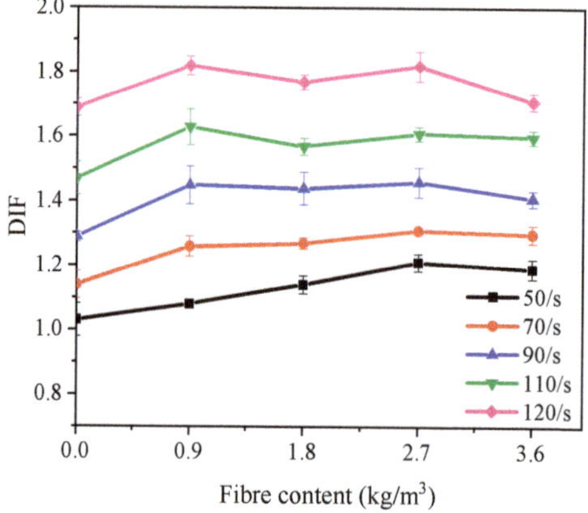

Figure 10. Effect of polypropylene fibre content on DIF.

DIF is essential for structural design and getting a reliable DIF is helpful for obtaining an appropriate and economical constructional design [9]. Therefore, it is of importance to obtain a reliable fitted DIF function for each concrete mixture in this study. It was reported by Tedesco et al. [44] that the relationship between DIF and logarithm of strain rate was approximately linear. Following this suggestion, the fitted DIF curves of all concrete mixtures are acquired and demonstrated in Figure 11 as well as Table 8, where the detailed fitted DIF equations for each mixture along with the correlation coefficient (R) are presented. As shown in Table 8, all R^2 values are greater than 0.9 suggesting that the fitted DIF functions are reliable. The results are consistent with previous study [40]. In addition, it is known that static compressive strength was mainly affected by fibre content, showing reducing tendency with increasing fibre content. On the other hand, for PP09 specimen, although static compressive strength decreased, dynamic compressive strength at relatively low loading rate reduced more remarkably, resulting in a decreased DIF. At high loading rate, DIF increased since with increasing fibre dosage, dynamic compressive strength increased while static compressive strength was reduced.

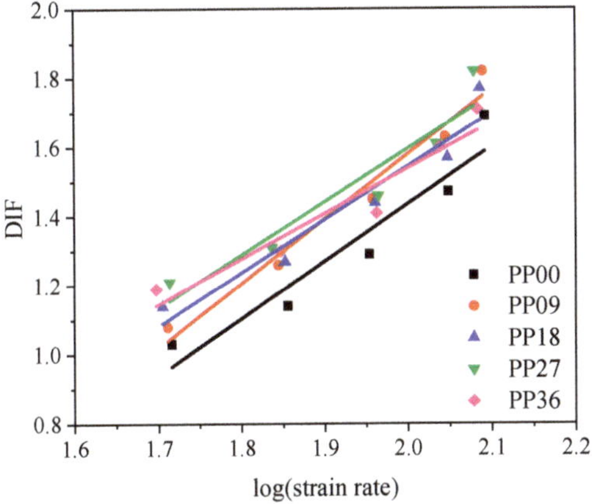

Figure 11. Fitted DIF curves of concrete reinforced with various polypropylene fibre content.

Table 8. Summary of the fitted DIF equations for different mixtures.

No.	Symbol	Fitted Equation of DIF	R^2
1	PP00	$\mathrm{DIF} = 1.656 \log \dot{\varepsilon} - 1.877$	0.911
2	PP09	$\mathrm{DIF} = 1.871 \log \dot{\varepsilon} - 2.161$	0.966
3	PP18	$\mathrm{DIF} = 1.541 \log \dot{\varepsilon} - 1.536$	0.928
4	PP27	$\mathrm{DIF} = 1.537 \log \dot{\varepsilon} - 1.477$	0.902
5	PP36	$\mathrm{DIF} = 1.312 \log \dot{\varepsilon} - 1.084$	0.922

Figure 12 shows the influence of PP fibre content on energy absorption capacity of concrete. The fracture energy presented is defined as the energy absorbed by the specimen up to the peak stress during the SHPB test, while the total energy reflects the energy absorption capacity in all regions. The fracture energy had an increase up to the PP fibre content of 0.9 kg/m^3 followed by a gradual decrease with further addition of PP fibre at different strain rates, except 110 s^{-1} strain rate (see Figure 12a). The fracture energy of PP09 was enhanced by 7.7%, 31.4%, 40.1%, −3.0% and 40.3%, respectively at the average strain rate around of 50 s^{-1}, 70 s^{-1}, 90 s^{-1}, 110 s^{-1} and 120 s^{-1} compared to that of the reference mixture (PP00). Figure 12b shows the development of total energy of PP fibre mixtures with varying content. A similar tendency can be seen that the PP fibre content of 0.9 kg/m^3 corresponds to the maximum total energy absorption capacity and thus it can be regarded as the optimal dosage regarding the energy absorption capacity.

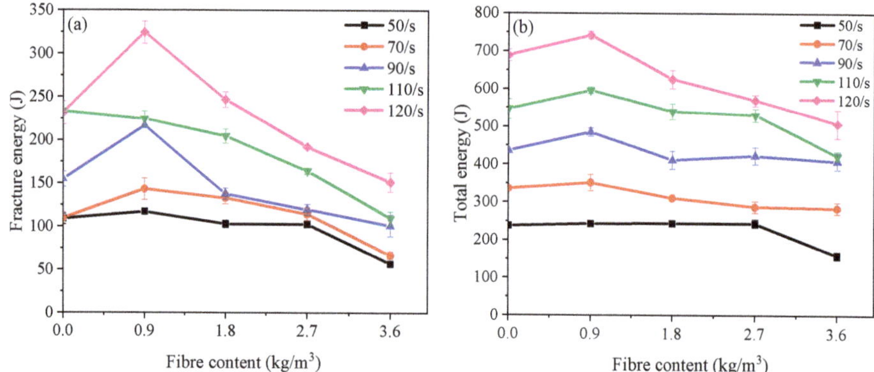

Figure 12. Energy absorption capacity of concrete reinforced with various polypropylene fibre content:
(**a**) Fracture energy; (**b**) Total energy.

The mechanical properties of FRC are highly dependent on its internal structure and the properties of fibres and cement matrix. The number of fibres inside the matrix can affect the spatial spacing of the fibres. It was reported that the static mechanical properties were influenced by space between fibres inside the matrix. For instance, the tensile strength decreased with increasing fibre spacing [45]. Based on the current experimental results, it is found that the static mechanical properties showed an exactly dissimilar trend with that of dynamic properties especially with relatively high strain rate. When the strain rate increases, the loading time is very short. Under such conditions, the FRC specimen may only exhibit the optimal dynamic compressive properties when its internal structure reaches the optimal state. In this work, the optimal fibre content in terms of workability, static and dynamic compressive properties was found to be 0.9 kg/m^3 (i.e., PP09), as PP09 presented the best workability among all mixtures and its static compressive strength was not affected significantly (i.e., only 5.8% reduction as compared to PP00). In addition, the DIF of PP09 increased by 7.7% at the strain rate of 120 s^{-1} as compared to PP00, which is the highest among all mixtures.

3.4. Dynamic Failure Mode

The effect of strain rate on the dynamic failure mode of all mixtures is illustrated in Figure 13. It is found that the failure pattern of all mixtures changed from splitting failure into pulverization failure with increasing strain rate, which is consistent with previous studies [2,7,9,12]. As shown in Figure 13a,d,g,j,m, several big broken fragments were observed for all mixtures when the strain rate is lower. However, when the strain rate is relatively high, all deformed specimens are broken into many tiny fragments (see Figure 13c,f,i,l,o). This phenomenon is in consistence with the "strain rate effect", which can also be attributed to the extremely short loading time under high-velocity impact loading.

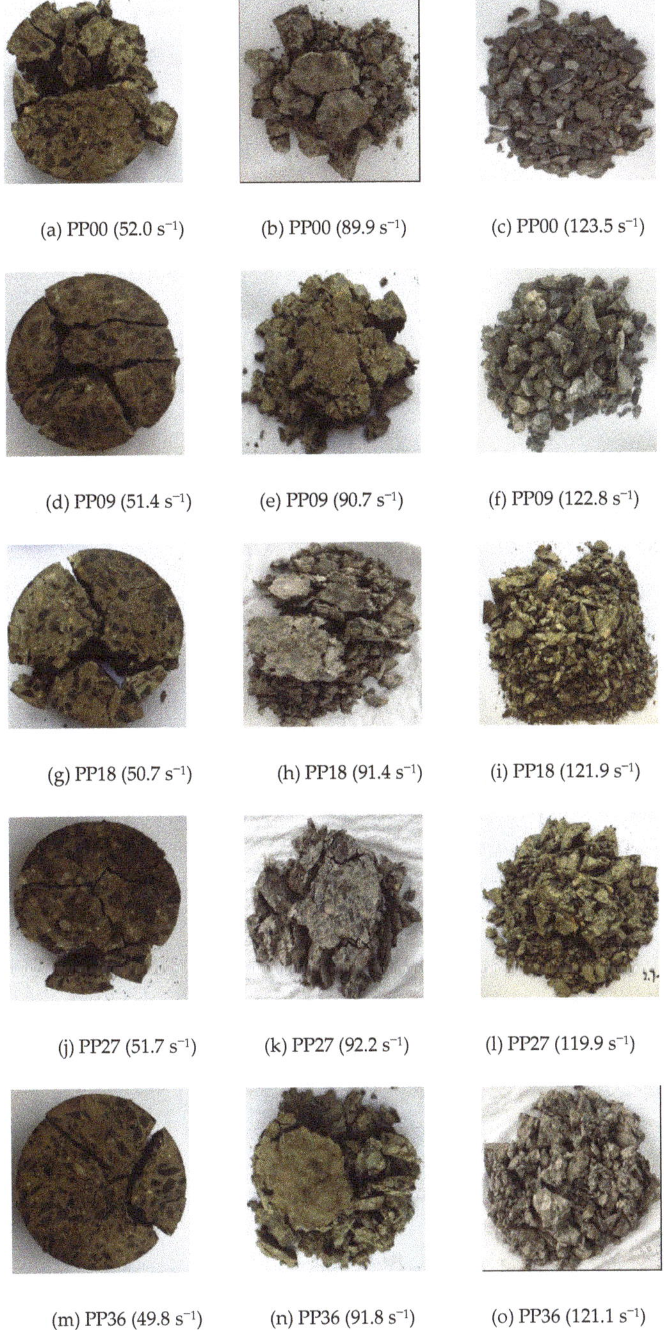

(a) PP00 (52.0 s^{-1})   (b) PP00 (89.9 s^{-1})   (c) PP00 (123.5 s^{-1})

(d) PP09 (51.4 s^{-1})   (e) PP09 (90.7 s^{-1})   (f) PP09 (122.8 s^{-1})

(g) PP18 (50.7 s^{-1})   (h) PP18 (91.4 s^{-1})   (i) PP18 (121.9 s^{-1})

(j) PP27 (51.7 s^{-1})   (k) PP27 (92.2 s^{-1})   (l) PP27 (119.9 s^{-1})

(m) PP36 (49.8 s^{-1})   (n) PP36 (91.8 s^{-1})   (o) PP36 (121.1 s^{-1})

Figure 13. Effects of strain rate and polypropylene fibre dosage on failure mode of all mixtures.

It can be clearly observed that the failure patterns of the mixtures were more complete with the PP fibre under low strain rate. As shown in Figure 13d,g,j,m, only several longitudinal splitting failures were observed. However, PP00 exhibited a brittle failure pattern. The results suggested that the PP fibre can restrict the growth of cracks under low strain rate. When the strain rate is relatively high (e.g., 120 s^{-1}), PP fibres were not so efficient to control cracking of mixtures as pulverization failure pattern was observed for all mixtures. It generally requires two parts of energy to break a fibre reinforced concrete, i.e., the energy to break the matrix and the energy to pull out or rupture fibre reinforcements [9]. The SEM images presented in Figures 14 and 15 can be used to compare the original morphology of PP fibres with the failure morphology in the specimen after impact test. As seen in Figure 15a, PP fibres were found to be pulled out of the concrete matrix, resulting in scratches and damage on the surface of the fibres. Figure 15b shows that PP fibres were ruptured in the process of specimen fracture. When the strain rate is relatively low, the generated micro-cracks had enough time to reach the transition zone between cement matrix and aggregate resulting in complete failure of the specimen. Thus, PP00 (without fibre reinforcement) failed quickly as compared to other PP fibre reinforced concrete mixtures. With the incorporation of PP fibres, many micro-cracks were restricted by the fibres and thus the ductility of the specimen enhanced. As a result, the failed specimens exhibited more ductile failure pattern compared to that of PP00. When the strain rate is relatively high, the impact loading time is extremely short and therefore the dynamic compressive strength of all mixtures was enhanced by increasing the energy to generate cracks. Although PP fibres can help to resist some cracks, the efficiency would be lower due to the extremely high impact velocity.

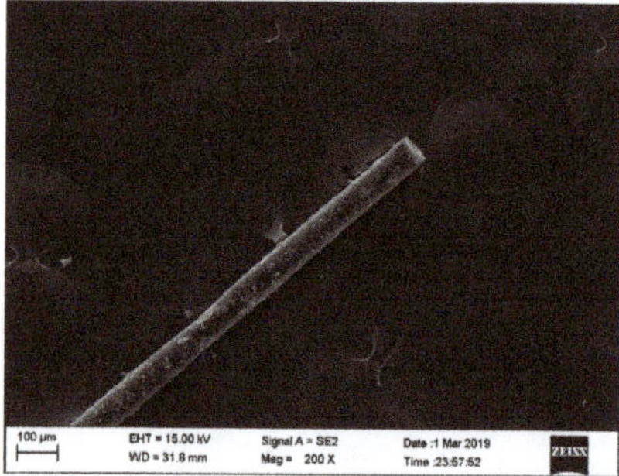

Figure 14. SEM micrographs showing the original morphology of PPF.

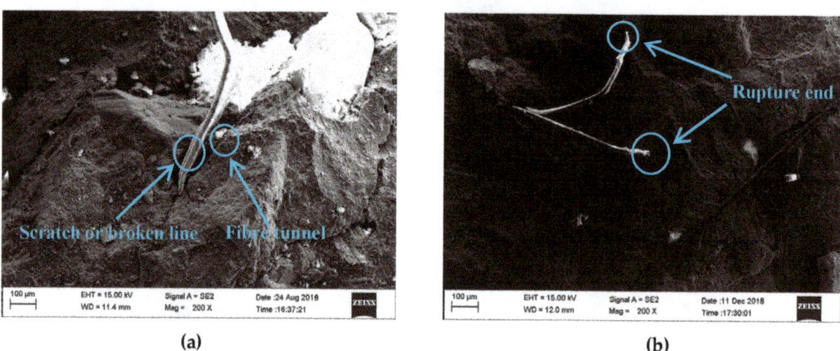

(**a**) (**b**)

Figure 15. SEM micrographs showing the morphology of PPF after impact: (**a**) Scratch or broken line and fibre tunnel; (**b**) Rupture end.

4. Conclusions

In this study, the workability, static compressive strength and dynamic compressive properties of concrete reinforced with PP fibres with various fibre content (i.e., 0.9, 1.8, 2.7 and 3.6 kg/m^3) were investigated and compared with those of the reference mixture without fibre (PP00). Compared with the previous researches concerning the dynamic properties of PP fibre reinforced normal strength concrete, the dynamic properties of high strength concrete with series content of PP fibres was emphasized here. Based on the experimental results, the main conclusions are drawn as follows:

- The incorporation of PP fibres enhanced the dynamic compressive behaviour but reduced the workability and static compressive strength of concrete mixtures. PP36 exhibited the lowest workability among all mixtures, where its workability reduced 14.6% as compared to PP00. The optimal PP fibre content was found 0.9 kg/m^3 (i.e., PP09) in terms of workability, and static and dynamic compressive properties as PP09 exhibited the best workability among all PP fibre reinforced concrete mixtures and showed the least reduction (5.8%) in static compressive strength and an increase of 7.7% in DIF when the strain rate was120 s^{-1} compared to PP00.

- The 'strain rate effect' can be observed for all mixtures regarding the dynamic compressive properties. As strain rate increased, dynamic compressive strength of all mixtures increased and the DIF of all mixtures was enhanced gradually.

- High correlation coefficients of the fitted DIF curves can be found for all mixtures, which indicates that the proposed equations can be used to describe well the relationship between the DIF and strain rate of PP fibre reinforced concrete under impact loading.

- When the strain rate is relatively low, PP fibres were efficient in controlling the growth of micro-cracks under impact loading. However, the effectiveness of PP fibres was reduced when the strain rate was increased, which can be mainly attributed to the extremely short impact loading time.

Author Contributions: Conceptualization, M.C.; Data curation, Y.Y. and X.L.; Funding acquisition, M.C.; Investigation, M.C.; Supervision, E.W.; Writing–original draft, C.R.; Writing–review & editing, Y.L.

Funding: The authors gratefully acknowledge the financial support of the Research Fund of Liaoning Natural Science Foundation (No. 20170540304), State Key Laboratory of Silicate Materials for Architectures (Wuhan University of Technology) (No. SYSJJ2017-08) and Fundamental Research Funds for the Central Universities (No. N170104023) and National Training Program of Innovation and Entrepreneurship for Undergraduates (No. 201810145049).

Acknowledgments: Thanks for the technical support provided by Design & Research Institute of Wuhan University of Technology.

Conflicts of Interest: The funders had no role in the design of the study; in the collection, analyses, or interpretation of data; in the writing of the manuscript, and in the decision to publish the results.

References

1. Parande, A.K.; Babu, B.R.; Karthik, M.A.; Deepak, K.-K.; Palaniswamy, N. Study on strength and corrosion performance for steel embedded in metakaolin blended concrete/mortar. *Constr. Build. Mater.* **2008**, *22*, 127–134. [CrossRef]

2. Fu, Q.; Niu, D.; Zhang, J.; Huang, D.; Wang, Y.; Hong, M.; Zhang, L. Dynamic compressive mechanical behaviour and modelling of basalt-polypropylene fibre-reinforced concrete. *Arch. Civ. Mech. Eng.* **2018**, *18*, 914–927. [CrossRef]

3. Song, P.S.; Hwang, S.; Sheu, B.C. Strength properties of nylon- and polypropylene-fiber-reinforced concretes. *Cem. Concr. Res.* **2005**, *35*, 1546–1550. [CrossRef]

4. Zhang, W.; Zhang, Y. Research on the static and dynamic compressive properties of high performance cementitious composite (HPCC) containing coarse aggregate. *Arch. Civ. Mech. Eng.* **2015**, *15*, 711–720. [CrossRef]

5. Fang, G.; Liu, Y.; Qin, S.; Ding, W.; Zhang, J.; Hong, S.; Xing, F.; Dong, B. Visualized tracing of crack self-healing features in cement/microcapsule system with X-ray microcomputed tomography. *Constr. Build. Mater.* **2018**, *179*, 336–347. [CrossRef]

6. Luccioni, B.; Isla, F.; Codina, R.; Ambrosini, D.; Zerbino, R.; Giaccio, G.; Torrijos, M.C. Effect of steel fibers on static and blast response of high strength concrete. *Int. J. Impact. Eng.* **2017**, *107*, 23–37. [CrossRef]

7. Li, W.; Xu, J. Mechanical properties of basalt fiber reinforced geopolymeric concrete under impact loading. *Mater. Sci. Eng. A* **2009**, *505*, 178–186. [CrossRef]

8. Nedeljkovic, M.; Lukovic, M.; van Breugel, K.; Hordijk, D.; Ey, G. Development and application of an environmentally friendly ductile alkali-activated composite. *J. Cleaner. Prod.* **2018**, *180*, 524–538. [CrossRef]

9. Wu, Z.; Shi, C.; He, W.; Wang, D. Static and dynamic compressive properties of ultra-high performance concrete (UHPC) with hybrid steel fiber reinforcements. *Cem. Concr. Compos.* **2017**, *79*, 148–157. [CrossRef]

10. Ali, M.A.E.M.; Nehdi, M.L. Innovative crack-healing hybrid fiber reinforced engineered cementitious composite. *Constr. Build. Mater.* **2017**, *150*, 689–702. [CrossRef]

11. Ali, M.A.E.M.; Soliman, A.M.; Nehdi, M.L. Hybrid-fiber reinforced engineered cementitious composite under tensile and impact loading. *Mater. Des.* **2017**, *117*, 139–149. [CrossRef]

12. Hao, Y.; Hao, H. Dynamic compressive behaviour of spiral steel fibre reinforced concrete in split Hopkinson pressure bar tests. *Constr. Build. Mater.* **2013**, *48*, 521–532. [CrossRef]

13. Romualdi, J.P.; Batson, G.B. Mechanics of crack arrest in concrete. *J. Eng. Mech. Div.* **1963**, *89*, 147–168.

14. Su, Y.; Li, J.; Wu, C.; Wu, P.; Li, Z. Effects of steel fibres on dynamic strength of UHPC. *Constr. Build. Mater.* **2016**, *114*, 708–718. [CrossRef]

15. Mao, L.; Barnett, S.J. Investigation of toughness of ultra high performance fibre reinforced concrete (UHPFRC) beam under impact loading. *Int. J. Impact. Eng.* **2017**, *99*, 26–38. [CrossRef]

16. Karimi, M.M.; Darabi, M.K.; Jahanbakhsh, H.; Jahangiri, B.; Rushing, J.F. Effect of steel wool fibers on mechanical and induction heating response of conductive asphalt concrete. *Int. J. Pav. Eng.* **2019**, 1–14. [CrossRef]

17. Fu, Q.; Niu, D.; Zhang, J.; Huang, D.; Hong, M. Impact response of concrete reinforced with hybrid basalt-polypropylene fibers. *Powder. Technol.* **2018**, *326*, 411–424. [CrossRef]

18. Mo, K.H.; Yap, S.P.; Alengaram, U.J.; Jumaat, M.Z.; Bu, C.H. Impact resistance of hybrid fibre-reinforced oil palm shell concrete. *Constr. Build. Mater.* **2014**, *50*, 499–507. [CrossRef]

19. Toutanji, H.; McNeil, S.; Bayasi, Z. Chloride permeability and impact resistance of polypropylene-fiber-reinforced silica fume concrete. *Cem. Concr. Res.* **1998**, *28*, 961–968. [CrossRef]

20. Zhang, H.; Liu, Y.; Sun, H.; Wu, S. Transient dynamic behavior of polypropylene fiber reinforced mortar under compressive impact loading. *Constr. Build. Mater.* **2016**, *111*, 30–42. [CrossRef]

21. Jahanbakhsh, H.; Karimi, M.M.; Tabatabaee, N. Experimental and numerical investigation of low-temperature performance of modified asphalt binders and mixtures. *Road. Mater. Pavement. Des.* **2016**, 1–22. [CrossRef]

22. Karimi, M.M.; Jahanbakhsh, H.; Jahangiri, B.; Nejad, F.M. Induced heating-healing characterization of activated carbon modified asphalt concrete under microwave radiation. *Constr. Build. Mater.* **2018**, *178*, 254–271. [CrossRef]

23. Jahanbakhsh, H.; Karimi, M.M.; Jahangiri, B.; Nejad, F.M. Induction heating and healing of carbon black modified asphalt concrete under microwave radiation. *Constr. Build. Mater.* **2018**, *174*, 656–666. [CrossRef]

24. Zhang, W.; Chen, S.; Liu, Y. Effect of weight and drop height of hammer on the flexural impact performance of fiber-reinforced concrete. *Constr. Build. Mater.* **2017**, *140*, 31–35. [CrossRef]

25. Bhutta, A.; Borges, P.H.R.; Zanotti, C.; Farooq, M.; Banthia, N. Flexural behavior of geopolymer composites reinforced with steel and polypropylene macro fibers. *Cem. Concr. Compos.* **2017**, *80*, 31–40. [CrossRef]

26. Wu, Y.; Song, W.; Zhao, W.; Tan, X. An Experimental Study on Dynamic Mechanical Properties of Fiber-Reinforced Concrete under Different Strain Rates. *Appl. Sci.* **2018**, *8*, 1904.

27. Wang, S.; Zhang, M.H.; Quek, S.T. Effect of high strain rate loading on compressive behaviour of fibre-reinforced high-strength concrete. *Mag. Concr. Res.* **2011**, *63*, 813–827. [CrossRef]

28. Ibrahim, S.M.; Almusallam, T.H.; Al-Salloum, Y.A.; Abadel, A.A.; Abbas, H. Strain rate dependent behavior and modeling for compression response of hybrid fiber reinforced concrete. *Lat. Am. J. Solids. Struct.* **2016**, *13*, 1695–1715. [CrossRef]

29. ASTM C128-15. *Standard Test Method for Relative Density (Specific Gravity) and Absorption of Fine Aggregate*; ASTM International: West Conshohocken, PA, USA, 2015.

30. ASTM C127-15. *Standard Test Method for Relative Density (Specific Gravity) and Absorption of Coarse Aggregate*; ASTM International: West Conshohocken, PA, USA, 2015.

31. Ranjbar, N.; Talebian, S.; Mehrali, M.; Kuenzel, C.; Cornelis Metselaar, H.-S.; Jumaat, M.-Z. Mechanisms of interfacial bond in steel and polypropylene fiber reinforced geopolymer composites. *Compos. Sci. Technol.* **2016**, *122*, 73–81. [CrossRef]

32. ASTM C143-15a. *Standard Test Method for Slump of Hydraulic-Cement Concrete*; ASTM International: West Conshohocken, PA, USA, 2015.

33. Ministry of Construction of the People's Republic of China. *Standard for Test MEthod of Mechanical Properties on Ordinary Concrete*; GB/T50081-2002; China Architecture & Building Press: Beijing, China, 2003.

34. Chen, M.; Chen, W.; Zhong, H.; Chi, D.; Wang, Y.; Zhang, M. Experimental study on dynamic compressive behaviour of recycled tyre polymer fibre reinforced concrete. *Cem. Concr. Compos.* **2019**, *98*, 95–112. [CrossRef]

35. Li, W.; Xu, J.; Zhai, Y.; Li, Q. Mechanical properties of carbon fiber reinforced concrete under impact loading. *China. Civ. Eng. J.* **2009**, *42*, 24–30. (In Chinese)

36. Kim, D.-J.; Sirijaroonchai, K.; El-Tawil, S.; Naaman, A.-E. Numerical simulation of the Split Hopkinson Pressure Bar test technique for concrete under compression. *Int. J. Impact. Eng.* **2010**, *37*, 141–149. [CrossRef]

37. Akca, K.-R.; Çakir, O.; Ipek, M. Properties of polypropylene fiber reinforced concrete using recycled aggregates. *Constr. Build. Mater.* **2015**, *98*, 620–630. [CrossRef]

38. Qian, C.; Stroeven, P. Development of hybrid polypropylene-steel fibre-reinforced concrete. *Cem. Concr. Res.* **2000**, *30*, 63–69. [CrossRef]

39. Jameran, A.; Ibrahim, I.-S.; Yazan, S.-H.; Rahim, S.-N. Mechanical Properties of Steel-polypropylene Fibre Reinforced Concrete Under Elevated Temperature. *Procedia Eng.* **2015**, *125*, 818–824. [CrossRef]

40. Zhang, H.; Wang, B.; Xie, A.; Qi, Y. Experimental study on dynamic mechanical properties and constitutive model of basalt fiber reinforced concrete. *Constr. Build. Mater.* **2017**, *152*, 154–167. [CrossRef]

41. Cotsovos, D.M.; Pavlovic, M.N. Numerical investigation of concrete subjected to compressive impact loading. Part 1: A fundamental explanation for the apparent strength gain at high loading rates. *Comput. Struct.* **2008**, *86*, 145–163. [CrossRef]

42. Cotsovos, D.M.; Pavlovic, M.N. Numerical investigation of concrete subjected to compressive impact loading. Part 2: Parametric investigation of factors affecting behaviour at high loading rates. *Comput. Struct.* **2008**, *86*, 164–180. [CrossRef]

43. Ross, C.-A.; Jerome, D.-M.; Tedesco, J.-W.; Hughes, M.-L. Moisture and strain rate effects on concrete strength. *ACI. Mater. J.* **1996**, *93*, 293–300.

44. Tedesco, J.-W.; Ross, C.-A. Strain-Rate-Dependent Constitutive Equations for Concrete. *J. Press. Vess-T.* **1998**, *120*, 398–405. [CrossRef]

45. Mangat, P.-S. Tensile strength of steel fiber reinforced concrete. *Cem. Concr. Res.* **1976**, *428*, 245–252. [CrossRef]

 *materials*

Article

Anti-Corrosion Property of Glass Flake Reinforced Chemically Bonded Phosphate Ceramic Coatings

Ge Yan [1], Mingyang Wang [2], Tao Sun [3,*], Xinping Li [1], Guiming Wang [3] and Weisong Yin [1]

[1] School of Civil Engineering and Architecture, Wuhan University of Technology, Wuhan 430070, China
[2] State Key Laboratory for Disaster Prevention & Mitigation of Explosion & Impact,
   Army Engineering University, Nanjing 210007, China
[3] State Key Laboratory of Silicate Materials for Architectures, Wuhan University of Technology,
   Wuhan 430070, China
* Correspondence: sunt@whut.edu.cn

Received: 23 May 2019; Accepted: 26 June 2019; Published: 28 June 2019

Abstract: Glass flake (GF) was used as the reinforcement in chemically bonded phosphate ceramic (CBPC) coatings to promote anti-corrosion property. The crystalline phase, curing behavior, micromorphology and electrochemical performance of the coatings were studied. The results indicate that with the addition of magnesia (MgO), a new bonding phase ($Mg_3(PO_4)_2$) can be formed, which can help the CBPCs achieve a more compact and denser structure. The effect of the magnesia and the GF additives on curing behavior is obvious: the heat of reaction of the phosphate ceramic materials increases with the addition of the magnesia and the GF, which emphasizes the higher crosslinking density in the phosphate ceramic materials. The phosphate ceramic coatings with the magnesia have a higher impedance value compared with the neat phosphate ceramic coating, while the highest impedance value is obtained with increased content of GF. The corrosion mechanism is mainly contributed by the new bonding phase and GF particles, which can hinder the permeation pathway and make the permeation more circuitous.

Keywords: anti-corrosion property; glass flake; chemically bonded phosphate ceramic (CBPC); corrosion mechanism

---

1. Introduction

Reinforced concrete (RC) is one of the most commonly used modern construction materials. The corrosion of the interface between concrete and steel rebar affects the service life of RC seriously. In the past decades, in order to increase the lifespan of RC, researchers have dedicated efforts towards the enhancement of bond strength and corrosion resistance in RC [1,2]. Among all the anti-corrosion technology, a most effective way to improve the property of corrosion resistance is to cover the rebar with a durable and adhesive coating [3], which can be divided into organic coatings and inorganic coatings. Epoxy [4,5] and modified resin [6] have been generally adopted as organic coatings, while inorganic coatings include galvanizing metal coatings [7], chemically reactive enamel (CRE) coatings [8], and chemically bonded phosphate ceramic (CBPC) coatings [9]. In the application of the epoxy coating in reinforced concrete, Brown MC [10] arrived at the conclusion that the durability of the steel with the epoxy coating was not ideal due to the moisture that penetrates beneath the coating, which means that the epoxy-coated steel has a higher corrosion rate than bare steel. What's more, the bone strength loss between the epoxy-coated steel and the concrete reached 20% compared to bare rebar [11]. For the inorganic coatings, a weakening effect on the bond strength between the galvanized reinforcing steel and the concrete was observed due to hydrogen evolution at the interface [12]. A high temperature of 800 °C needs to be applied to the chemically reactive enamel-coated steel, which can lead to a reduction in the creep limit and strength [1].

In recent years, more and more studies have focused on the utilization of CBPCs, which are derived from the reactions between base and acid, such as the metal oxide (Al_2O_3, MgO) and the soluble acid phosphate (KH_2PO_4, $Al(H_2PO_4)_3$) [13,14]. Due to their corrosion resistance, mechanical resistance, thermal conductivity and low temperature used in their processing, CBPC coatings have considerable technological importance [15]. CBPC coatings can be seen to use the phosphate matrixes as the binding phases and the metal oxides as suitable fillers. Researchers [16–18] have investigated the influence of ceramic oxides (AlN, MgO, and ZrO_2) on the improvement of the thermal properties of CBPC coatings. The abrasive filler of Al_2O_3-SiC in aluminum phosphate has been produced to enhance the abrasion resistance of coatings [19]. The ceramic coatings of aluminum phosphate formed by the reaction between soluble acid phosphates and alumina and alumina–sol–gel systems was studied [20]. H.M. Hawthorne et al. [21] prepared phosphate ceramic coatings on stainless steel substrates and analyzed the mechanical performance and electrochemical property. The application by thermal spraying for binders was carried out, and the microstructure defects of the coatings were improved [22]. A series of CBPC coatings were applied to steel surfaces to reduce or eliminate corrosion [23]. Da Bian et al. [24] paid attention to graphene-reinforced CBPC coatings. Zhu Ding et al. [25] studied the mechanical characterization and microstructure of CBPC composites reinforced with fiber, prepared at room temperature.

Glass flake (GF) particles are inorganic platelets with excellent resistance to chemicals and aging. Many efforts have been focused towards corrosion-resistant coatings with the advanced barrier properties of GF [26–30], and the parallel and overlapped arrangement of GF particles could form a compact impermeable layer in organic coatings. There are microstructure defects between GF particles and the organic coating matrix due to the coating matrix belonging to organic material, while GF belongs to inorganic materials; thus, we can only effectively improve the permeability resistance of organic coatings by surface treatment and functionalization of GF particles.

In all cases cited, there have been few reports on the application of GF particles as fillers in CBPC coatings. Meanwhile, the anticorrosion mechanism of the GF-reinforced CBPC coatings also needs to be illustrated. In this study, CBPC coatings reinforced with GF particles were prepared in order to protect the round steel. The crystalline phase, curing behavior, and micromorphology of the CBPC-based ceramic materials were analyzed, and the electrochemical characterization of the CBPC coatings was carried out in 3.5 wt.% NaCl solution using electrochemical measurement. Furthermore, the anticorrosion mechanism of the coating was also investigated.

2. Materials and Methods

2.1. Materials

The CBPC coatings reinforced with GF were applied onto round steel 8 mm in diameter and 20 mm in length by mixing raw materials of phosphate ceramic materials as shown in Table 1.

Table 1. Raw material of phosphate ceramic materials.

Name	Chemical Formula	Manufacturer
Monoaluminium phosphate	$Al(H_2PO_4)_3$	-
Chromium trioxide	CrO_3	Sinopharm Chemical Reagent Co., Ltd., Shanghai, China
Alumina	Al_2O_3	Aladdin Industrial Corporation Tech Co., Ltd., Shanghai, China
Magnesia	MgO	Sinopharm Chemical Reagent Co., Ltd., Shanghai, China
Glass flake (150 mesh)	SiO_2	Hebei Huawei Glass Flake Co., Ltd., Langfang, China

The preparation of monoaluminium phosphate (MAP) binder is based on the reaction between aluminum hydroxide and phosphoric acid at 120 °C under constant stirring for 60 min. The quantities were calculated according to Equation (1). For the preparation of MAP binder, 135.1 g of pure water were used to dilute 345.9 g of 85% phosphoric acid to 60%, and then 78.0 g of aluminum hydroxide were added.

$$3H_3PO_4 + Al(OH)_3 \rightarrow Al(H_2PO_4)_3 + 3H_2O \tag{1}$$

The mixture proportion of phosphate ceramic coating materials are shown in Table 2. The coated samples were prepared by brush coating with a clean bristle brush. Before preparing coating, 800 grit silicon carbide abrasive papers were used to polish the surface of the round steel, the round steel was degreased in acetone for 15 min (in ultrasonic bath) and then rinsed with deionized water. The coated round steel was placed at room temperature (25 ± 1 °C) for 10 h and then heated in an electric oven according to the curing process as shown in Figure 1.

Table 2. Mixture proportion of phosphate ceramic coating pastes.

Sample	MAP (g)	Powders (g)			H$_2$O (g)
		Al$_2$O$_3$	MgO	GF	
CBPC	10.0	10.0	-	-	5.0
GCBPC0	10.0	9.5	0.5	0	5.0
GCBPC5	10.0	9.0	0.5	0.5	5.0
GCBPC10	10.0	8.5	0.5	1.0	5.0
GCBPC15	10.0	8.0	0.5	1.5	5.0

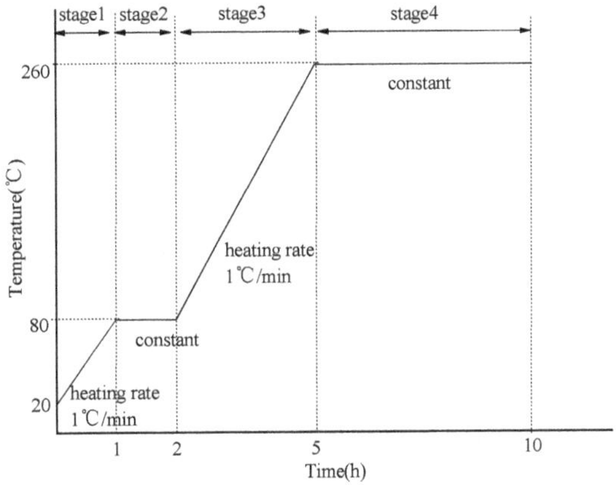

Figure 1. Curing process of phosphate ceramic coatings.

2.2. Characterization

Laser scanning confocal microscopy (LSCM, manufacture, Oberkochen, Germany) was used to characterize the thickness and surface topography of GF. The thickness of the coating was tested using a QNIX4500 coating thickness gauge measurement (QNIX, Oberkochen, Germany) at a precision of 1 μm. The crystalline phases of phosphate ceramic coatings after curing were measured by X-ray diffractometer (PANalytical Empyrean, Almelo, The Netherlands) using a CuKα source scanning from 5° to 70° in 2θ. A differential scanning calorimeter (STA instruments, Selb, Germany) was used to analyze the curing behavior of phosphate ceramic coating materials under N$_2$ atmosphere with gas flow of 30 mL/min, heat velocity of 10 °C/min, start temperature 25 °C and end temperature 400 °C. Furthermore, the SEM micrographs of the surface and cross-section of the coated samples were investigated on a JSM-IT300 (JEOL, Tokyo, Japan) under secondary electron mode, test voltage of 10 kV and surface treatment with platinum.

The potentiodynamic polarization was conducted using a workstation CS350 (Wuhan Corrtest Instrument Co., Ltd., Wuhan, China), as well as the electrochemical impedance spectroscopy (EIS) measurement, with a three-electrode system, containing the reference electrode (saturated calomel electrode), the counter electrode (platinum electrode), and the working electrode (sample).

The electrochemical experiment was conducted in 3.5 wt.% NaCl solution at 25 ± 1 °C. The exposed surface area was around 2.5 cm^2. After the samples were immersed in the NaCl solution for 10 h, the potentiodynamic polarization was performed at a speed of 2 mV·s^{-1}, from −100 mV to 100mV. EIS measurements were carried out with AC signals of 5 mV peak-to-peak amplitude in the frequency start at 100 kHz and end at 0.01 Hz. Z-View software was used to evaluate the EIS data. At least three repeated electrochemical tests were carried out to confirm the reliability of the measurement.

3. Results and Discussions

3.1. Characterization of GF

As shown in Figure 2, the GF, with 150 mesh, has an average thickness of 3–4 µm and an average particle size of 105 µm. Meanwhile, the GF is irregularly formed and has a smooth surface topography. The amorphous peaks of GF in the XRD patterns were observed in the range of 15–40°, as can be seen in Figure 3, which means the GF particles are silica-based materials with a wide typical diffraction peak [31].

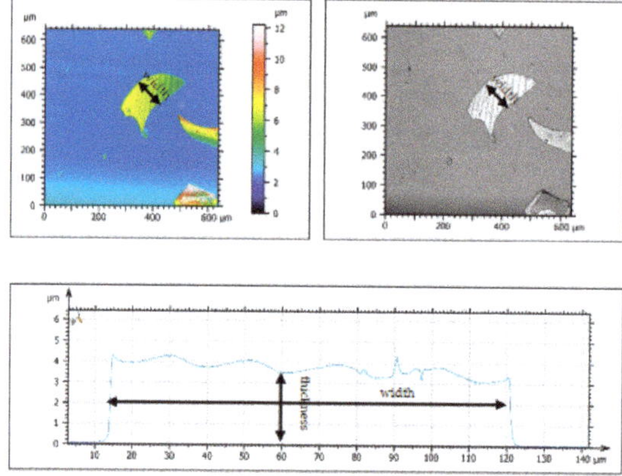

Figure 2. Surface topography of GF.

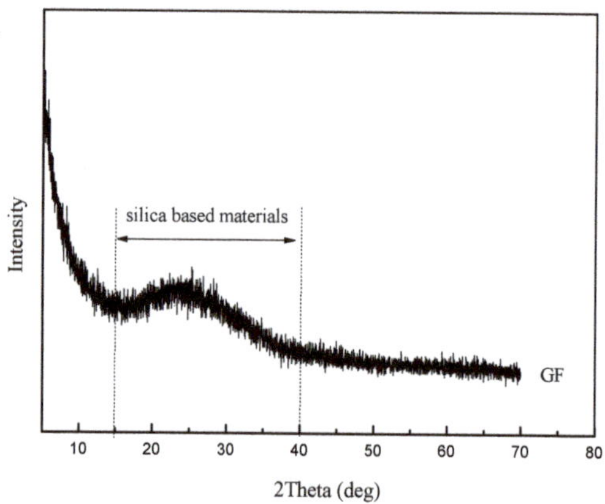

Figure 3. XRD patterns of GF.

3.2. Characterization of Composites and Coatings

3.2.1. Thickness of the Coatings

The thickness of the CBPC coatings is shown in Table 3. It is clear that the thickness of the different GCBPC coatings varied from 193 µm to 217 µm with the increase of GF particles, while the thickness of the CBPC coating was 186 µm. The coating thicknesses showed a little bit of increase, with regard to the standard deviation of the thickness measurement. The difference in thickness is not enough to affect the corrosion resistance of the coatings.

Table 3. Thickness of phosphate ceramic coatings after curing.

	Coatings				
	CBPC	GCBPC0	GCBPC5	GCBPC10	GCBPC15
Thickness (µm)	186	193	201	210	217
Standard deviation	0.3	0.2	0.2	0.3	0.2

3.2.2. XRD

The results obtained from the XRD patterns of the phosphate ceramic coatings are shown in Figure 4. In the case of the CBPC, only Al_2O_3 and $AlPO_4$ phase were present. While the unreacted Al_2O_3, MgO phase and the reacted $AlPO_4$, $Mg_3(PO_4)_2$ phase was identified in GCBPCs. A new bonding phase ($Mg_3(PO_4)_2$) can be found, which may benefit the structural and anti-corrosion properties of the phosphate ceramic coatings, as it has fine needle-like crystals [15]. The new bonding phase was the reaction product of magnesia and MAP solution. MgO can fully react with MAP solutions in the current pastes, so only the product of $Mg_3(PO_4)_2$ and its XRD peaks remain in the GCBPCs. The main binding phase in the phosphate ceramic coatings originates in these reactions of Equations (2) and (3) [15,32]. The reason for the higher peak of $Mg_3(PO_4)_2$ in GCBPCs is related to two aspects. Firstly, the product of CBPC materials is largely limited by the solubility of metal oxides, which means the dissolved magnesia can react faster with MAP due to the higher solubility of the magnesia [33]. Secondly, the peak of $AlPO_4$ phase may not be prominent in the presence of magnesia due to the covering of the $Mg_3(PO_4)_2$ [34]. What's more, the results indicate that the phase of the GF did not alter under the heat treatment of the GCBPCs, which did not affect the crystalline phase of others.

$$Al_2O_3 + Al(H_2PO_4)_3 \rightarrow 3AlPO_4 + 3H_2O \tag{2}$$

$$3MgO + Al(H_2PO_4)_3 \rightarrow AlPO_4 + Mg_3(PO_4)_2 + 3H_2O \tag{3}$$

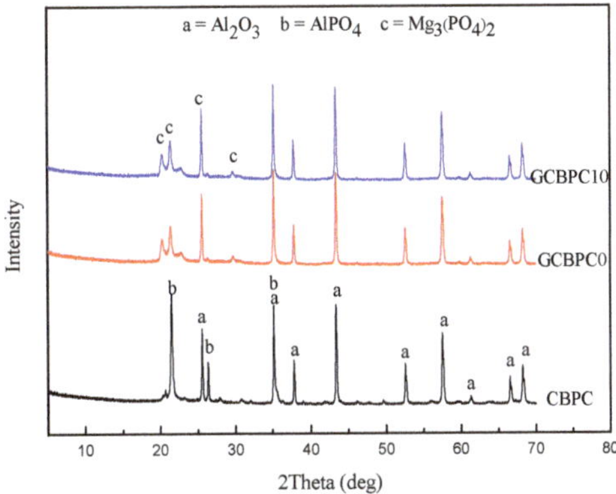

Figure 4. XRD patterns of phosphate ceramic coatings.

3.2.3. Thermal Characterization

To evaluate the effects of the magnesia and the GF additives in the phosphate ceramic materials on the curing behavior, the DSC analysis of the phosphate ceramic materials before curing was conducted, which are shown in Figure 5. The temperature values of the onset (T_{onset}), the peak (T_p), the endset (T_{endset}) and the curing enthalpy (ΔH) are available from the DSC curves, which are predicted in Table 4. All phosphate ceramic materials have only one significant endothermic peak. With the addition of magnesia, the endothermic peak temperature decreased from 228.0 °C (CBPC) to 126.9 °C (GCBPC0), which approaches the temperature of the maximum dissolution of the bauxite and magnesite [12]. Meanwhile, the endothermic peaks of coatings reinforced with GF are 126.3 °C (GCBPC5), 125.4 °C (GCBPC10), 125.2 °C (GCBPC15), respectively.

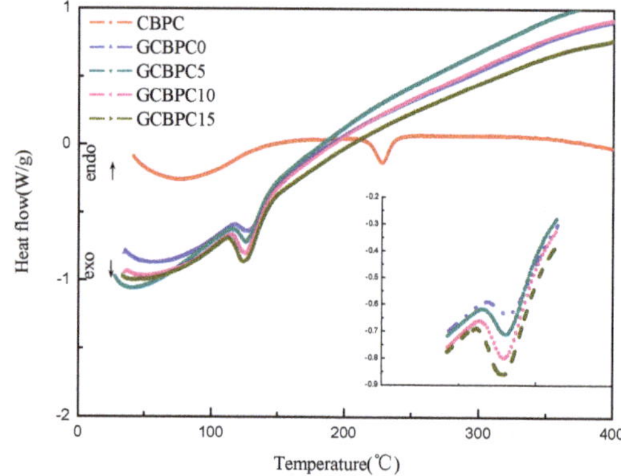

Figure 5. DSC curves of phosphate ceramic materials.

Table 4. Curing parameters from DSC curves.

Sample	T_{onset} (°C)	T_p (°C)	T_{endset} (°C)	ΔH (J/g)
CBPC	209.5	228.0	245.5	−30.32
GCBPC0	118.9	126.9	153.9	−37.65
GCBPC5	117.8	126.3	153.3	−48.30
GCBPC10	115.4	125.4	153.4	−54.38
GCBPC15	113.7	125.2	153.2	−56.39

where T_{onset}, T_p, T_{endset} represents the temperature values of the onset, the peak and the endset, respectively, ΔH is the curing enthalpy.

The great changes in curing temperature can be seen clearly with the addition of magnesia [17]. The exothermic peak temperature (T_p) of CBPC is 228 °C, while the GCBPCs are all around 125 °C. Besides, the ΔH are −30.32, −37.65, −48.30, −54.38, −56.39 J/g, respectively. Therefore, the curing enthalpy of reaction of the phosphate ceramic materials increases with the addition of the magnesia and the GF additives, which emphasizes the higher crosslinking density in the phosphate ceramic materials [35,36].

3.2.4. SEM

The typical SEM micrographs of different coating surfaces after curing are shown in Figure 6. The coatings of CBPCs consist of alumina distributed in the MAP binder. The binding phase is produced by the reaction between MAP and the surface of the particles during the curing process. As can be seen, the solidified binder fills the space between the particles. There were differences in the micrograph. The particles of the CBPC coating have an inhomogeneous morphology, the main component of which is Al_2O_3, while the particles spread more evenly in GCBPCs, which has a denser structure due to the new bonding phase. The result is consistent with the results of DSC. Furthermore, the GF particles on the surface and internal of the coating is shown in Figures 6 and 7. The GF particles and the GCBPC maintain a good adhesion. On the other hand, the GF particles achieve a homogeneous and parallel dispersion on the surface and internal of the coatings.

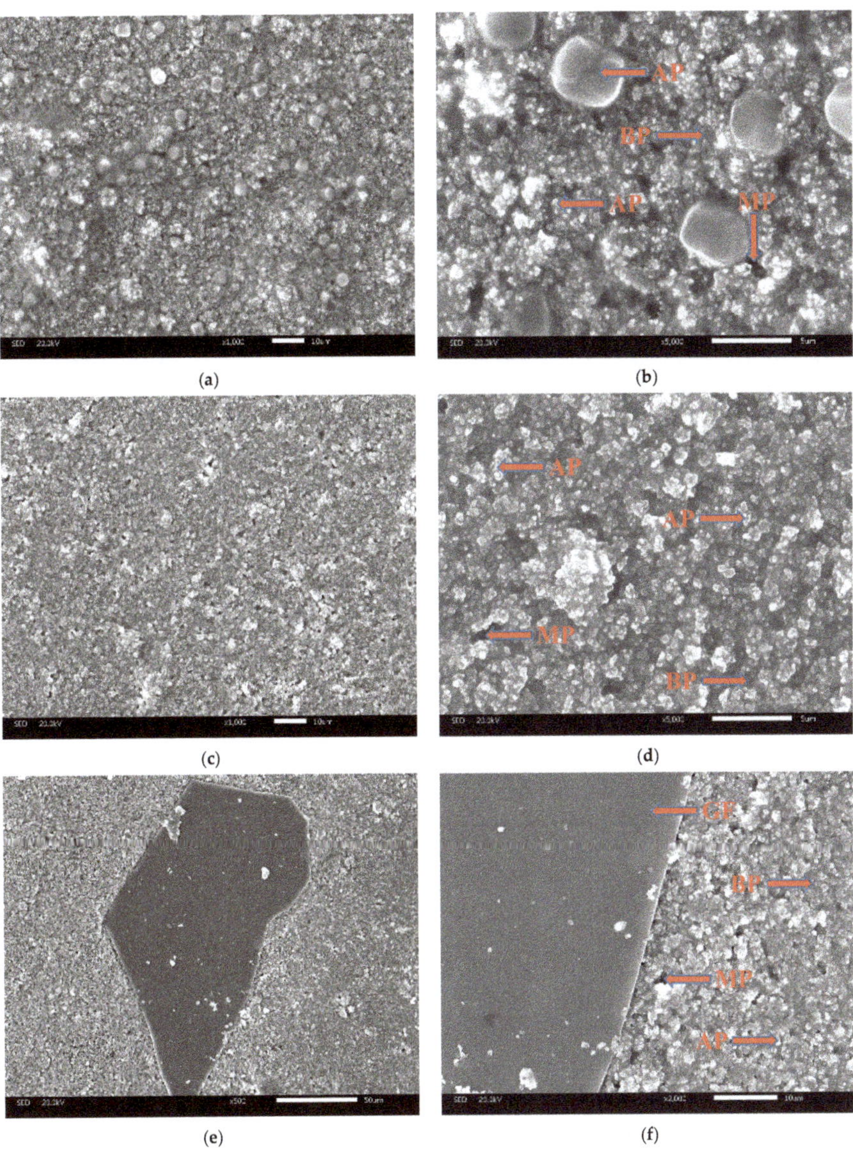

Figure 6. SEM micrographs of coating surfaces (after curing): (**a**) 1000× CBPC; (**b**) 5000× CBPC; (**c**) 1000× GCBPC0; (**d**) 5000× GCBPC0; (**e**) 500× GCBPC10; (**f**) 2000× GCBPC10. Notes: AP (alumina particles), BP (binding phases), MP (micro-porous).

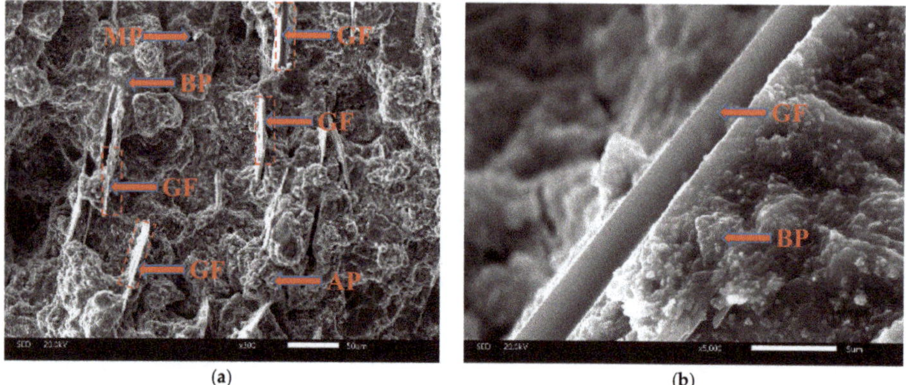

Figure 7. SEM images of the cross-section in GCBPCs: (**a**) 300×; (**b**) 5000×. Notes: AP (alumina particles), BP (binding phases), MP (micro-porous).

3.3. Anti-Corrosion Performance

3.3.1. Potentiodynamic Polarization

The effect of the magnesia and the GF particles on the corrosion resistance of CBPCs samples was evaluated by using the polarization measurement in 3.5 wt.% NaCl solution. Figure 8 illustrates the potentiodynamic polarization data of the uncoated round steel, and the coated different phosphate ceramic samples.

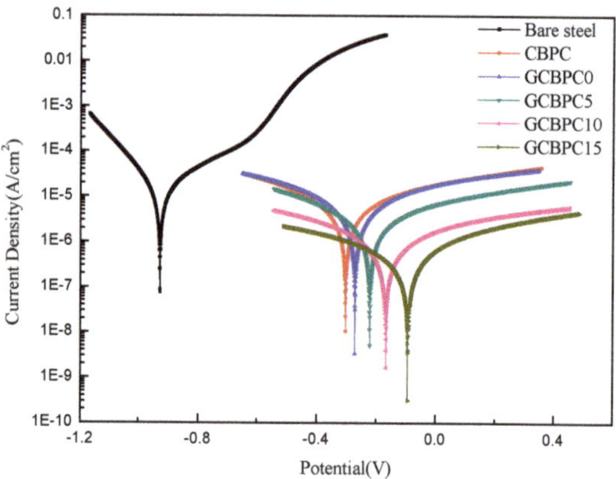

Figure 8. Potentiodynamic polarization curves of phosphate ceramic coatings.

The corresponding electrochemical parameters are shown in Table 5, derived using the Tafel extrapolation method. The results show that the GCBPC0 has a higher E_{corr} and a lower i_{corr} compared to CBPC, which means the GCBPC0 has a better corrosion resistance than the CBPC. Meanwhile, with the increase of the GF in the GCBPCs, the value of E_{corr} increases, and the value of i_{corr} and corrosion rate decreases, indicating that the presence of GF can improve the anti-corrosion performance of the CBPCs [37]. In the GCBPC coating samples, the value of E_{corr} of GCBPC15 increases from −0.272 V of GCBPC0 to −0.094 V. Additionally, the value of i_{corr} and corrosion rate decreases from 7.396×10^{-6} A/cm^2, 0.087 mm/a to 4.522×10^{-7} A/cm^2, 0.005 mm/a, respectively. The reason for this may attributed to the most compact microstructure and the strongest resistance to electrolytes penetration being in GCBPC15. The effective inhibition η is calculated from Equation (4) [38].

$$\eta = \left(1 - \frac{i_{corr}}{i_{corr,blank}}\right) \times 100\% \tag{4}$$

where $i_{corr,blank}$ is the etching current density of the uncoated sample, while i_{corr} represents the corrosion current density of the coated samples. As is well known, the higher the value of η, the better the anti-corrosion performance may be.

Table 5. Curing parameters based on the potentiodynamic polarization curves.

Sample	Electrochemical Parameter			η (%)
	E_{corr} (V)	i_{corr} (A/cm^2)	Corrosion Rate (mm/a)	
Bare Steel	−0.928	1.503×10^{-5}	0.177	-
CBPC	−0.303	8.312×10^{-6}	0.098	44.69
GCBPC0	−0.272	7.396×10^{-6}	0.087	50.79
GCBPC5	−0.222	3.134×10^{-6}	0.037	79.15
GCBPC10	−0.157	1.082×10^{-6}	0.013	92.80
GCBPC15	−0.094	4.522×10^{-7}	0.005	96.99

where E_{corr}, i_{corr} represent the corrosion potential and corrosion current density, η is the effective inhibition, which is calculated from Equation (4).

3.3.2. Electrochemical Impedance Spectroscopy

The EIS method is an effective way to access the anticorrosion property of the CBPCs samples. The Nyquist and Bode plots of the CBPC coated samples immersed for 10 h in 3.5 wt.% NaCl solution are shown in Figure 9. EIS parameters (Table 6) from Nyquist and Bode plots were extracted by using an electrical equivalent circuit (R(RQ(RQ))) to model the experiment. R_s, R_c, CPE_c, R_{ct} and CPE_{ct} represent resistance of solution, resistance of the CBPC coating, nonideal capacity of the CBPC coating, resistance of the charge transfer and nonideal capacity of the double layer, respectively.

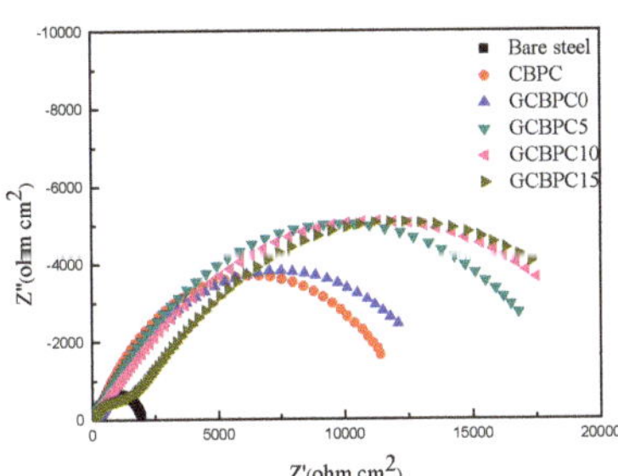
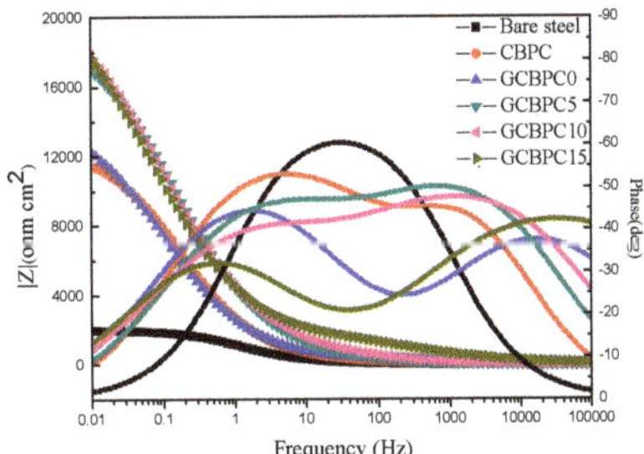

Figure 9. Nyquist and Bode plots of CBPCs.

Table 6. EIS parameters from Nyquist and Bode plots.

Sample	R_s ($\Omega \cdot cm^2$)	R_c ($\Omega \cdot cm^2$)	CPE_c ($\Omega \cdot cm^2$)	N_c	R_{ct} ($\Omega \cdot cm^2$)	CPE_{ct} (F/cm^2)	N_{ct}
Bare Steel	9.46	-	-	-	1989	1.4327×10^{-4}	0.75
CBPC	11.43	267	4.1701×10^{-5}	0.69	12406	5.4811×10^{-5}	0.67
GCBPC0	11.27	371	2.0892×10^{-5}	0.57	14187	9.5405×10^{-5}	0.64
GCBPC5	10.98	1128	3.1458×10^{-5}	0.65	18370	3.6515×10^{-5}	0.58
GCBPC10	10.90	1509	2.9823×10^{-5}	0.61	20619	4.785×10^{-5}	0.53
GCBPC15	10.88	1784	9.6835×10^{-6}	0.54	20950	8.5690×10^{-5}	0.57

where R_s, R_c, CPE_c, R_{ct} and CPE_{ct} represent the resistance of the solution, resistance of the CBPC coating, the nonideal capacity of the CBPC coating, the resistance of charge transfer and the nonideal capacity of the double layer, respectively.

According to EIS parameters from Table 6, the resistance of the solution (R_s) remains stable, the addition of magnesia improves the anti-corrosion performance of CBPC to a certain extent, as the coating resistance (R_c) and charge transfer resistance (R_{ct}) increase. Meanwhile, the more GF added within the range of experimental dosage, the better anti-corrosion performance achieved. In addition, the GCBPC15 obtains the strongest anticorrosion property. According to the previous results (DSC, SEM and XRD), two main reasons can be noted for this phenomenon. Firstly, the presence of magnesia contributes to the formation of new phase and increases the compactness of GCBPCs to shield the round steels from the erosion of aggressive electrolyte. Secondly, the parallel distribution of GF particles on the surface and internal of the coating make the electrolyte diffusion path more tortuous.

The comparison of the EIS parameters from different coatings between the previous literature and the results obtained in this research is shown in Table 7. As can be seen, the anti-corrosion property of the epoxy coating is significantly higher than other species of coatings due to the excellent protection against aggressive substances of the polymer matrix. For other types of coatings, GCBPC15 in this paper has better corrosion resistance performance than other coatings, which can mainly be attributed to the compactness of the matrix and the parallel distribution of GF particles.

Table 7. Comparison of the EIS parameters from different coatings.

Species	R_s ($\Omega \cdot cm^2$)	R_c ($\Omega \cdot cm^2$)	CPE_c ($\Omega \cdot cm^2$)	N_c	R_{ct} ($\Omega \cdot cm^2$)	CPE_{ct} (F/cm^2)	N_{ct}
CBPC	11.43	267	4.1701×10^{-5}	0.69	12406	5.4811×10^{-5}	0.67
GCBPC15	10.88	1784	9.6835×10^{-6}	0.54	20950	8.5690×10^{-5}	0.57
Epoxy [39]	100	28935	4.3417×10^{-6}	0.75	44523	3.8667×10^{-6}	0.22
Galvanized [40]	10	491.6	4.0417×10^{-5}	0.80	2000	3.2378×10^{-5}	0.66
GCBPC0-ZnO [24]	-	817.2	7.2100×10^{-5}	0.64	911.7	3.8300×10^{-5}	0.79

where R_s, R_c, CPE_c, R_{ct} and CPE_{ct} represent the resistance of the solution, the resistance of the coating, the nonideal capacity of the coating, the resistance of the charge transfer and the nonideal capacity of the double layer, respectively.

3.4. Corrosion Protection Mechanism

As is widely known, the initial corrosion is commonly due to the penetration of sufficient H_2O, O_2, Cl^-, or their combined effect on the steel [2,41]. The corrosion will cease to exist until the penetration is inhibited. For CBPCs, the substrate is protected by the compact coatings, which can retard the substrate corrosion under in cases with H_2O, O_2 and Cl^-. The surface microstructures of the CBPC coatings are presented in Figure 7. Compared to GCBPC0, the GF particles are found in phosphate ceramics coatings, as shown in Figures 7 and 8. The reason the GCBPC10 has better anticorrosion performance than GCBPC0 is that the new bonding phase and GF can hinder the permeation pathway of H_2O, O_2 and Cl^- and make the permeation more circuitous.

These results indicate that GF-reinforced chemically bonded phosphate ceramic coatings are potentially superior compared to the neat chemically bonded phosphate ceramics in terms of anti-corrosion property. To clarify the anticorrosion performance, a model of GF reinforced chemically bonded phosphate ceramic coating was established. As shown in Figure 10, for CBPC or GCBPC0, aggressive substances such as H_2O, O_2 and Cl^- can be passed to the substrate through the diffusion route between the binding phase and the particle. However, the binding phase and the GF can hinder the aggressive substances permeation pathway and make the diffusion route more tortuous in GCBPCs, preventing H_2O, O_2 and Cl^- from engaging the substrate effectively.

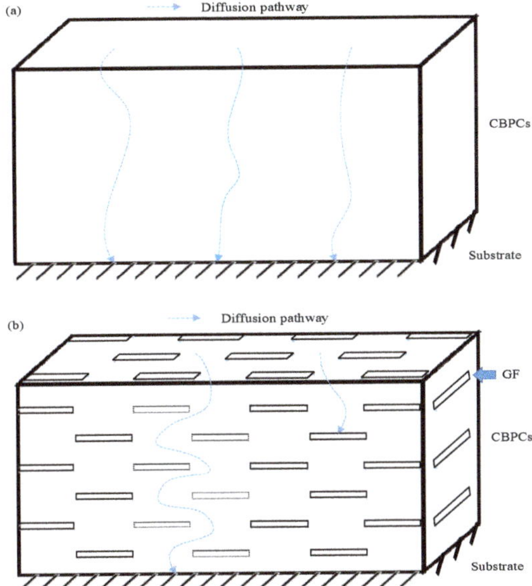

Figure 10. Schematic diagram of the anti-corrosion mechanism in (**a**) CBPCs and (**b**) GCBPCs.

4. Conclusions

The glass flake-reinforced chemically bonded phosphate ceramic coatings were prepared on round steel. The crystalline phase, curing behavior, micromorphology and electrochemical performance of the coating were studied. The main conclusions can be obtained as follows:

(1) With the addition of magnesia, a new bonding phase ($Mg_3(PO_4)_2$) can be formed, which can help the GCBPCs obtain a more compact and denser structure. Meanwhile, The GF particles have a good adhesion with GCBPC and achieve a homogeneous and parallel dispersion on the surface and internal of the coatings.

(2) The effect of the magnesia and the GF additives on curing behavior is obvious; the heat of reaction of the phosphate ceramic materials increases, which emphasizes the higher crosslinking density in the phosphate ceramic materials.

(3) The phosphate ceramic coatings with the magnesia have a higher impedance value compared with the neat phosphate ceramic coating, while the highest impedance value is acquired with increase content of GF. It is found that GCBPC0 has a smaller particle size and a denser structure due to the new bonding phase compared with CBPC, and GF is distributed parallel on the surface and internal of the GCBPCs homogeneously. The corrosion mechanism is mainly contributed by the new bonding phase and GF, which can hinder the permeation pathway and make the permeation more circuitous.

Author Contributions: Conceptualization and Methodology, G.Y. and W.Y.; Validation, G.Y. and T.S.; Resources, T.S.; Data Curation, G.W.; Writing—Original Draft Preparation, G.Y.; Writing—Review & Editing, M.W. and X.L.

Funding: This study was financially supported by the National Key Research and Development Program of China (No. 2017YFB0310905), Yang Fan Innovative & Entrepreneurial Research Team Project (No. 201312C12), and the 111 Project (No. B18038).

References

1. Yan, D.; Reis, S.; Tao, X.; Chen, G.; Brow, R.K.; Koenigstein, M. Effect of chemically reactive enamel coating on bonding strength at steel/mortar interface. *Constr. Build. Mater.* **2012**, *28*, 512–518. [CrossRef]
2. Zhu, X.; Chen, Z.; Wang, H.; Chen, Y.; Xu, L. Probabilistic Generalization of a Comprehensive Model for the Deterioration Prediction of RC Structure under Extreme Corrosion Environments. *Sustainability* **2018**, *10*, 3051. [CrossRef]
3. Selvaraj, R.; Selvaraj, M.; Iyer, S. Studies on the evaluation of the performance of organic coatings used for the prevention of corrosion of steel rebars in concrete structures. *Prog. Org. Coat.* **2009**, *64*, 454–459. [CrossRef]
4. Kobayashi, K.; Takewaka, K. Experimental studies on epoxy coated reinforcing steel for corrosion protection. *Int. J. Cem. Compos. Light. Concr.* **1984**, *6*, 99–116. [CrossRef]
5. Končan Volmajer, N.; Steinbücher, M.; Berce, P.; Venturini, P.; Gaberšček, M. Electrochemical Impedance Spectroscopy Study of Waterborne Epoxy Coating Film Formation. *Coatings* **2019**, *9*, 254. [CrossRef]
6. Pour-Ali, S.; Dehghanian, C.; Kosari, A. Corrosion protection of the reinforcing steels in chloride-laden concrete environment through epoxy/polyaniline–camphorsulfonate nanocomposite coating. *Corros. Sci.* **2015**, *90*, 239–247. [CrossRef]
7. Cheng, A.; Huang, R.; Wu, J.; Chen, C. Effect of rebar coating on corrosion resistance and bond strength of reinforced concrete. *Constr. Build. Mater.* **2005**, *19*, 404–412. [CrossRef]
8. Tang, F.; Chen, G.; Brow, R.K.; Volz, J.S.; Koenigstein, M.L. Corrosion resistance and mechanism of steel rebar coated with three types of enamel. *Corros. Sci.* **2012**, *59*, 157–168. [CrossRef]
9. Liu, Y.; Bian, D.; Zhao, Y.; Wang, Y. Influence of curing temperature on corrosion protection property of chemically bonded phosphate ceramic coatings with nano-titanium dioxide reinforcement. *Ceram. Int.* **2019**, *45*, 1595–1604. [CrossRef]
10. Brown, M.C. Corrosion protection service life of epoxy coated reinforcing steel in virginia bridge decks. *Diss. Abstr. Int.* **2002**, *63*, 3825.
11. Kayali, O.; Yeomans, S.R. Bond of ribbed galvanized reinforcing steel in concrete. *Cem. Concr. Compos.* **2000**, *22*, 459–467. [CrossRef]
12. Arenas, M.A.; Casado, C.; Nobel-Pujol, V.; De Damborenea, J. Influence of the conversion coating on the corrosion of galvanized reinforcing steel. *Cem. Concr. Compos.* **2006**, *28*, 267–275. [CrossRef]
13. Wagh, A.S.; Jeong, S.Y. Chemically Bonded Phosphate Ceramics: I, A Dissolution Model of Formation. *J. Am. Ceram. Soc.* **2003**, *86*, 1838–1844. [CrossRef]
14. Formosa, J.; Chimenos, J.M.; Lacasta, A.M.; Niubó, M. Interaction between low-grade magnesium oxide and boric acid in chemically bonded phosphate ceramics formulation. *Ceram. Int.* **2012**, *38*, 2483–2493. [CrossRef]
15. Wagh, A.S. *Chemically Bonded Phosphate Ceramics, Twenty-First Century Materials with Diverse Applications*, 2nd ed.; Elsevier: Naperville, IL, USA, 2016.
16. Han, H.-J.; Kim, D.-P. Studies on Curing Chemistry of Aluminum-Chromium-Phosphates as Low Temperature Curable Binders. *J. Sol-Gel Sci. Technol.* **2003**, *26*, 223–228. [CrossRef]
17. Hong, L.Y.; Han, H.J.; Ha, H.; Lee, J.Y.; Kim, D.P. Development of Cr-free aluminum phosphate binders and their composite applications. *Compos. Sci. Technol.* **2007**, *67*, 1195–1201. [CrossRef]
18. Yang, X.; Huang, Y.; Zhang, J.; Cao, H. Preparation and properties of phosphate base heat-resisting composites. *Chem. Adhes.* **2005**, *27*, 67–70.
19. Chen, D.; He, L.; Shang, S. Study on aluminum phosphate binder and related Al_2O_3–SiC ceramic coating. *Mater. Sci. Eng. A* **2003**, *348*, 29–35. [CrossRef]
20. Moorlag, C.; Yang, Q.; Troczynski, T.; Bretherton, J.; Fyfe, C. Aluminum Phosphates Derived from Alumina and Alumina-Sol-Gel Systems. *J. Am. Ceram. Soc.* **2004**, *87*, 2064–2071. [CrossRef]
21. Hawthorne, H.; Neville, A.; Troczynski, T.; Hu, X.; Thammachart, M.; Xie, Y.; Fu, J.; Yang, Q. Characterization of chemically bonded composite sol–gel based alumina coatings on steel substrates. *Surf. Coat. Technol.* **2004**, *176*, 243–252. [CrossRef]
22. Vippola, M.; Ahmaniemi, S.; Keränen, J.; Vuoristo, P.; Lepistö, T.; Mäntylä, T.; Olsson, E. Aluminum phosphate sealed alumina coating: characterization of microstructure. *Mater. Sci. Eng. A* **2002**, *323*, 1–8. [CrossRef]
23. Wagh, A.; Drozd, V. Inorganic Phosphate Corrosion Resistant Coatings. U.S. Patent US20130139930A1, 6 June 2013.

24. Bian, D.; Zhao, Y. Preparation and corrosion mechanism of graphene-reinforced chemically bonded phosphate ceramics. *J. Sol-Gel Sci. Technol.* **2016**, *80*, 30–37. [CrossRef]

25. Ding, Z.; Li, Y.Y.; Lu, C.; Liu, J. An Investigation of Fiber Reinforced Chemically Bonded Phosphate Ceramic Composites at Room Temperature. *Materials* **2018**, *11*, 858. [CrossRef] [PubMed]

26. Lokuge, W.; Aravinthan, T. Effect of fly ash on the behaviour of polymer concrete with different types of resin. *Mater. Des.* **2013**, *51*, 175–181. [CrossRef]

27. Momber, A.; Irmer, M.; Glück, N. Performance characteristics of protective coatings under low-temperature offshore conditions. Part 1: Experimental set-up and corrosion protection performance. *Cold Reg. Sci. Technol.* **2016**, *127*, 76–82. [CrossRef]

28. Chi, S.; Park, J.; Shon, M. Study on cavitation erosion resistance and surface topologies of various coating materials used in shipbuilding industry. *J. Ind. Eng. Chem.* **2015**, *26*, 384–389. [CrossRef]

29. Du, Y.; Sun, J.; Yang, H.; Yan, L. Recovery of gallium in the alumina production process from high-alumina coal fly ash. *Rare Met. Mater. Eng.* **2016**, *45*, 1893–1897.

30. Barbhuiya, S.; Choudhury, M.I. Nanoscale Characterization of Glass Flake Filled Vinyl Ester Anti-Corrosion Coatings. *Coatings* **2017**, *7*, 116. [CrossRef]

31. Manna, J.; Roy, B.; Sharma, P. Efficient hydrogen generation from sodium borohydride hydrolysis using silica sulfuric acid catalyst. *J. Power Sources* **2015**, *275*, 727–733. [CrossRef]

32. Deng, S.; Wang, C.; Zhou, Y.; Huang, F.; Du, L. Preparation and Characterization of Fiber-Reinforced Aluminum Phosphate/Silica Composites with Interpenetrating Phase Structures. *Int. J. Appl. Ceram. Technol.* **2010**, *8*, 360–365. [CrossRef]

33. Soudée, E.; Pera, J. Mechanism of setting reaction in magnesia-phosphate cements. *Cem. Concr. Res.* **2000**, *30*, 315–321. [CrossRef]

34. Will, G. *Powder Diffraction: The Rietveld Method and the Two Stage Method to Determine and Refine Crystal Structures from Powder Diffraction Data*; Springer Science & Business Media: Bonn, Germany, 2006.

35. Karasinski, E.; Da Luz, M.; Lepienski, C.M.; Coelho, L. Nanostructured coating based on epoxy/metal oxides: Kinetic curing and mechanical properties. *Thermochim. Acta* **2013**, *569*, 167–176. [CrossRef]

36. Sari, M.G.; Saeb, M.R.; Shabanian, M.; Khaleghi, M.; Vahabi, H.; Vagner, C.; Zarrintaj, P.; Khalili, R.; Paran, S.M.R.; Ramezanzadeh, B.; et al. Epoxy/starch-modified nano-zinc oxide transparent nanocomposite coatings: A showcase of superior curing behavior. *Prog. Org. Coat.* **2018**, *115*, 143–150. [CrossRef]

37. Romano, A.-P.; Olivier, M.-G.; Vandermiers, C.; Poelman, M. Influence of the curing temperature of a cataphoretic coating on the development of filiform corrosion of aluminium. *Prog. Org. Coat.* **2006**, *57*, 400–407. [CrossRef]

38. Shao, F.; Yang, K.; Zhao, H.; Liu, C.; Wang, L.; Tao, S. Effects of inorganic sealant and brief heat treatments on corrosion behavior of plasma sprayed Cr_2O_3–Al_2O_3 composite ceramic coatings. *Surf. Coat. Technol.* **2015**, *276*, 8–15. [CrossRef]

39. Golabadi, M.; Aliofkhazraei, M.; Toorani, M.; Rouhaghdam, A.S. Corrosion and cathodic disbondment resistance of epoxy coating on zinc phosphate conversion coating containing Ni^{2+} and Co^{2+}. *J. Ind. Eng. Chem.* **2017**, *47*, 154–168. [CrossRef]

40. Hosseini, M.; Ashassi-Sorkhabi, H.; Ghiasvand, H.A.Y. Corrosion Protection of Electro-Galvanized Steel by Green Conversion Coatings. *J. Rare Earths* **2007**, 537–543. [CrossRef]

41. Chang, C.H.; Huang, T.C.; Peng, C.W.; Yeh, T.C.; Lu, H.I.; Hung, W.I.; Weng, C.J.; Yang, T.I.; Yeh, J.M. Novel anticorrosion coatings prepared from polyaniline/graphene composites. *Carbon* **2012**, *50*, 5044–5051. [CrossRef]

MDPI

St. Alban-Anlage 66

4052 Basel

Switzerland

Tel. +41 61 683 77 34

Fax +41 61 302 89 18

www.mdpi.com

Materials Editorial Office

E-mail: materials@mdpi.com

www.mdpi.com/journal/materials